Advances in
ATOMIC AND MOLECULAR PHYSICS

VOLUME 23

CONTRIBUTORS TO THIS VOLUME

J. BAUCHE

C. BAUCHE-ARNOULT

D. L. EDERER

IAN P. GRANT

M. KLAPISCH

J. L. PICQUE

HARRY M. QUINEY

C. R. VIDAL

D. E. WILLIAMS

F. J. WUILLEMEUMIER

JI-MIN YAN

ADVANCES IN
ATOMIC AND MOLECULAR PHYSICS

Edited by

Sir David Bates

DEPARTMENT OF APPLIED MATHEMATICS AND THEORETICAL PHYSICS
THE QUEEN'S UNIVERSITY OF BELFAST
BELFAST, NORTHERN IRELAND

BENJAMIN BEDERSON

DEPARTMENT OF PHYSICS
NEW YORK UNIVERSITY
NEW YORK, NEW YORK

VOLUME 23

 ACADEMIC PRESS, INC.

Harcourt Brace Jovanovich, Publishers

Boston   San Diego   New York
Berkeley   London   Sydney
Tokyo   Toronto

Copyright © 1988 by Academic Press, Inc.
All rights reserved.
No part of this publication may be reproduced or
transmitted in any form or by any means, electronic
or mechanical, including photocopy, recording, or
any information storage and retrieval system, without
permission in writing from the publisher.

ACADEMIC PRESS, INC.
1250 Sixth Avenue, San Diego, CA 92101

United Kingdom Edition published by
ACADEMIC PRESS INC. (LONDON) LTD.
24–28 Oval Road, London NW1 7DX

LIBRARY OF CONGRESS CATALOG CARD NUMBER: 65-18423

ISBN 0-12-003823-4

PRINTED IN THE UNITED STATES OF AMERICA
88 89 90 91   9 8 7 6 5 4 3 2 1

Contents

Vacuum Ultraviolet Laser Spectroscopy of Small Molecules

C. R. Vidal

I. Introduction	1
II. Multiphoton Spectroscopy and Harmonic Generation	2
III. Coherent VUV Sources	6
IV. Absorption and Excitation Spectrosopy	8
V. Fluorescence Spectroscopy	15
VI. Photodissociation Spectroscopy	22
VII. Ionization Spectroscopy	23
VIII. Two-Step Excitation Spectroscopy	24
IX. Vacuum UV Multiphoton Spectroscopy	31
X. Summary	32
References	32

Foundations of the Relativistic Theory of Atomic and Molecular Structure

Ian P. Grant and Harry M. Quiney

I. Introduction	37
II. Preliminaries	39
III. From QED to Atomic Structure Theory	46
IV. New Developments—Approximation by Finite Basis Sets	64
V. Outlook and Conclusions	81
References	83

Point-Charge Models for Molecules Derived from Least-Squares Fitting of the Electric Potential

D. E. Williams and Ji-Min Yan

I. Introduction	87
II. Calculation of the Electric Potential	90
III. Calculation of the PD/LSF Point Charges in Molecules	94

IV. Examples	101
V. Conclusion	128
References	129

Transition Arrays in the Spectra of Ionized Atoms

J. Bauche, C. Bauche-Arnoult, and M. Klapisch

I. Introduction	132
II. Energy Distribution of Configuration States	137
III. Transition Arrays	142
IV. Comparisons with Experiment	155
V. Level Emissivity	164
VI. Extension to More Physical Situations	171
VII. Level and Line Statistics	179
VIII. Conclusion	186
IX. Appendix	192
References	192

Photoionization and Collisional Ionization of Excited Atoms Using Synchrotron and Laser Radiations

F. J. Wuilleumier, D. L. Ederer, and J. L. Picqué

I. Introduction	198
II. Experimental Techniques	201
III. Theoretical Background	209
IV. Photoionization of an Outer Electron in Excited Atoms	210
V. Results from Synchrotron Radiation Ionization of Laser-Excited Atoms	229
VI. Collisional Ionization of Laser-Excited Atoms	261
VII. Conclusion	278
References	279

Index 287

Contributors

Numbers in parentheses indicate the pages on which the authors contributions begin.

J. BAUCHE, Laboratoire Aime Cotton, Centre National de la Recherche Scientifique, Orsay, France (131)

C. BAUCHE-ARNOULT, Laboratoire Aime Cotton, Centre National de la Recherche Scientifique, Orsay, France (131)

D. L. EDERER, Radiation Physics Division, National Bureau of Standards, Gaithersberg, Maryland 20899 (197)

IAN P. GRANT, Department of Theoretical Chemistry, 1 South Parks Road, Oxford OX1 3TG, England (37)

M. KLAPISCH, Racah Institute of Physics, The Hebrew University, Jerusalem, Israel (131)

J. L. PICQUE, Laboratoire Aime Cotton, Centre National de la Recherche Scientifique, Université Paris Sud, B505, Orsay Cedex, France 91405 (197)

HARRY M. QUINEY, Department of Theoretical Chemistry, 1 South Parks Road, Oxford OX1 3TG, England (37)

C. R. VIDAL, Max-Planck-Institut für Extraterrestrische Physik, D-8046 Garching, Federal Republic of Germany (1)

D. E. WILLIAMS, Department of Chemistry, University of Louisville, Louisville, Kentucky (87)

F. J. WUILLEMEUMIER, Laboratoire de Spectroscopie Atomique et Ionique *and* Laboratoire pour l'Utilisation du Rayonnement Electromagnetique (LURE), Université de Paris Sud, B350, Orsay Cedex, France 91405 (197)

JI-MIN YAN, Institute of Chemistry, Academic Sinica, Beijing, China (87)

VACUUM ULTRAVIOLET LASER SPECTROSCOPY OF SMALL MOLECULES

C. R. VIDAL*

Max-Planck-Institut für Extraterrestrische Physik,
D-8046 Garching, Federal Republic of Germany

I. Introduction	1
II. Multiphoton Spectroscopy and Harmonic Generation	2
III. Coherent VUV Sources	6
IV. Absorption and Excitation Spectroscopy	8
A. Nonselective Detection Methods	8
B. Frequency-Selective Excitation Spectroscopy	11
V. Fluorescence Spectroscopy	15
A. Wavelength-Selective Fluorescence Spectroscopy	15
B. Time-Resolved Fluorescence Spectroscopy	16
C. Lifetime Measurements	18
VI. Photodissociation Spectroscopy	22
VII. Ionization Spectroscopy	23
VIII. Two-Step Excitation Spectroscopy	24
A. Principle	24
B. Analysis of Perturbations	26
C. Predissociation	27
D. Collision-Induced Transitions	30
E. Fine Structure Measurements	30
IX. Vacuum UV Multiphoton Spectroscopy	31
X. Summary	32
Acknowledgements	32
References	32

I. Introduction

The vacuum uv (VUV) spectral region has long been the domain of conventional spectroscopy where the synchrotron or laboratory discharges have served as broad band continuum light sources (Kunz, 1979; Radler and Berkowitz, 1978) and where spectrometers have been used to disperse the radiation. In a few applications synchrotrons have also served as light sources for low resolution fluorescence spectroscopy. In the visible and infrared parts of the spectrum, on the other hand, lasers have greatly advanced the

* also: *Institut für Theoretische Physik, Universität Innsbruck, A-6020 Innsbruck, Austria.*

spectroscopic techniques. With the advent of coherent VUV sources the situation has changed quite significantly and by now several laboratories have started to explore the different possible applications of these new light sources in the VUV spectral region.

The review starts with a brief summary of laser multiphoton techniques which were exclusively employed for some time before the advent of suitable VUV sources to get access to the spectral region below 200 nm. It continues with a short summary describing the technical parameters of typical, state-of-the-art VUV sources. It then describes the different methods of VUV laser spectroscopy which have so far been successfully demonstrated. In order to limit the scope of the present review, the various examples are primarily taken from the gas phase spectroscopy of small diatomic molecules such as H_2, CO, NO and others, and they cover the spectral region mostly below 200 nm (VUV) and in some instances also below 100 nm (XUV). It is therefore not the intent of this review to achieve completeness, but to illustrate the upcoming techniques of high-resolution laser spectroscopy in the vacuum ultraviolet spectral region as they have already been demonstrated in a number of examples and to indicate further possible applications, for example, in laser photochemistry.

II. Multiphoton Spectroscopy and Harmonic Generation

VUV laser spectroscopy was first performed by means of multiphoton spectroscopy. Using tunable lasers of sufficient intensity in the visible part of the spectrum, highly excited states of atoms or molecules can be pumped. The particular transitions were detected either by the subsequent fluorescence (Friedrich and McClain, 1980) or by an additional ionization process (Johnson and Otis, 1981) requiring typically one or two extra photons. Combining the latter method with a mass spectrometer also mass selective detection becomes feasible. Using lasers with a sufficiently small pulse duration, time-of-flight spectrometers can also be particularly advantageous. Among all the possible multiphoton processes of different order the most frequently applied two-photon excitation is described by a pump rate from the atomic or molecular ground state $|g\rangle$ to some excited level $|p\rangle$ which is given by

$$P \sim \Re e \left\{ \frac{\hbar^{-3}}{(\Omega_{pg} - 2\omega)} \sum_a \left(\frac{\mu_{ga}\mu_{ap}}{(\Omega_{ag} - \omega)} \right)^2 \right\} \cdot I^2. \tag{1}$$

μ_{ga} is the dipole matrix element between levels $|g\rangle$ and $|a\rangle$ and the complex transition frequency Ω_{ag} is given by

$$\Omega_{ag} = \omega_{ag} - i\Gamma_{ag}. \tag{2}$$

ω_{ag} is the real transition frequency and Γ_{ag} is the corresponding homogeneous damping constant.

Due to the selection rules multiphoton spectroscopy provides access to states which cannot be reached directly by electric dipole allowed transitions, and for the case of two-photon absorption O-, P-, Q-, R- and S-lines ($\Delta J = J' - J'' = -2, -1, 0, +1, +2$) can be observed. In the VUV multiphoton fluorescence (MPF) or multiphoton ionization (MPI) has been demonstrated for several molecular species such as NO,[1] CO,[2] N_2,[3] H_2,[4] and NH_3[5] as summarized in Table I. In addition, several atomic species such as hydrogen (Bokor et al., 1981; Zacharias et al., 1981b) and others have been investigated using MPF and/or MPI. According to Eq. (1) multiphoton processes can be rather efficient due to a resonant enhancement originating from the two-photon resonant denominator.[6] In addition, high resolution, Doppler free spectra can be obtained using counterpropagating waves[7] in which the ultimate resolution is limited by the homogeneous linewidth of the atomic or molecular system.

However, there is a major drawback of multiphoton spectroscopy for high-resolution spectroscopy. An efficient multiphoton excitation is inevitably associated with a corresponding sizable AC Stark effect since the lowest order contributions to both effects originate from the same order in a perturbation approach (Armstrong et al., 1962; Puell and Vidal, 1976a). Since the sample is in general exposed to a focused beam with a Gaussian profile in space and time, a quantitative interpretation may require an extensive average over electric fields. Hence the levels are not only shifted, but also broadened, limiting the resolution of multiphoton spectra. Girard et al. (1983) observed

---

[1] Bray et al., 1974; Johnson et al., 1975; Bray et al., 1975; Gelbwachs et al., 1975; Zacharias et al., 1976; Freedman, 1977; Asscher and Haas, 1978; Wallenstein and Zacharias, 1978; Zandee and Bernstein, 1979; Zacharias et al., 1980a; Halpern et al., 1980; Zacharias et al., 1981a; Miller and Compton, 1981; Miller et al., 1982; Ebata et al., 1982; Sirkin et al., 1982; Cheung et al., 1983; Ebata et al., 1983; Ebata et al., 1984; Anezaki et al., 1985; Verschuur et al., 1986.

[2] Halpern et al., 1980; Filseth et al., 1977; Faisal et al., 1977; Bernheim et al., 1978; Jones et al., 1982; Girard et al., 1982; Pratt et al., 1983a; Kittrell et al., 1983; Loge et al., 1983; Ferrell et al., 1983; Pratt et al., 1983b; Sha et al., 1984a; Sha et al., 1984b; Rottke and Zacharias, 1985a.

[3] Halpern et al., 1980; Filseth et al., 1977.

[4] Marinero et al., 1982; Marinero et al., 1983a.

[5] Glownia et al., 1980a; Glownia et al., 1980b; Hellner et al., 1984.

[6] Jones et al., 1982; Pratt et al., 1983a; Pratt et al., 1983b; Sha et al., 1984a.

[7] Gelbwachs et al., 1975; Wallenstein and Zacharias, 1978; Filseth et al., 1977; Bernheim et al., 1978.

TABLE I

SUMMARY OF LASER MULTIPHOTON EXPERIMENTS ON SMALL MOLECULES.
FOR THE DETECTION OF THE INDIVIDUAL TRANSITIONS EITHER
MULTIPHOTON FLUORESCENCE (MPF) OR MULTIPHOTON IONIZATION (MPI) HAS
BEEN EMPLOYED. FOR ADDITIONAL MULTIPHOTON EXPERIMENTS ON
ATOMIC OR OTHER MOLECULAR SYSTEMS SEE FRIEDRICH AND MCCLAIN (1980),
JOHNSON AND OTIS (1981).

Observed electronic states	Measurements	Detection	Footnotes
NO: $A^2\Sigma^+(v = 0)$	Excitation spectra	MPF	(8)
NO: $A^2\Sigma^+(v = 0)$	Doppler free spectra	MPF	(9)
NO: $A^2\Sigma^+(v = 0 - 2)$	Lifetimes, trans. moments	MPF	(10)
NO: $C^2\Pi(v = 0)$	Predissociation	MPF	(11)
NO: $A^2\Sigma^+(v = 0 - 3)$	Dissociation limit	MPF	
$C^2\Pi(v = 0, 1) \, D^2\Sigma^+(v = 0)$	Fluorescence spectra	MPF	(12)
NO: $A^2\Sigma^+(v = 0 - 2)$	Double resonance	MPF, MPI	(13)
NO: Rydberg states	Opt. opt. double res.	MPI, MPF	(14)
NO: $A^2\Sigma^+(v = 0, 1)$	Rot. energy transfer	MPF, MPI	(15)
CO: $A^1\Pi$	Doppler free spectra	MPF	(16)
CO: $A^1\Pi(v = 2)$	Excitation spectra	MPF	(17)
CO: $A^1\Pi$	Jet cooled samples	MPI	(18)
CO: $A^1\Pi(v = 0)$	Lifetimes, trans. moments	MPF	(19)
CO: $A^1\Pi(v = 1 - 3)$	Photoelectron study	MPI	(20)
CO: $B^1\Sigma^+$	Fluorescence spectra	MPF, MPI	(21)
CO: $D^1\Delta(v = 7 - 12)$	Spectral assignment	MPF	(22)
CO: $A^1\Pi, B^1\Sigma^+$	Opt. opt. double res.	MPI	(23)
$N_2: a^1\Pi_g - X^1\Sigma_g^+$	Trans. moments	MPF	(24)
$H_2: E, F^1\Sigma_g^+$	Ion spectra	MPI	(25)
$NH_3: \tilde{C}, \tilde{D}, \tilde{E}$	Spectral assignment	MPF, MPI	(26)

[8] Bray et al., 1974; Bray et al., 1975.
[9] Gelbwachs et al., 1975; Wallenstein and Zacharias, 1978; Halpern et al., 1980.
[10] Zacharias et al., 1976; Halpern et al., 1980.
[11] Freedman, 1977.
[12] Asscher and Haas, 1978.
[13] Zacharias et al., 1981a; Ebata et al., 1982; Sirkin et al., 1982; Anezaki et al., 1985; Verschuur et al., 1986.
[14] Miller et al., 1982; Cheung et al., 1983; Ebata et al., 1983; Anezaki et al., 1985; Verschuur et al., 1986.
[15] Ebata et al., 1984.
[16] Halpern et al., 1980; Filseth et al., 1977; Bernheim et al., 1978.
[17] Faisal et al., 1977.
[18] Jones et al., 1982.
[19] Halpern et al., 1980; Girard et al., 1982.
[20] Pratt et al., 1983a; Pratt et al., 1983b.
[21] Loge et al., 1983; Rottke and Zacharias, 1985a.
[22] Kittrell et al., 1983.
[23] Ferrell et al., 1983; Sha et al., 1984a; Sha et al., 1984b.
[24] Halpern et al., 1980; Filseth et al., 1977.
[25] Marinero et al., 1982; Marinero et al., 1983a.
[26] Glownia et al., 1980a; Glownia et al., 1980b.

that for an intensity of 10^{10} W/cm^2 the $A^1\Pi$ state of the CO molecule exhibits a linewidth of 0.6 cm^{-1} and a lineshift of 1.0 cm^{-1} due to an interaction of the $A^1\Pi$ state with the $B^1\Sigma^+$ and $C^1\Sigma^+$ states. Using larger electric field amplitudes even larger Stark shifts have been seen in molecular hydrogen (Pummer et al., 1983). Similar or even larger values will occur for electronic systems where the interacting states are closely spaced. This has to be kept in mind, for example, if one- and two-photon spectra which are taken at very different intensities, are compared for measuring small splittings between energy levels of different symmetry such as the Λ-type doubling and the spin rotation splitting of NO (Wallenstein and Zachrias, 1978). In the latter case a Doppler-free two-photon spectrum was compared with a Doppler-limited one-photon spectrum. Furthermore, field-induced interactions between different electronic states may become important modifying the transition moments and making the lifetimes of individual levels field-dependent.

In order to overcome the problems associated with the AC Stark effect, one can in some situations delay the detection process after the end of the multiphoton excitation process. This, however, only provides unperturbed, small signal spectra for a subsequent fluorescence or excitation process, but not for the delayed detection of the initial multiphoton process which was subject to the Stark-shifted transition.

Other multiphoton techniques such as resonant and nonresonant harmonic generation or frequency mixing, have also been successfully used for analyzing high-lying states of atoms or molecules. In this case the excitation is detected not by the subsequent fluorescence, but by the resulting sum or difference frequency signal. Sum frequency signals in the VUV spectral region have been observed on several molecules such as H_2,[27] CO,[28] NO,[29] and N_2[30] as summarized in Table II. Further experiments reaching the XUV spectral region where also the fundamental, incident waves are in the VUV, will be mentioned below in Section IX. Similar investigations have been carried out on numerous atomic systems. A summary is given in Table 3.1 of Vidal (1986).

Because of the Stark-effect-induced changes of the spectrum it is generally better for high-resolution applications to separate the nonlinear process from the atomic or molecular system to be investigated, and to expose the sample only to moderate VUV intensities. This is best done by generating the VUV radiation before entering the sample, exploiting, for example, four wave sum frequency mixing in atomic systems, as reviewed in the subsequent section.

[27] Srinivasan et al., 1983.
[28] Vallee et al., 1982; Vallee and Lukasik, 1982; Glownia and Sander, 1982; Hellner and Lukasik, 1984.
[29] Innes et al., 1976; Wallace and Innes, 1980.
[30] Hellner and Lukasik, 1984.

TABLE II

Sum Frequency and Third Harmonic Generation of Small Molecules. Electronic States in Brackets Give the Two-photon Resonant Intermediate States. Measurements Have Been Performed for Generating Vacuum Ultraviolet Radiation (VUV), for Performing Multiphoton Ionization (MPI) and for Measuring the Linear, Complex Susceptibility (SUSC).

Observed electronic system	Measurements	Wavelength [nm]	Footnotes
H_2: $X^1\Sigma_g^+ - (E, F^1\Sigma_g^+)$ — Rydberg	VUV	≈ 79	(31)
CO: $X^1\Sigma^+ - (A^1\Pi) - B^1\Sigma^+$	SUSC	114–116	(32)
CO: $X^1\Sigma^+ - (A^1\Pi)$ — continuum	MPI	147.4	(33)
CO: $X^1\Sigma^+ - (A^1\Pi)$ — Rydberg	VUV	93.7–96.7	(34)
NO: $X^2\Pi - (A^2\Sigma^+) - M^2\Pi$	VUV	130–150	(35)
N_2: $X^1\Sigma_g^+ - (a^1\Pi_g) - o, b^1\Pi_u, b^l{}^1\Sigma_u^+$	VUV	93.5–96.7	(36)

III. Coherent VUV Sources

Coherent VUV sources can be generated basically in two different ways. Several systems like the H_2, CO and F_2 lasers and the inert gas halogen excimer lasers (Brau, 1979) have been electronically excited to produce stimulated emission. Some of these systems may approach output powers even as high as a few GWatts. However, only a few of these systems, like the inert gas excimer lasers (McCusker, 1979), are tunable over a limited region of the spectrum. A further disadvantage is a rather large linewidth of typically 10 cm^{-1}.

For the purposes of high-resolution spectroscopy, widely tunable sources with a typical linewidth of 0.1 cm^{-1} and better are desirable. These kinds of sources can be generated using the methods of four wave sum frequency mixing in gases.[37] Also higher-order anti-Stokes Raman sources have been successfully applied (Schomburg et al., 1982). The latter sources, however, have not yet achieved a similar spectral quality and flexibility. Recently, nonlinear systems using the methods of sum frequency mixing in gases have been amplified by means of high-power excimer lasers (Egger et al., 1981;

---

[31] Srinivasan et al., 1983.
[32] Vallee et al., 1982; Vallee and Lukasik, 1982.
[33] Glownia and Sander, 1982.
[34] Hellner and Lukasik, 1984.
[35] Innes et al., 1976; Wallace and Innes, 1980.
[36] Hellner and Lukasik, 1984.
[37] Vidal, 1986; Vidal, 1980; Jamroz and Stoicheff, 1983.

Döbele et al., 1984). In this manner peak powers as high as 1 GWatt (pulse duration: 10 psec) have been demonstrated with a synchronously pumped, mode-locked dye laser in the frequency mixing system (Egger et al., 1982).

Most VUV systems have so far been based on two-photon resonant sum frequency mixing in gases.[38] Mostly dye lasers have been used which are pumped by excimer laser, nitrogen laser or frequency-doubled Nd laser. The dye laser generally consists of an oscillator amplifier arrangement, in order to achieve the small linewidth and the beam quality required for high-resolution spectroscopy. In this case, one of the dye lasers is tuned to a suitable two-photon resonance of the nonlinear medium, whereas another dye laser is continuously tunable. For high-resolution work the dye lasers may be equipped with air-spaced etalons which allow a pressure scan covering the spectral region of interest (Wallenstein and Hänsch, 1974; Wallenstein and Zacharias, 1980). For the nonlinear medium a gaseous two-component system is generally employed. It is generated either inside a heat pipe oven (Vidal and Haller, 1971; Scheingraber and Vidal, 1982b) containing a metal vapor inert gas mixture or inside a cell containing a mixture of two different inert gases.

For efficient sum frequency mixing in gaseous two-component systems, one component of the nonlinear medium is generally selected to provide a large non-linear susceptibility which may be resonantly enhanced by a suitable two-photon resonance. The other component is selected to adjust the overall refractive index. In this manner the phase-matching condition (Vidal, 1986) is satisfied. As a useful spin-off the latter condition has also been successfully exploited for accurate measurements of transition moments (Puell and Vidal, 1976b; Wynne and Beigang, 1981) by measuring the refractive index at a given wavelength.

The most efficient nonlinear media so far have employed phase-matched metal vapor inert gas mixtures. Laser systems with this kind of nonlinear medium have achieved a pulse duration of the order of a few nsec and can provide an energy of typically 10^{11} to 10^{13} photons per shot at a repetition rate of a few Hertz and a spectral linewidth as small as 0.1 cm^{-1} in the VUV.[39] In this manner Doppler-limited spectra can be obtained. The spectral brightness of these sources, which is defined by the number of photons per time and wavelength interval and per unit solid angle, greatly exceeds the one of the synchrotron or of laboratory plasma discharges (Kunz, 1979; Radler and Berkowitz, 1978). This is to some extent a result of the diffraction-limited beam profile. At shorter wavelengths below typically Lyman α the spectral brightness decreases significantly to something like 10^6

---

[38] Vidal, 1986; Vidal, 1980; Jamroz and Stoicheff, 1983.
[39] Wallenstein, 1980; Hutchinson and Thomas, 1983; Klopotek and Vidal, 1984; Scheingraber and Vidal, 1985a.

photons per shot because a number of other competing nonlinear processes start to limit the efficiency of four wave mixing in gases (Scheingraber and Vidal, 1983, 1987).

Third harmonic generation or sum frequency mixing offers, in addition, the great advantage of allowing an accurate wavelength calibration in the visible part of the spectrum where secondary length standards are more reliable than in the VUV spectral region. For this purpose one may use, for example, the molecular iodine or tellurium spectrum as a calibration spectrum. Figure 1 shows a small portion of the VUV excitation spectrum taken on the (1, 0) band of the NO $D^2\Sigma^+ - X^2\Pi$ system (Scheingraber and Vidal, 1985a), which was recorded together with the corresponding iodine absorption spectrum (Gerstenkorn and Luc, 1978, 1979) in the visible part of the spectrum. In order to simplify the experiment the excitation spectrum was taken with the third harmonic radiation generated near the $5s^2\ ^1S$-$5p^2\ ^1S$ two-photon resonance of strontium (Vidal, 1986) requiring the calibration of only one dye laser. The spectra in Fig. 1 were taken by pressure tuning a single dye laser (Scheingraber and Vidal, 1985a). Figure 1 gives the quantum numbers of the individual electronic transitions in NO and demonstrates the resolving power of the VUV source. As an illustration of the technical progress it is worth noting that the closely spaced doublets in Fig. 1 have not been resolved before. The wavelength of the fundamental wave was evaluated from a least squares fit of the iodine spectrum with a typical standard error of 0.003 cm^{-1} corresponding to an accuracy of 0.01 cm^{-1}, or 2 parts in 10^7, in the VUV. This accuracy exceeds that which can be achieved with conventional grating spectrometers of any size and any order.

IV. Absorption and Excitation Spectroscopy

A. NONSELECTIVE DETECTION METHODS

The simplest method of applying tunable coherent VUV sources is by means of absorption or excitation spectroscopy. Straight absorption spectra have been taken in CO[40] as well as in H_2.[41] In a related technique known as excitation spectroscopy, the VUV source can be tuned across individual electronic transitions of an atom or molecule and a photodetector collects the total subsequent fluorescence. Excitation spectra of this kind closely resemble absorption spectra if the quantum efficiency is similar for all excited states investigated. Serious changes of the quantum efficiency may occur, for

---

[40] Miller et al., 1982; Provorov et al., 1977; Maeda and Stoicheff, 1984.
[41] Rothschild et al., 1980; Rothschild et al., 1981; Marinero et al., 1983b.

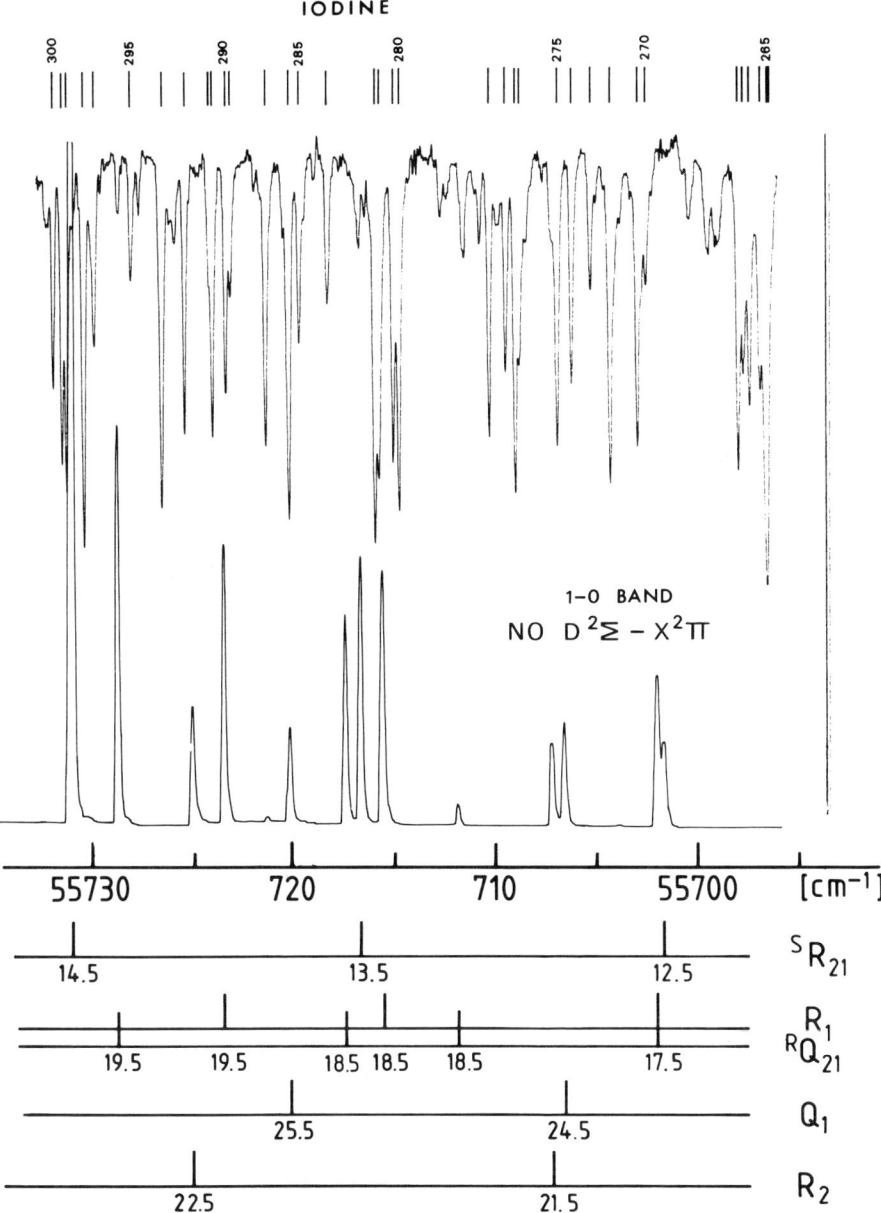

FIG. 1. Small section of the (1, 0) band of the NO $D^2\Sigma^+ - X^2\Pi$ excitation spectrum. The excitation spectrum was taken with the third harmonic of a visible dye laser which was calibrated by the iodine spectrum shown above. The spectral resolution and the accuracy of the wavelength measurement in the VUV spectral region are 0.12 cm^{-1} and 0.01 cm^{-1}, respectively.

example, in case of perturbations or in case of a strong predissociation or autoionization.

With a sufficiently small linewidth of the VUV source, absorption and excitation spectra can provide a Doppler limited resolution. Excitation spectra have been obtained on several molecules such as CO,[42] NO,[43] H_2,[44] Kr_2,[45] and Xe_2.[46] By using a pulsed nozzle beam, rotationally cold noble gas

TABLE III

SUMMARY OF VACUUM ULTRAVIOLET LASER ABSORPTION AND EXCITATION EXPERIMENTS. FOR THE DETECTION OF THE INDIVIDUAL TRANSITIONS EITHER ABSORPTION (AS), EXCITATION (ES), FLUORESCENCE (FS) OR FREQUENCY SELECTIVE EXCITATION (FSES) SPECTRA HAVE BEEN TAKEN.

Observed electronic states	Measurements	Detection	Footnotes
NO: $A^2\Sigma^+$, $B^2\Pi$, $C^2\Pi$, $D^2\Sigma^+$	Franck Condon factors	FS	(47)
NO: $A^2\Sigma^+$, $D^2\Sigma^+$	Hönl London factors	FSES	(48)
NO: $B'\ ^2\Delta(v = 0 - 8)$	Lifetimes	ES	(49)
CO: $A^1\Pi - X^1\Sigma^+$	Photoionization	AS	(50)
CO: $A^1\Pi$	Lifetimes	AS, ES	(51)
CO: $A^1\Pi$	Fluorescence	ES	(52)
CO: $a'\ ^3\Sigma^+$, $e^3\Sigma^-$, $d^3\Delta_{1,2}$	Spectral assignment	FSES	(53)
HD, H_2: $B''^1\Sigma_u^+ - X^1\Sigma^+$	Line shapes	AS	(54)
H_2, HD, D_2: $D^1\Pi_u$, $D''^1\Pi_u B''^1\Sigma_u^+$	Line shapes	AS	(55)
H_2: $B^1\Sigma_u^+(v = 7, 8)$, $C^1\Pi_u(v = 0)$		AS, ES	(56)
H_2, HD, D_2: $B^1\Sigma_u^+$		ES	(57)
Kr_2: $X^1\Sigma_g^+$, $A^3\Sigma_u^+$, $B^1\Sigma_u^+$, $C^1\Sigma_u^+$	Spectral assignment	ES	(58)
Xe_2: $A1_u$, $B0_u^+$, $C0_u^+$	Spectral assignment	ES	(59)

[42] Klopotek and Vidal, 1984; Hilbig and Wallenstein, 1981.
[43] Scheingraber and Vidal, 1985a; Banic et al., 1981.
[44] Marinero et al., 1983b; Northrup et al., 1984.
[45] LaRocque et al., 1986.
[46] Lipson et al., 1984; Lipson et al., 1985.
[47] Scheingraber and Vidal, 1985a.
[48] Scheingraber and Vidal, 1985b.
[49] Banic et al., 1981.
[50] Miller et al., 1982.
[51] Provorov et al., 1977; Maeda and Stoicheff, 1984.
[52] Klopotek and Vidal, 1984; Hilbig and Wallenstein, 1981.
[53] Klopotek and Vidal, 1984.
[54] Rothschild et al., 1980; Rothschild et al., 1981.
[55] Rothschild et al., 1980; Rothschild et al., 1981.
[56] Marinero et al., 1983b.
[57] Northrup et al., 1984.
[58] LaRocque et al., 1986.
[59] Lipson et al., 1984; Lipson et al., 1985.

molecules have been generated. For the Kr_2 and Xe_2 molecules, rotationally resolved spectra and vibrational isotope shifts have been unambiguously measured, the quantum numbers have been assigned and the corresponding molecular constants have been determined. A summary of all applications is given in Table III.

As an example Fig. 2 shows an excitation spectrum of the (2, 0) band of the CO $A^1\Pi - X^1\Sigma^+$ system (Klopotek and Vidal, 1984) also known as the "fourth positive system". The spectrum gives an impression of the achievable resolution. Several lines like the $R(9)$ and the $Q(2)$ lines, for example, have not been resolved in previous VUV spectra and are listed with identical linepositions in existing tables of the CO spectrum (Tilford and Simmons, 1972). The same spectral region is shown again in Fig. 3 with a greatly enhanced sensitivity. In this spectrum the (2, 0) band of the $A^1\Pi - X^1\Sigma^+$ system for the less abundant isotopic species $^{13}C\ ^{16}O$ (1% in a sample of natural abundance) is indicated. The lines for the most abundant isotopic species $^{12}C\ ^{16}O$ are off scale. The latter two spectra clearly show that individual levels even of rare isotopic species can be excited and detected. This particular option is very valuable for some of the more sophisticated techniques such as the two-step excitation spectroscopy to be discussed below in Section VIII.

At this point it is interesting to compare laser-induced VUV spectra with representative absorption and fluorescence measurements as taken with a synchrotron on the NO (Guest and Lee, 1981) and the CO molecule (Lee and Guest, 1981). The latter spectra do not reveal the resolution of the rotational lines and clearly demonstrate the advances in VUV spectroscopy which have been achieved using tunable coherent VUV sources.

B. FREQUENCY-SELECTIVE EXCITATION SPECTROSCOPY

In case of severely overlapping spectroscopic features a modification of the previous nonselective excitation spectroscopy, namely the method of frequency selective excitation spectroscopy, can be particularly valuable. In the preceding nonselective excitation spectroscopy the initial absorption process is detected by collecting the *total* subsequent fluorescence from a series of excited levels. By looking only at the fluorescence in a *small* selected spectral region one can exploit the fact that different electronic states have their dominant fluorescence in distinctly different spectral regions depending on the Franck–Condon factors and the different electronic transition moments.

In scanning the spectral region shown in Fig. 2 and Fig. 3, one primarily pumps electronic transitions of the (2, 0) band of the CO $A^1\Pi - X^1\Sigma^+$ system. In the same spectral region one can also pump transitions of the intercombination bands whose transition moments due to perturbations are

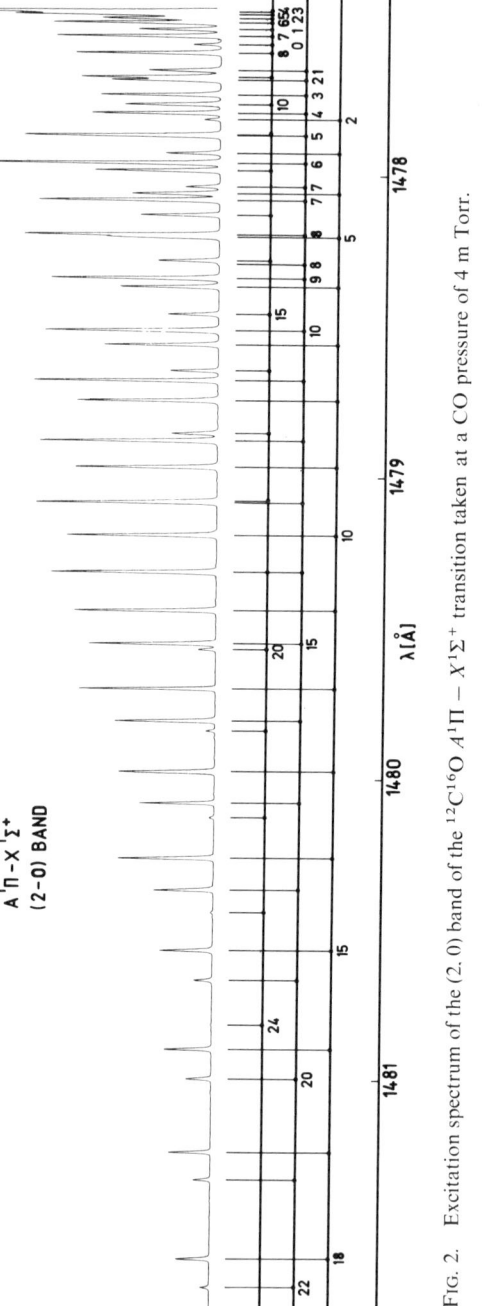

FIG. 2. Excitation spectrum of the (2, 0) band of the $^{12}C^{16}O$ $A^1\Pi - X^1\Sigma^+$ transition taken at a CO pressure of 4 m Torr.

EXCITATION SPECTRUM OF CO
THE $A^1\Pi - X^1\Sigma^+$ (2-0) BAND

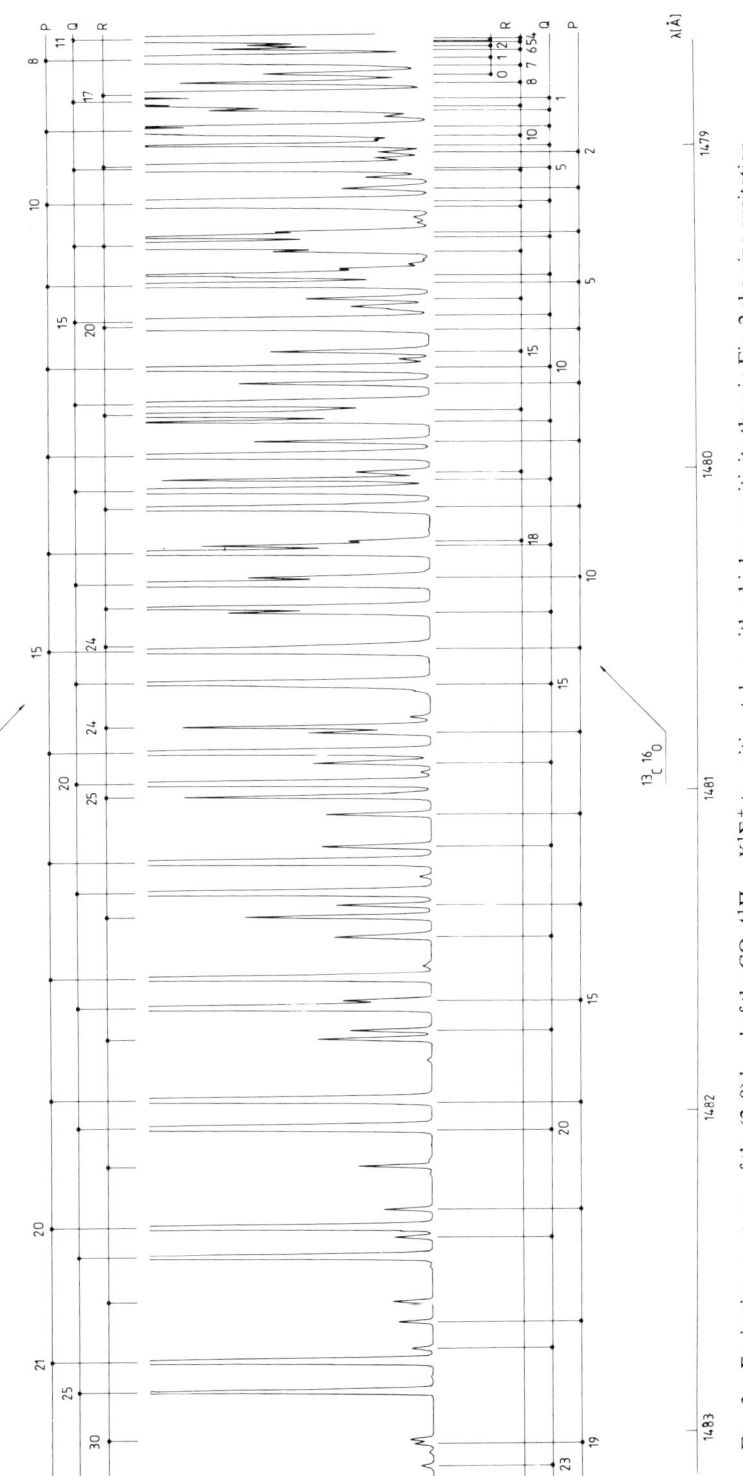

FIG. 3. Excitation spectrum of the (2, 0) band of the CO $A^1\Pi - X^1\Sigma^+$ transition taken with a higher sensitivity than in Fig. 2 showing excitation lines of the less abundant isotopic molecule $^{13}C\,^{16}O$.

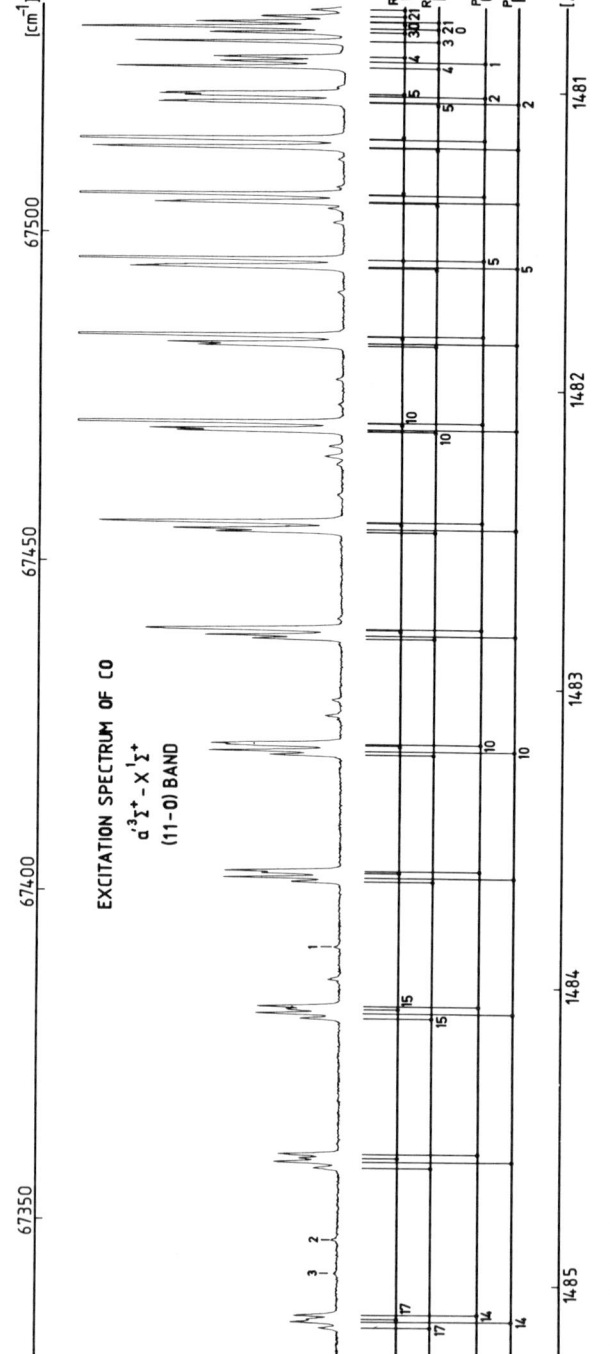

FIG. 4. Excitation spectrum of the (11, 0) band of the CO $a'\,^3\Sigma^+ - X\,^1\Sigma^+$ transition taken at a CO pressure of 0.35 Torr. The lines marked by numerals belong to the (4, 0) band of the $e\,^3\Sigma^- - X\,^1\Sigma^+$ transition and show up due to an interaction of the $e\,^3\Sigma^-$ state with the $A\,^1\Pi$ state.

typically three to four orders of magnitude smaller. In pumping the $A^1\Pi$ state, the dominant fluorescence occurs in the VUV spectral region back to the $X^1\Sigma^+$ ground state. However, in pumping the triplet states through one of the intercombination lines in the same spectral region, the fluorescence occurs primarily in the green to infrared spectral region and is dominated by transitions to the $a^3\Pi$ state. The latter fluorescence can easily be separated from the VUV fluorescence by an appropriate filter and does not interfere at all with the spectrum of the $A^1\Pi - X^1\Sigma^+$ system. Figure 4 shows a frequency selective excitation spectrum of the (11, 0) band of the CO $a'\ ^3\Sigma^+ - X^1\Sigma^+$ transition (Klopotek and Vidal, 1984). Note that the wavelength range of Fig. 4 partially overlaps with that of Figs. 2 and 3. The lines marked by numerals belong to the (4, 0) band of the CO $e\ ^3\Sigma^- - X^1\Sigma^+$ system and show up due to an interaction of the $e\ ^3\Sigma^-$ state with the $A^1\Pi$ state.

Similar frequency selective excitation spectra have also been taken on several other intercombination bands of the CO molecule (Klopotek and Vidal, 1984). On the $A^2\Sigma^+ - X^2\Pi$ and the $B^2\Pi - X^2\Pi$ systems of the NO molecule (Scheingraber and Vidal, 1985a) and similarly on the $A^2\Sigma^+ - X^2\Pi$ and the $D^2\Sigma^+ - X^2\Pi$ systems of the same molecule further frequency selective excitation spectra were taken (Scheingraber and Vidal, 1985b). In the latter two cases of the NO molecule a spectrometer had to be used as a filter for separating the closely spaced bands of different overlapping electronic systems.

V. Fluorescence Spectroscopy

In this review a clear distinction between excitation and fluorescence spectroscopy is made. In excitation spectroscopy one generally scans the pump source and collects the total or a large part of the resulting fluorescence. For fluorescence spectroscopy, however, the pump source is tuned into resonance with an individual line of an electronic transition exciting an individual level of a particular state and the subsequent fluorescence is dispersed by means of a spectrometer. This is a well-known technique (Demtröder, 1981) which has been widely used for measuring Franck–Condon factors in the visible part of the spectrum.

A. Wavelength-Selective Fluorescence Spectroscopy

The first state-selective experiments in the VUV spectral region have exploited accidental coincidences, for example, with atomic resonance lines of the inert gases or of mercury. Experiments of this kind have been performed, for example, in NO.[60] Recently, accidental coincidences of VUV lasers have also been employed to carry out laser induced fluorescence on the NO molecule using an ArF laser (Shibuya and Stuhl, 1982) or an F_2 laser (Taherian and Slanger, 1984).

Fluorescence experiments using continuously tunable VUV sources have been carried out on the NO molecule.[61] Figure 5 shows an interesting example where the fluorescence of the individual lines in the (3, 1) band of the NO $A^2\Sigma^+ - X^2\Pi$ system has been resolved.[62] It has been obtained after pumping a line of the (3, 0) band of the $A^2\Sigma^+ - X^2\Pi$ system. The line intensities provide the Hönl London factors for the (3, 1) band. The triangles in Fig. 5 originate from a relation given by Earls (1935). The measurements are in good agreement with the calculations. The results clearly illustrate the transition from Hund's coupling case (a) to case (b) which is manifested by the J-dependence of the line intensities and depends on the ratio B/A where B is the rotational constant and A the fine structure splitting of the system (Hougen, 1970).

An interesting extension of the fluorescence technique was demonstrated on the CO molecule (Hancock and Zacharias, 1981), which was initially formed in highly excited vibrational levels of the ground state by means of a chemical reaction between atomic oxygen and CS radicals. In this manner the absorption spectrum was effectively extended to longer wavelengths and became accessible to a Raman-shifted, frequency-doubled dye laser. The subsequent fluorescence was dominated again by transitions in the VUV spectral region.

B. TIME-RESOLVED FLUORESCENCE SPECTROSCOPY

As an extension of the preceding wavelength selective technique, different overlapping electronic transitions can also be separated, exploiting the different lifetimes of the participating electronic states. This method is particularly valuable if the fluorescence of highly excited states is investigated where an excitation of lower electronic states cannot be avoided, because of subsequent cascade processes from the initially populated state which compete through their additional fluorescence. Provided the lifetimes are sufficiently different, the fluorescence signals can be separated by looking only at a small preselected interval in time. This time interval has to be

[60] Bergeman and Zare, 1974; Hikida et al., 1975; Hikida et al., 1980.
[61] Scheingraber and Vidal, 1985a; Scheingraber and Vidal, 1985b; Scheingraber and Vidal, 1982a.
[62] Scheingraber and Vidal, 1985b; Scheingraber and Vidal, 1982b.

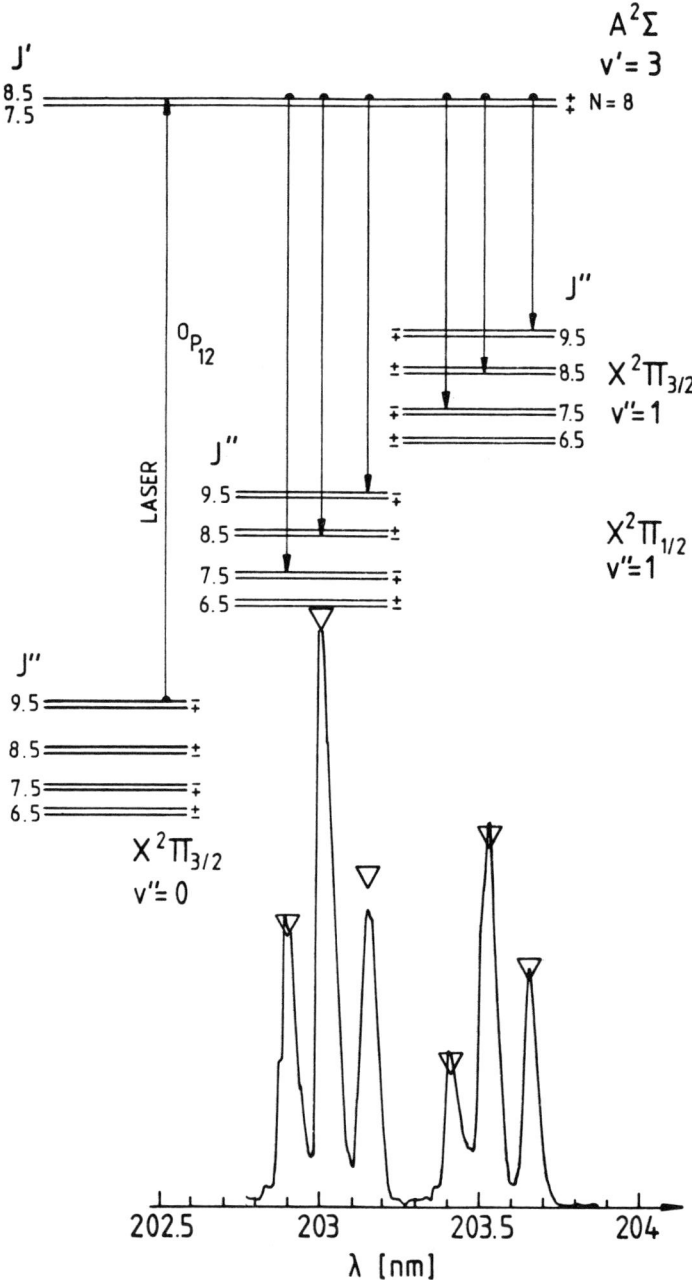

FIG. 5. Measurements of the Hönl-London factors obtained from fluorescence spectra of the (3, 0) band of the NO $A^2\Sigma^+ - X^2\Pi$ transition after state selective excitation.

chosen in such a manner that the transition of interest has still sufficient intensity and that the disturbing fluorescence signal either has already disappeared or has not yet been built up.

Figure 6 shows an example of a time resolved fluorescence spectrum which was taken on the $D^2\Sigma^+ - X^2\Pi$ system of the NO molecule (Scheingraber and Vidal, 1985a). Figure 6 contains three fluorescence spectra after pumping a $v = 0$ level of the $D^2\Sigma^+$ state. A gate width of 20 nsec was used. The upper trace shows the fast $D^2\Sigma^+ - X^2\Pi$ fluorescence ($D^2\Sigma^+$ state lifetime ≈ 20 nsec) which was measured without any delay of the gate. The middle trace shows the fluorescence taken with a delay of 160 nsec after the excitation pulse of the VUV laser. At that time the fluorescence is dominated by the $A^2\Sigma^+ - X^2\Pi$ transition ($A^2\Sigma^+$ state lifetime ≈ 200 nsec). The $A^2\Sigma^+$ state was populated by a fluorescence from the $D^2\Sigma^+$ state, whereas the $D^2\Sigma^+$ state has relaxed almost completely as indicated by the small remaining signal amplitude in the middle trace. A very similar technique was recently demonstrated for the time-delayed photoionization of NO (Rottke and Zacharias, 1985b) where again states of widely different lifetimes could easily be separated.

The lower trace in Fig. 6, finally is included as a warning. It was taken with the same time window as the upper trace. However, the pump intensity was raised by a factor of 4 which was enough for causing stimulated emission from the $D^2\Sigma^+$ to the $A^2\Sigma^+$ state. In this case the population density in the $A^2\Sigma^+$ state builds up almost instantaneously. Hence, in this situation the fluorescence from the $D^2\Sigma^+$ state can no longer be separated from that of the $A^2\Sigma^+$ state by selecting an appropriate time window.

C. Lifetime Measurements

As a natural extension of the preceding time dependent measurements it is obvious that lifetime measurements can also be carried out in the VUV spectral region on individual levels of a particular electronic state. For sufficiently small intensities the radiative lifetime due to spontaneous emission from some level $|p\rangle$ is defined by

$$\frac{1}{\tau_p} = \sum_{q(<p)} A_{pq} \tag{3}$$

where A_{pq} is the Einstein A-coefficient for all transitions between level $|p\rangle$ and the levels $|q\rangle$. In general, the lifetimes for all levels of an unperturbed electronic state are identical except for a weak v^{-3} dependence which in most cases can be neglected. However, in case of perturbed states the lifetimes may

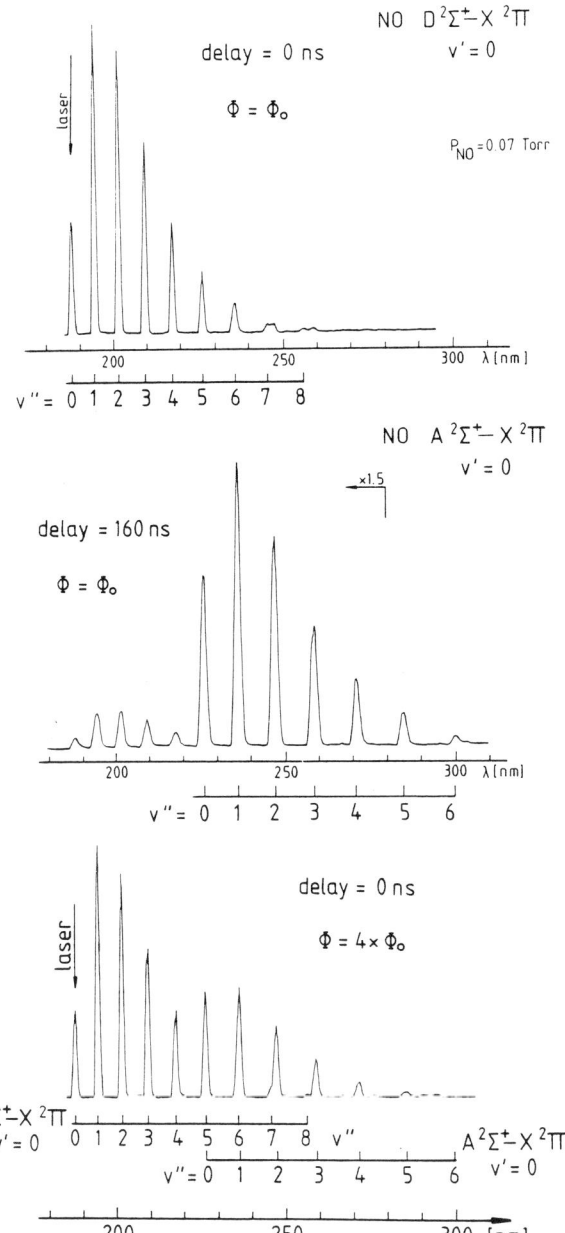

FIG. 6. Time-resolved fluorescence spectra following the excitation of a $v = 0$ level of the $D^2\Sigma^+$ state. All spectra were taken with a gate width of 20 nsec. The upper trace shows the $D^2\Sigma^+ - X^2\Pi$ fluorescence taken with no delay, the middle trace the $A^2\Sigma^+ - X^2\Pi$ fluorescence with a delay of 160 nsec. The lower trace was taken with an enhanced pump intensity giving rise to stimulated emission from the $D^2\Sigma^+$ state to the $A^2\Sigma^+$ state.

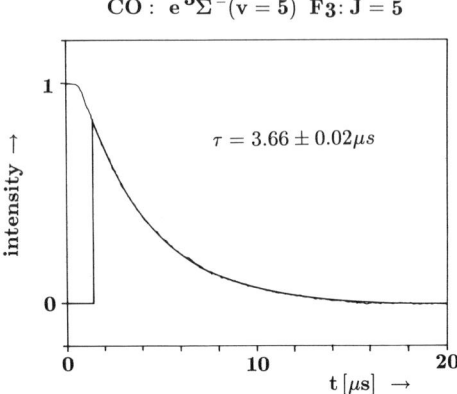

FIG. 7. Fluorescence decay from an individual $v = 5$, $J = 5$ level of the CO $e\,^3\Sigma^-$ state as measured by means of a transient digitizer after averaging over 500 pulses.

differ widely from one level to the next, depending on the interaction between different electronic states. As a specific example of an interaction between singlet (S) and triplet (T) states the lifetime of a perturbed singlet or triplet state is given by

$$\frac{1}{\tau_S} = \frac{1-\beta_S^2}{\tau_S^0} + \frac{\beta_S^2}{\tau_T^0} \quad \text{and} \quad \frac{1}{\tau_T} = \frac{1-\beta_T^2}{\tau_T^0} + \frac{\beta_T^2}{\tau_S^0} \qquad (4)$$

where $\tau_{S,T}^0$ is the lifetime of the unperturbed singlet or triplet state and $|\beta_{S,T}|^2$ the interaction or mixing coefficient between the two electronic states.

One of the first experiments in the VUV spectral range pumping individual levels, was carried out on the $A^1\Pi$ state of the CO molecule.[63] Perturbations between the singlet and the triplet states of the CO molecule could be detected by the J-dependence of the lifetimes which according to Eq. (4) show variations by as much as a factor of 2 for singlet states (Provorov et al., 1977; Maeda and Stoicheff 1984). These perturbations have recently been observed more clearly by measuring the lifetime of individual levels, not of the singlet states, but of the interacting triplet states after an excitation through one of the weaker intercombination lines (see Fig. 4 (Strobl and Vidal, 1987). Variations in the lifetime by as much as two orders of magnitude have been detected. Figure 7 shows a decay signal obtained with a charge sensitive amplifier and a transient digitizer after averaging over 500 laser pulses. These measurements reveal a very pronounced dependence of the lifetimes on the vibrational quantum number v and the total angular momentum J (Strobl and Vidal, 1987). This is shown in Fig. 8. The large variations between the lifetimes of neighboring rotational vibrational levels of the triplet states has

---

[63] Girard et al., 1982; Provorov et al., 1977; Maeda and Stoicheff, 1984.

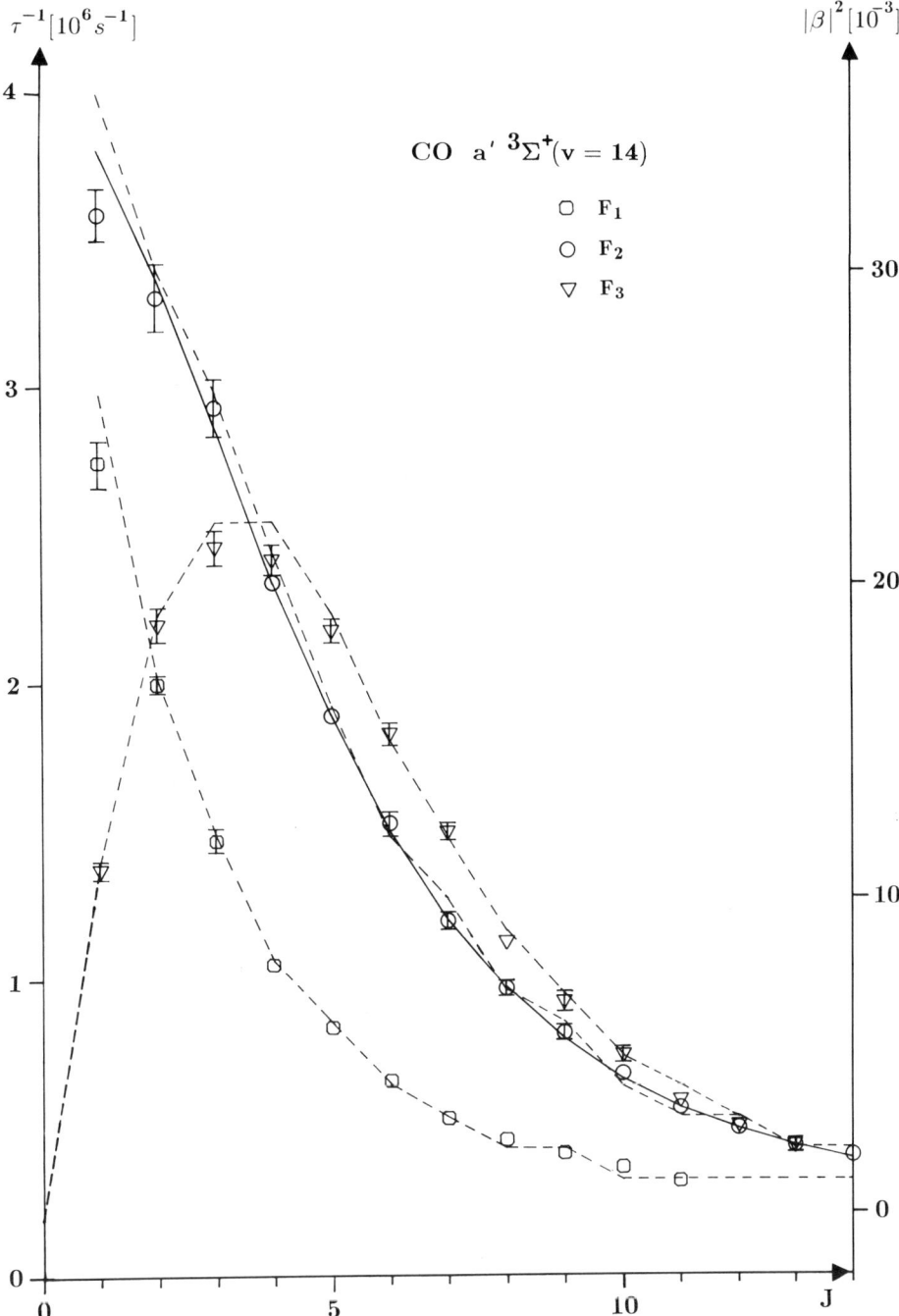

FIG. 8. Measured and calculated radiative lifetimes for the three different fine structure components of the $a'\ ^3\Sigma^+(v=14)$ state which is perturbed by the $A\ ^1\Pi\ (v=4)$ state. The solid line is from a least squares fit of the measured radiative lifetimes.

not been observed previously. It gives rise to a very intricate pressure dependence because neighboring levels of different lifetimes act either as reservoirs or as sinks for the level of interest which is coupled by collisions. Hence, for more complicated situations accurate lifetimes can only be obtained under truely *collisionfree* conditions and by pumping *single* levels of a particular electronic state. This can be tested experimentally by observing a single exponential decay as shown in Fig. 7. Measurements which show more than one exponential decay, therefore, have to be regarded with great caution and generally indicate a more complicated radiative and/or collisional coupling.

Further lifetime measurements have also been performed on several levels of the NO $B'^2\Delta$ state (Banic *et al.*, 1981) which differed significantly from the results obtained after electron excitation. For a meaningful measurement of the Einstein A-coefficients from lifetime measurements a careful study of the branching ratios and an extensive test of the intensity and pressure dependence is required. By adding various gases at well-defined pressures quenching cross sections have been obtained for NO (Zacharias *et al.*, 1976; Shibuya and Stuhl, 1983).

VI. Photodissociation Spectroscopy

The VUV excitation of an atomic or molecular system can also be detected by its laser-induced photogragments, which result either from the neutral molecule or the subsequent ion. A typical case is the photofragmentation of NH_3 using an ArF (Donnelly *et al.*, 1979) or a xenon excimer laser (Hellner *et al.*, 1984) generating NH_2 and NH fragments.

Using an ArF laser, two-photon dissociation has been studied in CO (Bokor *et al.*, 1980) where $C(2^1D)$ atoms have been detected in a subsequent absorption of a third laser photon. The free carbon atoms then gave rise to an association process, forming C_2 molecules which have been detected by their Swan band emission.

Investigations of this kind are certainly key experiments in the understanding of photochemistry, where tunable VUV sources allow a state selective analysis of the reactants and reveal the dependence of the individual rates on the quantum numbers of the laser excited levels.

In this context it is useful to refer to a general review of Jortner and Leach (1980) on the perspectives of synchrotron radiation for the investigation of molecular dynamics and photochemistry. The review contains a comparison with other laboratory light sources. It is clear that lasers will soon excel over these light sources in almost any application. The biggest advantage of the synchrotron with respect to existing VUV laser sources still appears to be the wide and easy tunability down to a spectral region not yet accessible to VUV lasers. This option is particularly useful for low resolution survey spectra. For high-resolution, Doppler-limited spectroscopy in the near VUV spectral region, however, existing laser sources have already an adequate tuning range and are by far superior to conventional technology.

VII. Ionization Spectroscopy

Absorption of VUV radiation can also be observed through the formation of atomic or molecular ions which are amenable to mass-spectrometric detection and hence to isotopic analysis. As mentioned above, ionization has frequently been used for the detection of multiphoton excitation and experiments have been carried out on H_2,[64] CO,[65] NO,[66] and NH_3.[67] In a double-resonance experiment NO was excited in the infrared and subsequently ionized by means of multiphoton ionization (Esherick and Anderson, 1980).

Time-delayed ionization can provide the excited state lifetimes if they are longer than the exciting laser pulse. Very short lifetimes, on the other hand, can be determined in the frequency domain by tuning the excitation laser across the excitation line while keeping the ionization laser fixed in frequency and intensity. Intermediate ionization rates have to be determined from a detailed comparison with the complete system of rate equations.

An experiment in which the excitation was not done by multiphoton excitation, but by a direct VUV single-photon process, was reported for CO (Zacharias et al., 1980b) and NO (Zacharias et al., 1980b; Rottke and Zacharias, 1985b) where the third or fourth harmonic of a Nd:YAG laser was used to subsequently ionize the excited states. By saturating the ionization step, photoionization cross sections of individual rotational levels of excited states have been measured. A similar experiment has recently been carried out on H_2 (Meier et al., 1985) in which excitation spectra and autoionizing resonances have been observed. In another experiment, rotationally cooled

---

[64] Marinero et al., 1982; Marinero et al., 1983a.
[65] Jones et al., 1982; Pratt et al., 1983a; Loge et al., 1983; Sha et al., 1984a; Rottke and Zacharias, 1985a.
[66] Johnson et al., 1975; Zacharias et al., 1980a; Zacharias et al., 1981a; Ebata et al., 1982; Sirkin et al., 1982; Ebata et al., 1984; Anezaki et al., 1985; Verschuur et al., 1986.
[67] Glownia et al., 1980b.

NO from a pulsed nozzle beam was ionized by means of the third harmonic of an excimer laser-pumped dye laser (Miller et al., 1985).

VIII. Two-Step Excitation Spectroscopy

With the advent of suitable coherent VUV sources the two-step excitation technique has become a very attractive method for the detailed analysis of transitions between highly excited states which has already been exploited in a series of optical-optical double resonance experiments in the visible and the infrared part of the spectrum (Demtröder 1981).

A. PRINCIPLE

For some molecular systems Fig. 9 illustrates schematically such a two-step excitation process. In the first step the molecule is excited by the VUV

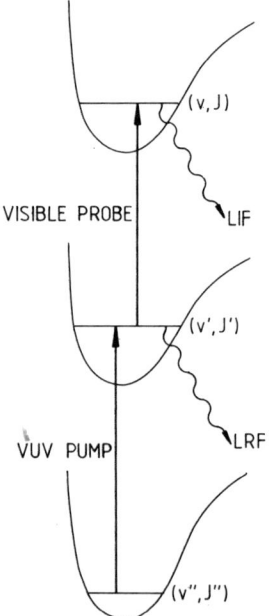

FIG. 9. Principle of the two-step excitation spectroscopy indicating the laser induced fluorescence (LIF) from the final upper level and the laser reduced fluorescence (LRF) from the intermediate level.

radiation from its ground state to some intermediate state. In a second step radiation which can be again in the visible part of the spectrum, excites the molecule from the intermediate state to some higher excited electronic state. The selection rules for such a process turn the two-step excitation into a highly selective method. The first two-step excitation in the VUV spectral region was performed on the NO molecule (Cheung et al., 1983) where the initial step was carried out by means of a two-photon excitation and where the second step gave access to high-lying Rydberg states of the NO molecule. The first experiment in which the first step was carried out by a tunable VUV source in a one-photon excitation, was a two-step excitation on the CO molecule (Klopotek and Vidal, 1985).

As indicated in Fig. 9, the intermediate as well as the final level are both capable of radiating back to a lower electronic state. However, there is a significant difference between the fluorescence from the two levels. Keeping the wavelength of the VUV pump laser fixed, a steady fluorescence occurs from the intermediate level. In this case the second excitation step of the probe laser can be detected in two different ways. First of all, the probe laser will reduce the population density of the intermediate level giving rise to a laser-reduced fluorescence (LRF). Secondly, the probe laser will cause significant population in the final upper level which can be observed by the subsequent laser-induced fluorescence (LIF). The sensitivity of both kinds of fluorescence signals depends strongly on the quantum efficiency of the particular excited level, which can be very low, for example, in case of severe predissociation. The idea of the laser-reduced fluorescence was first shown in the visible part of the spectrum (Feinberg et al., 1977; Bergmann et al., 1980) and was first demonstrated in the VUV spectral region on the CO molecule (Klopotek and Vidal, 1985). A laser-reduced fluorescence in the VUV was also reported in a double resonance experiment (Moutard et al., 1985) in which a BRV broad band continuum source was used for the initial excitation step. For comparison with the properties of the coherent VUV sources mentioned above, the BRV source was operated in connection with a monochromator giving a linewidth of 2000 cm^{-1} at 120 nm with 10^9 photons per shot.

Figure 10 shows a two-step excitation spectrum in which the VUV laser pumps a single line of the CO $A^1\Pi - X^1\Sigma^+$ transition. Depending on the parity of the intermediate level the visible probe laser pumps either a doublet of P- and R-lines or a single Q-line. By pumping individual lines of different isotopic species (see Fig. 3), it has been possible to measure the isotope shifts for transitions between excited electronic states (Klopotek and Vidal, 1985) which were already investigated in a more indirect manner by Kepa et al. (1975, 1978).

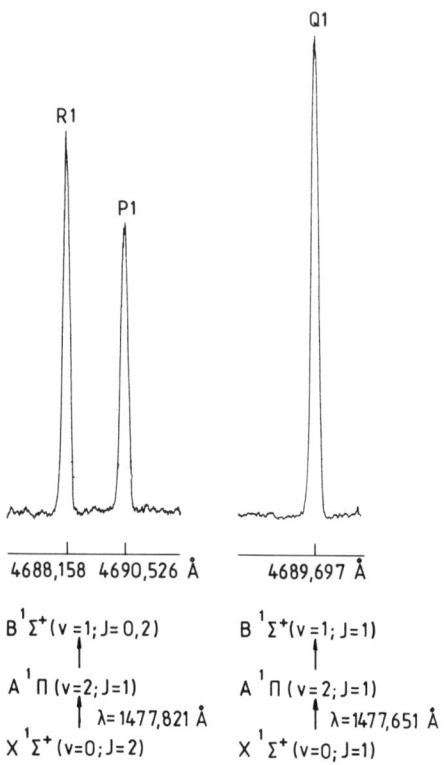

FIG. 10. Two-step excitation spectrum of the CO $X^1\Sigma^+ - A^1\Pi - B^1\Sigma^+$ system showing the LIF signal from the final $B^1\Sigma^+$ state.

B. ANALYSIS OF PERTURBATIONS

A powerful application of the two-step excitation spectroscopy has been demonstrated in the detailed analysis of perturbations in the CO $E^1\Pi$ state (Klopotek and Vidal, 1985). Figure 11 shows part of a two-step excitation spectrum for the $X^1\Sigma^+ - A^1\Pi - E^1\Pi$ system. By pumping in the first excitation step either the $R(29)$ or the $Q(30)$ line, it was possible to alternately pump either one of the two Λ-type components of the intermediate level with $v = 4$ and $J' = 30$ which have opposite parity. The probe signals of the second excitation step are shown in Fig. 11 and reveal a very interesting behavior. In pumping the $J' = 30$ level of negative parity, normal LIF and LRF signals are observed where only one component of the complete triplet of lines for a $^1\Pi - {}^1\Pi$ transition is shown in Fig. 11. A completely different

VACUUM-ULTRAVIOLET LASER SPECTROSCOPY 27

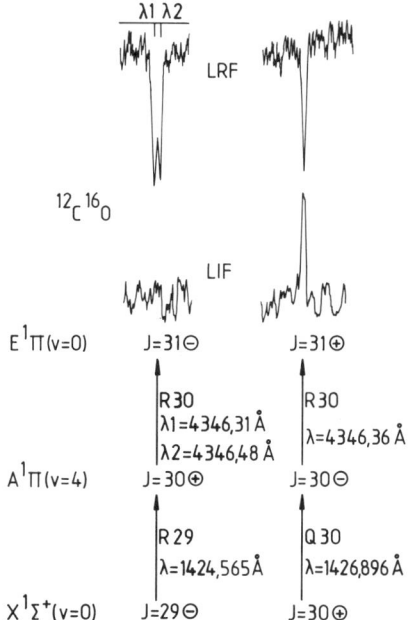

FIG. 11. Two-step excitation spectrum of the CO $X^1\Sigma^+ - A^1\Pi - E^1\Pi$ system showing a perturbation of the $|v = 0, J = 31\rangle$ level of negative parity in the $E^1\Pi$ state as manifested by the vanishing LIF signal and the doublet in the LRF signal originating from the perturbed and the perturbing level. For comparison, the other component of positive parity shows normal LIF and LRF signals.

situation is encountered in pumping the $J' = 30$ level of positive parity. In this case the LIF signal from the final $E^1\Pi$ state has vanished completely due to a severe accidental predissociation (Simmons and Tilford, 1974), whereas the LRF signal shows a closely spaced doublet originating from the perturbed level and the perturbing level. Due to the selection rules for perturbations it was possible to identify for the first time the perturbing state as a $^1\Sigma^+$ state (Klopotek and Vidal, 1985).

C. PREDISSOCIATION

Another interesting application of the two-step excitation is shown in Fig. 12 for the $X^1\Sigma^+ - A^1\Pi - B^1\Sigma^+$ system. By measuring the LIF and LRF signals as a function of J one notices a strong enhancement of the LRF signal and a strong reduction of the LIF signal at $J > 17$ for the $^{12}C^{16}O$ isotope.

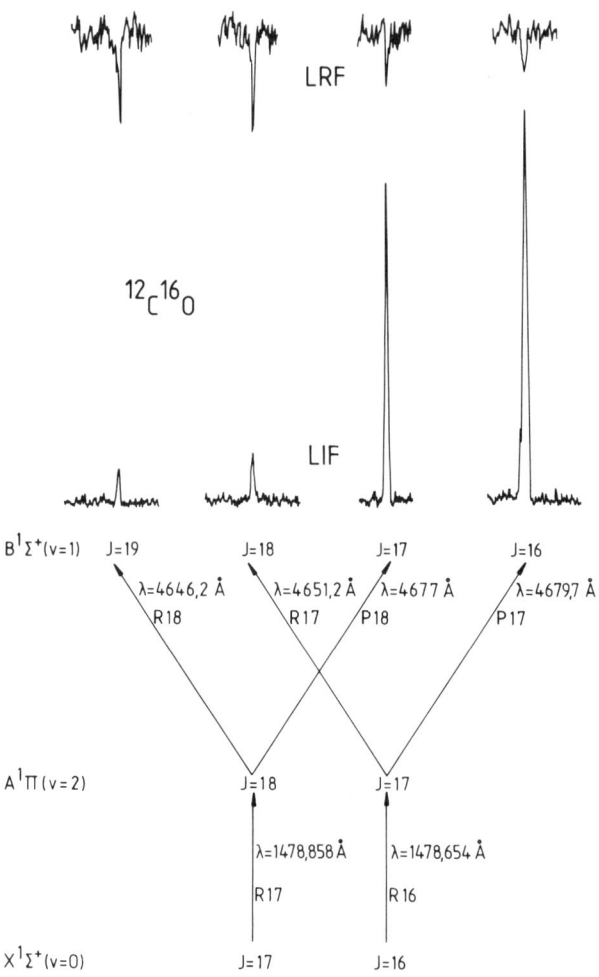

FIG. 12. Two-step excitation spectrum of the CO $X^1\Sigma^+ - A^1\Pi - B^1\Sigma^+$ system showing the predissociation of the $B^1\Sigma^+$ state with $v = 1$ for $J > 17$ as manifested by the vanishing LIF signal and the enhanced LRF signal.

This is a clear indication of a predissociation which reduces the quantum efficiency of the final state and hence the LIF signal. The LRF signal, however, is increased because the probe transition can no longer be saturated. As a result, the population density of the intermediate state is depleted more rapidly and the fluorescence signal from the intermediate level is reduced more effectively. Similar experiments on other isotopic species (Klopotek and Vidal, 1985) have shown a predissociation at the same energy

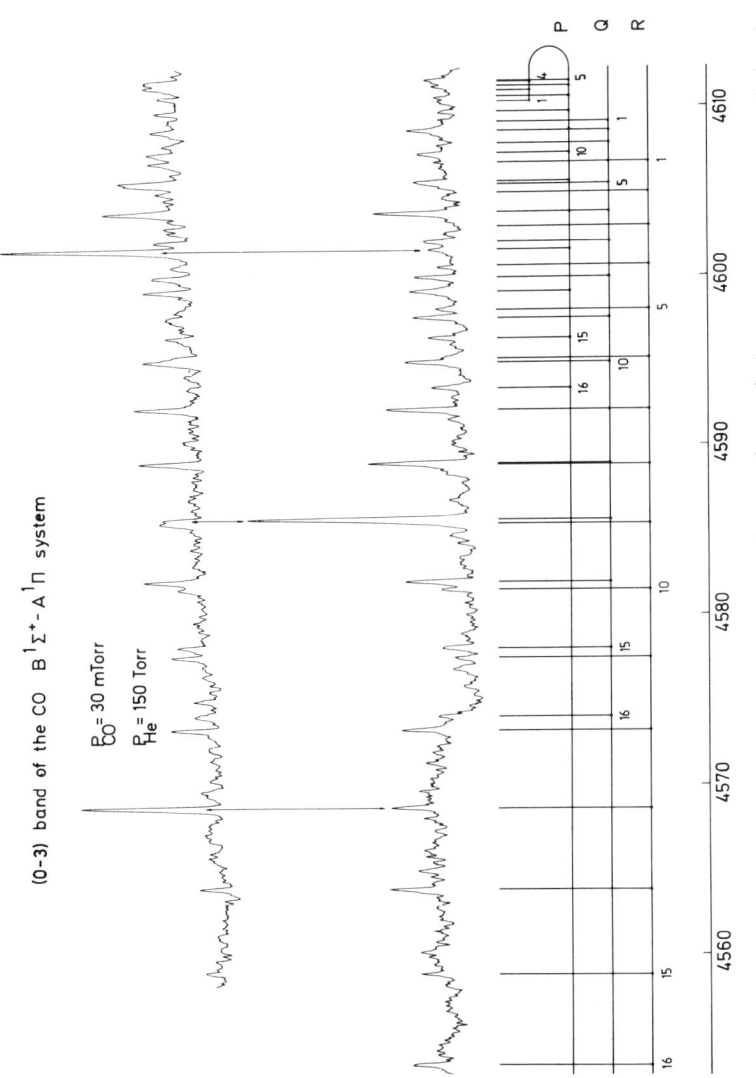

FIG. 13. Two-step excitation spectrum of the CO $X^1\Sigma^+ - A^1\Pi - B^1\Sigma^+$ system showing collision induced transitions of the (0, 3) band of the $B^1\Sigma^+ - A^1\Pi$ system.

for different values of J in agreement with the Born-Oppenheimer approximation. Related predissociation experiments have led in the past to the most accurate determination of the dissociation energy of the CO ground state (Douglas and Møller, 1955). A similar kind of predissociation was recently observed on the $B^2\Pi$ state of the NO molecule (Rottke and Zacharias, 1985b) by the breaking-off of the laser induced fluorescence and the absence of lines in the time-delayed photoionization spectrum.

D. COLLISION-INDUCED TRANSITIONS

In carrying out two-step experiments on the $X^1\Sigma^+ - A^1\Pi - B^1\Sigma^+$ system at different pressures, collision-induced double resonances were observed (Klopotek and Vidal, 1985). An example is shown in Fig. 13. The relative line intensities provide relative population densities which can be associated with individual rate constants in a system of rate equations (Vidal, 1978). In this manner the corresponding cross sections can be extracted. Similar experiments have been performed on the CO molecule (Sha et al., 1984b) and on the NO molecule (Ebata et al., 1984) where the two-step excitation was detected by means of the final ionization.

Another interesting process of laser-assisted collisions, which has already been seen on several atomic transitions in the visible (Green et al., 1979a, 1979b), has also been observed in the VUV spectral region on the CO molecule (Lukasik and Wallace, 1981). This effect was seen for laser intensities as high as 10^{10} W/cm^2 and larger, and it illustrates very clearly how cautious one has to be in interpreting high-intensity measurements. This was pointed out already in connection with the interpretation of multiphoton excitation and third harmonic experiments.

E. FINE STRUCTURE MEASUREMENTS

A further useful application of the two-step excitation is shown in Fig. 14 where the VUV source pumps a line of an intercombination band (see Fig. 4). In this manner a two-step excitation can be performed on the $X^1\Sigma^+ - a'\,^3\Sigma^+ - c\,^3\Pi$ system of the CO molecule (Klopotek and Vidal, 1985). This measurement gave the first accurate determination of the fine structure splitting of the CO $c\,^3\Pi$ state, which was difficult to get from conventional emission or absorption spectroscopy (Siwiec-Rytel, 1983). In the two-step excitation, however, it was possible to select the level of the $c\,^3\Pi$ state with the largest fine structures splitting.

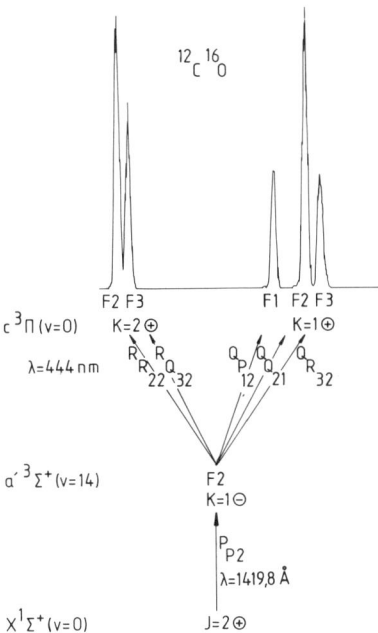

FIG. 14. Two-step excitation spectrum of the CO $X^1\Sigma^+ - a'\,^3\Sigma^+ - c\,^3\Pi$ system showing the fine structure splitting of the $c\,^3\Pi$ state for $N = 1$.

IX. Vacuum UV Multiphoton Spectroscopy

Finally, multiphoton experiments should be mentioned in which the wavelength of the incident high power laser beam lies also in the VUV spectral region and which are different from the experiments mentioned above in Sect. II where the pump laser is in the visible part of the spectrum. This method was mainly explored by Rhodes and coworkers in Chicago, who used high-power lasers of small linewidth by combining the methods of nonlinear optics with excimer laser amplifiers. With these sources of very high spectral brightness (up to 10^{15} W/cm^2) multiply charged ions of different atoms (Luk et al., 1983; Boyer et al., 1984) have been generated under collision free conditions. These results are not yet fully understood. By comparing the different laser-induced stages of ionization it is suspected that the ionization involves a collective response of the atom (Boyer et al., 1984). Similar high-power experiments in molecular hydrogen (Pummer et al., 1983; Luk et al., 1985) gave rise to strong stimulated emission with an energy conversion efficiency of 0.5%. As already mentioned above in Section II, it was noticed that at an intensity of $6 \cdot 10^{11}$ W/cm^2 the Stark effect plays an important role and lineshifts as large as 45 cm^{-1} were observed. Similarly, these high intensities give rise, to strong harmonic signals which have been seen in various atomic and molecular gases (Srinivasan et al., 1983; Plummer et al., 1980).

X. Summary

This review presents a short coverage of the various methods which have recently emerged with the advent of tunable coherent vacuum ultraviolet sources. In view of the large variety of methods developed in laser spectroscopy in the visible and infrared part of the spectrum (Demtröder, 1981), it is easy to anticipate a wealth of new information in the vacuum ultraviolet part of the spectrum. Besides detailed high-resolution spectroscopy, which has been the main subject of this review, interesting results are expected in different fields such as state-selective photochemistry and plasma diagnostics, which are both of particular interest in the vacuum ultraviolet spectral region.

ACKNOWLEDGEMENTS

The author is very grateful to his collaborators, in particular Drs. Scheingraber, Klopotek and Strobl who provided numerous discussions and without whom many measurements covered in this review, would not exist.

REFERENCES

Anezaki, Y., Ebata, T., Mikami, N., and Ito, M. (1985). *Chem. Phys.* **97**, 153.
Armstrong, J. A., Bloembergen, N., Ducuing, J., and Pershan, P. S. (1962). *Phys. Rev.* **127**, 1918.
Asscher, M., and Haas, Y. (1978). *Chem. Phys. Lett.* **59**, 231.
Banic, J. R., Lipson, R. H., Efthimiopoulos, T., and Stoicheff, B. P. (1981). *Opt. Lett.* **6**, 461.
Bergeman, T., and Zare, R. N. (1974). *J. Chem. Phys.* **61**, 4500.
Bergmann, K., Hefter, U., and Witt, J. (1980). *J. Chem. Phys.* **72**, 4777.
Bernheim, R. A., Kittrell, C., and Veirs, D. K. (1978). *J. Chem. Phys.* **69**, 1308.
Bokor, J., Freeman, R. R., White J. C., and Storz, R. H. (1981). *Phys. Rev. A* **24**, 612.
Bokor, J., Zavelovich, J., and Rhodes, C. K. (1980). *J. Chem. Phys.* **72**, 965.

Boyer, K., Egger, H., Luk, T. S., Pummer, H., and Rhodes, C. K. (1984). *J. Opt. Soc. Am. B* **1**, 3.
Brau, C. A. (1979). "Rare gas halogen excimers." *In* "Excimer Lasers" (C. K. Rhodes, ed.), *Topics in Applied Physics* **30**, pp. 87–134. Springer, Heidelberg.
Bray, R. G., Hochstraser, R. M., and Sung, H. N. (1975). *Chem. Phys. Lett.* **33**, 1.
Bray, R. G., Hochstrasser, R. M., and Wessel, J. E. (1974). *Chem. Phys. Lett.* **27**, 167.
Cheung, W. Y., Chupka, W. A., Colson, S. D., Gauyacq, D., Avouris, P., and Wynne, J. J. (1983). *J. Chem. Phys.* **78**, 3625.
Demtröder, W. (1981). "Laser Spectroscopy." *Springer Series in Chemical Physics*, **5**. Springer, Heidelberg.
Döbele, H. F., Röwekamp, M., and Rückle, B. (1984). *IEEE J. Quant. Electron.* **Q E-20**, 1284.
Donnelly, V. M., Baronavsky, A. P., and McDonald, J. R. (1979). *Chem. Phys. Lett.* **43**, 271.
Douglas, A. E., and Møller, C. K. (1955). *Can J. Phys.* **33**, 125.
Earls, L. T. (1935). *Phys. Rev.* **48**, 423.
Ebata, T., Abe, H., Mikami, N., and Ito, M. (1982). *Chem. Phys. Lett.* **86**, 445.
Ebata, T., Anezaki, Y., Fujii, M., Mikami, N., and Ito, M. (1984). *Chem. Phys. Lett.* **84**, 151.
Ebata, T., Mikami, N., and Ito, M. (1983). *J. Chem. Phys.* **78**, 1132.
Egger, H., Luk, T. S., Boyer, K., Muller, D. F., Plummer, H., Srinivasan, T., and Rhodes, C. K. (1982). *Appl. Phys. Lett.* **41**, 1032.
Egger, H., Srinivasan, T., Hohla, K., Scheingraber, H., and Vidal, C. R. (1981). *Appl. Phys. Lett.* **39**, 37.
Esherick, P., and Anderson, R. J. M. (1980). *Chem. Phys. Lett.* **70**, 621.
Faisal, F. H. M., Wallenstein, R., and Zacharias, H. (1977). *Phys. Rev. Lett.* **39**, 1138.
Feinberg, R., Teets, R. E., Rubbmark, J., and Schawlow, A. L. (1977). *J. Chem. Phys.* **66**, 4330.
Ferrell, W. R., Chen, C. H., Payne, M. G., and Willis, R. D. (1983). *Chem. Phys. Lett.* **97**, 460.
Filseth, S. V., Wallenstein, R., and Zacharias, H. (1977). *Opt. Commun.* **23**, 231.
Freedman, P. A. (1977). *Can. J. Phys.* **55**, 1387.
Friederich, D. M., and McClain, W. M. (1980). *Ann. Rev. Phys. Chem.* **31**, 559.
Gelbwachs, J. A., Jones, P. F., and Wessel, J. E. (1975). *Appl. Phys. Lett.* **27**, 551.
Gerstenkorn, S, and Luc, P. (1979). *Rev. Phys. Appl.* **14**, 791.
Gerstenkorn, S., and Luc, P. (1978). "Atlas du spectre d' absorption de la molécule d'iode." CNRS, Paris.
Girard, B., Billy, N., Vigue, J., and Lehmann, J. C. (1982). *Chem. Phys. Lett.* **92**, 615.
Girard, B., Billy, N., Vigue, J., and Lehmann, J. C. (1983). *Chem. Phys. Lett.* **102**, 168.
Glownia, J. H., and Sander, R. K. (1982). *Appl. Phys. Lett.* **40**, 648.
Glownia, J. H., Riley, S. J., Colson, S. D., and Nieman, G. C. (1980a). *J. Chem. Phys.* **72**, 5958.
Glownia, J. H., Riley, S. J., Colson, S. D., and Nieman, G. C. (1980b). *J. Chem. Phys.* **73**, 4296.
Green, W. R., Lukasik, J., Willison, J. R., Wright, M. D., Young, J. F., and Harris, S. E. (1979a). *Phys. Rev. Lett.* **42**, 970.
Green, W. R., Wright, M. D., Lukasik, J., Young, J. F., and Harris, S. E. (1979b). *Opt. Lett.* **4**, 265.
Guest, J. A., and Lee, L. C. (1981). *J. Phys. B* **14**, 3401.
Halpern, J. B., Zacharias, H., and Wallenstein, R. (1980). *J. Mol. Spectrosc.* **79**, 1.
Hancock, G., and Zacharias, H. (1981). *Chem. Phys. Lett.* **82**, 402.
Hellner, L., and Lukasik, J. (1984). *Opt. Commun.* **51**, 347.
Hellner, L., Grattan, K. T. V., and Hutchinson, M. H. R. (1984). *J. Chem. Phys.* **81**, 4389.
Hikida, T., Washida, N., Nakajima, S., Yagi, S., Ichimura, T., and Mori, Y. (1975). *J. Chem. Phys.* **63**, 5470.
Hikida, T., Yagi, S., and Mori, Y. (1980). *Chem. Phys.* **52**, 399.
Hilbig, R., and Wallenstein, R. (1981). *IEEE J. Quant. Electron.* **Q E 17**, 1566.
Hougen, J. T. (1970), *Natl. Bur. Stand. Monogr.* **115**.
Hutchinson, H. R., and Thomas, K. J. (1983). *IEEE J. Quant. Electron.* **Q E-19**, 1823.
Innes, K. K., Stoicheff, B. P., and Wallace, S. C. (1976). *Appl. Phys. Lett.* **29**, 715.

Jamroz, W., and Stoicheff, B. P. (1983). "Generation of tunable coherent vacuum ultraviolet radiation." *In* "Progress in Optics" (E. Wolf, ed.), vol. 20, pp. 326–380. North Holland, Amsterdam.
Johnson, P. M., and Otis, C. E. (1981). *Ann. Rev. Phys. Chem.* **32**, 139.
Johnson, P. M., Berman, M. R., and Zakheim, D. (1975). *J. Chem. Phys.* **62**, 2500.
Jones, W., Sivakumar, N., Rockney, B. H., Houston, B. H., and Grant, E. R. (1982). *Chem. Phys. Lett.* **91**, 271.
Jortner, J., and Leach, S. (1980). *H. de Chim. Physique* **77**, 7.
Kepa, R., Knot-Wisniewska, M., and Rytel, M. (1975). *Act. Phys. Pol. A* **48**, 819.
Kepa, R., Rytel, M., and Rzeszut, Z. (1978). *Act. Phys. Pol. A* **54**, 355.
Kittrell, C., Cameron, S., Butler, L., Field, R. W., and Barrow, R. F. (1983). *J. Chem. Phys.* **78**, 3623.
Klopotek, P., and Vidal, C. R. (1984). *Can J. Phys.* **62**, 1426.
Klopotek, P., and Vidal, C. R. (1985). *J. Opt. Soc. Am. B* **2**, 869.
Kunz, C. (1979). "Synchrotron Radiation," Topics in Current Physics, vol. 10. Springer, Heidelberg.
LaRocque, P. E., Lipson, R. H., Herman, P. R., and Stoicheff, B. P. (1986). *J. Chem. Phys.* **84**, 6627.
Lee, L. C., and Guest, J. A. (1981). *J. Phys. B* **14**, 3415.
Lipson, R. H., LaRocque, P. E., and Stoicheff, B. P. (1984). *Opt. Lett.* **9**, 402.
Lipson, R. H., LaRocque, P. E., and Stoicheff, B. P. (1985). *J. Chem. Phys.* **82**, 4470.
Loge, G. W., Tiee, J. J., and Wampler, F. B. (1983). *J. Chem. Phys.* **79**, 196.
Luk, T. S., Egger, H., Müller, W., Pummer, H., and Rhodes, C. K. (1985). *J. Chem. Phys.* **82**, 4479.
Luk, T. S., Pummer, H., Boyer, K., Shahidi, M., Egger, H., and Rhodes, C. K. (1983). *Phys. Rev. Lett.* **51**, 110.
Lukasik, J., and Wallace, S. C. (1981). *Phys. Rev. Lett.* **47**, 240.
Maeda, M., and Stoicheff, B. P, (1984). *AIP Conf. Proc.* **119**, 162.
Marinero, E. E., Rettner, C. T., and Zare, R. N. (1982). *Phys. Rev. Lett.* **48**, 1323.
Marinero, E. E., Vasudev, R., and Zare, R. N. (1983a). *J. Chem. Phys.* **78**, 692.
Marinero, E. E., Rettner, C. T., and Zare, R. N. (1983b). *Chem. Phys. Lett.* **95**, 486.
McCusker, M. V. (1979). "The rare gas excimers." *In* "Excimer Lasers" (C. K. Rhodes, ed.), *Topics in Applied Physics* **30**, pp. 47–86. Springer, Heidelberg.
Meier, W., Zacharias, H., and Welge, K. H. (1985). *J. Chem. Phys.* **83**, 4360.
Miller, J. C., and Compton, R. N. (1981). *J. Chem. Phys.* **75**, 22.
Miller, J. C., Compton, R. N., and Cooper, R. W. (1982). *J. Chem. Phys.* **76**, 3967.
Miller, P. J., Chen, P., and Chupka, W. A. (1985). *Chem. Phys. Lett.* **120**, 217.
Moutard, P., Laporte, P., Bon, M., Damany, N., and Damany, H. (1985). *Opt. Lett.* **10**, 538.
Northrup, F. J., Polanyi, J. C., Wallace, S. C., and Williamson, J. M. (1984). *Chem. Phys. Lett.* **105**, 34.
Pratt, S. T., Poliakoff, E. D., Dehmer, P. M., and Dehmer, J. L. (1983a). *J. Chem. Phys.* **78**, 65.
Pratt, S. T., Dehmer, P. M., and Dehmer, J. L. (1983b). *J. Chem. Phys.* **79**, 3234.
Provorov, A. C., Stoicheff, B. P., and Wallace, S. (1977). *J. Chem. Phys.* **67**, 5393.
Puell, H., and Vidal, C. R. (1976a). *Phys. Rev. A* **14**, 2225.
Puell, H., and Vidal, C. R. (1976b). *Opt. Commun.* **19**, 279.
Pummer, H., Egger, H., Luk, T. S., Srinivasan, T., and Rhodes, C. K. (1983). *Phys. Rev. A* **28**, 795.
Pummer, H., Srinivasan, T., Egger, H., Boyer, K., Luk, T. S., and Rhodes, C. K. (1980). *Opt. Lett.* **5**, 282.
Radler, K., and Berkowitz, J. (1978). *J. Opt. Soc. Am.* **68**, 1181.
Rothschild, M., Egger, H., Hawkins, R. T., Bokor, J., Pummer, H., and Rhodes, C. K. (1981). *Phys. Rev. A* **23**, 206.

Rothschild, M., Egger, H., Hawkins, R. T., Pummer, H., and Rhodes, C. K. (1980). *Chem. Phys. Lett.* **72**, 404.
Rottke, H., and Zacharias, H. (1985a). *Opt. Commun.* **55**, 87.
Rottke, H., and Zacharias, H. (1985b). *J. Chem. Phys.* **83**, 4831.
Scheingraber, H., and Vidal, C. R. (1982a). "Vacuum uv laser induced fluorescence of the NO molecule." *In* "Laser Techniques for Extreme Ultraviolet Spectroscopy" (T. J. McIlrath and R. R. Fremann, eds.), *AIP Conf. Proc.* **90**, 95.
Scheingraber, H., and Vidal, C. R. (1982b). *Rev. Scient. Instr.* **52**, 1010.
Scheingraber, H., and Vidal, C. R. (1983). *IEEE J. Quant. Electron.* Q **E-19**, 1747.
Scheingraber, H., and Vidal, C. R. (1985a). *J. Opt. Soc. Am. B* **2**, 343.
Scheingraber, H., and Vidal, C. R. (1985b). *J. Chem. Phys.* **83**, 3873.
Scheingraber, H., and Vidal, C. R. (1987). *AIP Conf. Proc.* **160**, 164.
Schomburg, H., Döbele, F., and Rückle, B. (1982). *Appl. Phys. B* **28**, 201.
Sha, G., Zhong, X., Zhao, S., and Zhang, C. (1984a). *Chem. Phys. Lett.* **110**, 405.
Sha, G., Zhong, X., Zhao, S., and Zhang, C. (1984b). *Chem. Phys. Lett.* **110**, 410.
Shibuya, K., and Stuhl, F. (1982). *J. Chem. Phys.* **76**, 1184.
Shibuya, K., and Stuhl, F. (1983). *Chem. Phys.* **79**, 367.
Simmons, J. D., and Tilford, S. G. (1974). *J. Mol. Spectrosc.* **49**, 167.
Sirkin, E. R., Asscher, M., and Haas, Y. (1982). *Chem. Phys. Lett.* **86**, 265.
Siwiec-Rytel, T. (1983). *J. Mol. Spectrosc.* **97**, 234.
Srinivasan, T., Egger, H., Pummer, H., and Rhodes, C. K. (1983). *IEEE J. Quant. Electron.* Q **E-19**, 1270 (1983).
Strobl, K. H., and Vidal, C. R. (1987). *J. Chem. Phys.* **86**, 62.
Taherian, M. R., and Slanger, T. G. (1984). *J. Chem. Phys.* **81**, 3796.
Tilford, S. G., and Simmons, J. D. (1972). *J. Phys. Chem. Ref. Data* **1**, 147.
Vallee, F., and Lukasik, J. (1982). *Opt. Commun.* **43**, 287.
Vallee, F., Wallace, S. C., and Lukasik, J. (1982). *Opt. Commun.* **42**, 148.
Verschuur, J. W. J., Kimman, J., Van Linden van den Neuvell, H. B., and Van der Wiel, M. J. (1986). *Chem. Phys.* **103**, 359.
Vidal, C. R. (1978). *Chem. Phys.* **35**, 215.
Vidal, C. R. (1980). *Appl. Opt.* **19**, 3897.
Vidal, C. R. (1986). "Four wave frequency mixing in gases." *In* "Tunable Lasers" (L. F. Mollenauer and J. C. White, eds.), *Topics in Applied Physics* **59**, pp. 19–75. Springer, Heidelberg.
Vidal, C. R., and Haller, F. B. (1971). *Rev. Scient. Instr.* **42**, 1779.
Wallace, S. C., and Innes, K. K. (1980). *J. Chem. Phys.* **72**, 4805.
Wallenstein, R. (1980). *Opt. Commun.* **33**, 119.
Wallenstein, R., and Hänsch, T. W. (1974). *Appl. Opt.* **13**, 1625.
Wallenstein, R., and Zacharias, H. (1978). *Opt. Commun.* **25**, 363.
Wallenstein, R., and Zacharias, H. (1980). *Opt. Commun.* **32**, 429.
Wynne, J. J., and Beigang, R. (1981). *Phys. Rev. A* **23**, 2736.
Zacharias, H., Halpern, J. B., and Welge, K. H. (1976). *Chem. Phys. Lett.* **43**, 41.
Zacharias, H., Schmiedl, R., and Welge, K. H. (1980a). *Appl. Phys.* **21**, 127.
Zacharias, H., Rottke, H., and Welge, K. H. (1980b). *Opt. Commun.* **35**, 185.
Zacharias, H., Rottke, H., and Welge, K. H. (1981a). *Appl. Phys.* **24**, 23.
Zacharias, H., Rottke, H., Danon, J., and Welge, K. H. (1981b). *Opt. Commun.* **37**, 15.
Zandee, L., and Bernstein, R. B. (1979). *J. Chem. Phys.* **71**, 1359.

FOUNDATIONS OF THE RELATIVISTIC THEORY OF ATOMIC AND MOLECULAR STRUCTURE

IAN P. GRANT and HARRY M. QUINEY

Department of Theoretical Chemistry
Oxford University
Oxford OX1 3TG, England

I. Introduction	37
II. Preliminaries	39
A. The Dirac Operator	39
B. Boundary Conditions as $r \to \infty$	43
C. Boundary Conditions at $r = 0$	44
III. From QED to Atomic Structure Theory	46
A. The Furry Bound Interaction Picture of QED	47
B. Perturbation Theory	49
C. The Two-Body Interaction Kernel	52
D. The "Standard" Model of Relativistic Atomic Structure Theory and the Hartree–Fock Approximation	54
E. The Open Shell Problem. MCDF and Other Approaches	56
F. Corrections to the Standard Model	59
IV. New Developments—Approximation by Finite Basis Sets	64
A. Principles of Basis Set Calculations in Atomic and Molecular Structure	65
B. The Matrix Dirac–Fock Equations Including the Breit Interaction	69
C. Choice of Basis Sets	71
D. Comparison of Methods	73
V. Outlook and Conclusions	81
Acknowledgements	83
References	83

I. Introduction

The use of relativistic quantum theory to treat problems in atomic physics has become commonplace in the last few years now that computer packages such as the multi-configuration Dirac–Fock systems of Desclaux (1975) and Grant *et al.* (1980) have become generally available. At the same time, critics such as Sucher (1980) have thrown doubt upon the whole basis of such calculations, claiming that relativistic many-electron Hamiltonians can *in principle* have no bound state solutions. In his opinion, this makes it essential to project out the "negative energy" states which inevitably appear in relativistic quantum theory (Foldy, 1956; Shirokov, 1958) to obtain "healthy Hamiltonians" (Sucher, 1985). The first object of this review, §III, is to show that the methods of relativistic quantum theory in current use are correct implementations of quantum electrodynamics in the Furry (1951) bound state interaction picture. As such, there are *no problems of principle* in accounting for the use of self-consistent field methods, in the treatment of many-electron correlation effects, or of renormalization of divergent radiative corrections.

There seem to be no problems of principle in computing the structure of many-electron atoms to arbitrary accuracy in relativistic quantum mechanics. Difficulties arise from purely practical considerations: truncation errors of finite difference or finite element schemes; rounding errors of arithmetic processes; and the number of configurational states that it is economically feasible to include in our wavefunctions. The dominant role of the atomic nucleus in the isolated atom means that the directional dependence of the wavefunction can be handled algebraically, so that only the dependence on the radial coordinate need be treated by numerical methods. In general, this simplification is not available for molecules, and the expansion of orbitals as linear combinations of suitable basis functions (Roothaan, 1951) has been the most popular way of dealing with the problem. Kim (1967) was the first to do this for atoms with but moderate success. His calculations were the first to be plagued by the phenomenon of "variational collapse," so-called because it was thought to be associated with the fact that the Dirac operator has an unbounded spectrum of negative energies. The need to overcome this and allied problems has dominated research in this field for a number of years.

We have shown in a series of articles that the unboundedness of the Dirac operator has only the most tenuous connection with these numerical problems. The second objective of this article is to summarize the conclusions of these articles which show that the difficulties disappear when the basis sets are chosen to reflect the analytic dependence of admissible solutions in the neighbourhood of the singular points of Dirac operators. The most important neighbourhood is the high field region around each nucleus; the behaviour at large distances from the nuclei is less critical. This is exactly what we should expect from our experience of simple boundary value problems: for example, the spectrum of an ideal plucked string depends on whether the endpoints are fixed or free (Lamb, 1960). We set out the mathematical elements required in §II, and make use of them to explain the principles of successful numerical calculation with finite basis sets in §IV.

The discussion of §III provides a link between the formalism of quantum electrodynamics and the simpler equations of relativistic atomic and molecular structure on which we focus in §IV. We also see how to include radiative corrections omitted from the relativistic atomic structure formalism, and how to make allowance for nuclear motions.

§IV gives a brief account of the software currently available for general-purpose calculation of atomic structure and properties before concentrating on a discussion of finite basis set methods. We explain the pathological behaviour experienced in many attempts to use finite basis sets, and describe highly accurate calculations that we have made for one-electron and many-electron systems at the Dirac–Fock level. The outlook for further work is the subject of §V.

II. Preliminaries

An understanding of the analytic behaviour of eigensolutions of the Dirac operator is fundamental to the construction of a rigorous description of the relativistic quantum mechanics of atoms and molecules. We shall assume that the reader is familiar with elementary accounts of Dirac's wave equation such as occur in books like Schiff (1968) or Messiah (1961). We shall use sign conventions and notation from the review *Relativistic Calculation of Atomic Structures* (Grant, 1970). Our equations throughout will be given in Hartree atomic units for which $m = 1$, $\hbar = 1$, $e^2/4\pi\varepsilon_0 = 1$, $c = \alpha^{-1} = 137.0360$; we shall sometimes insert these quantities explicitly if it helps to bring out the dimensionality.

A. The Dirac Operator

Although we shall work almost entirely with a nonrelativistic notation, it is helpful to remember that the natural setting for calculations in special relativity is Minkowski space. In some suitable reference frame we write the Minkowski space coordinates x^μ: $x^0 = ct$, $(x^1, x^2, x^3) = \mathbf{x}$; the scalar product of two such vectors, $x \cdot y$, is given by $x \cdot y = g_{\mu\nu} x^\mu y^\nu$ (summation over repeated μ, ν indices being understood) where the metric $g_{\mu\nu}$ has nonzero elements only for $\mu = \nu$ such that $g_{00} = 1 = -g_{11} = -g_{22} = -g_{33}$. Four-vectors x^μ with superscript indices are said to be contravariant; we define associated covariant four-vectors $x_\mu = g_{\mu\nu} x^\nu$, and in this notation, $x \cdot y = x_\mu y^\mu$; a scalar product will always be written in this form so that the metric $g_{\mu\nu}$ does not appear explicitly.

Let $A^\mu(x)$ be the four-potential of a classical electromagnetic field, so that

$$A^\mu(x) = \left(\frac{\Phi(x)}{c}, \mathbf{A}(x)\right) \tag{1}$$

where $\Phi(x)$ is the familiar scalar potential and $\mathbf{A}(x)$ the vector potential at the space-time point x. Dirac's wave equation for an electron, charge $-e$, in Hartree atomic units, can be written in the covariant form

$$[\gamma_\mu(p^\mu + eA^\mu) - mc]\psi(x) = 0. \tag{2}$$

Here the γ_μ are 4×4 matrices satisfying the conditions

$$\{\gamma_\mu, \gamma_\nu\} = \gamma_\mu\gamma_\nu + \gamma_\nu\gamma_\mu = 2g_{\mu\nu}, \tag{3}$$

and $p_\mu = i\partial/\partial x^\mu$. Premultiplying (2) with $c\gamma_0$ enables us to rewrite (2) in the more familiar nonrelativistic notation

$$\{i\partial/\partial t + e\Phi(x) - c\boldsymbol{\alpha} \cdot (\mathbf{p} + e\mathbf{A}(x)) - \beta mc^2\}\psi(x) = 0 \tag{4}$$

where we shall use the standard representation of the α and β matrices

$$\beta = \begin{bmatrix} 1 & 0 \\ 0 & -1 \end{bmatrix} \quad \boldsymbol{\alpha} = \begin{bmatrix} 0 & \boldsymbol{\sigma} \\ \boldsymbol{\sigma} & 0 \end{bmatrix}; \tag{5}$$

here $\boldsymbol{\sigma}$ are the usual 2×2 Pauli spin matrices, and 1 denotes the 2×2 identity matrix (Schiff 1968).

One-particle models provide a remarkably good first approximation for many processes in atoms and it is no coincidence that most many-body theories are built from one-body wavefunctions. In quantum electrodynamics textbooks, plane wave solutions of Dirac's equation for noninteracting free electrons are the commonest building blocks. However, solutions of Dirac's equation in some suitable classical potential are a better starting point for problems involving real atoms and molecules. For free atoms and molecules, it is natural to choose a model in which there is no magnetic field, $\mathbf{A}(x) = 0$, and the scalar potential $\Phi(\mathbf{x})$ does not depend upon time, and satisfies Poisson's equation

$$\nabla^2\Phi(\mathbf{x}) = -\frac{\rho_{ext}(\mathbf{x})}{\varepsilon_0} \tag{6}$$

where $\rho_{ext}(\mathbf{x})$ is a classical static charge density distribution. In practice, such an assumption implies that our preferred reference frame must be that of the nuclear centre of mass; that $\rho_{ext}(\mathbf{x})$ is the nuclear charge density distribution (either taken as a point charge or as a structure having finite size), possibly augmented by some screening potential representing the average effect of other electrons of the system. In the case of an atom in a molecule, this implies the Born–Oppenheimer approximation in which we neglect the nuclear motion. Naturally, we shall have to consider how to relax these restrictions later on.

The assumption of a static potential allows us to define stationary states of the form

$$\psi(x) = \psi_E(\mathbf{x})\exp(-iEt) \tag{7}$$

where E represents the energy of the electron. The wavefunction factorizes into a pure harmonic time-dependent factor multiplied by a function of the (three-space) coordinates only. Clearly,

$$[c\boldsymbol{\alpha}\cdot\mathbf{p} + \beta mc^2 - e\Phi(\mathbf{x})]\psi_E(\mathbf{x}) = E\psi_E(\mathbf{x}). \tag{8}$$

In general, a factorization of this type is destroyed if we make a Lorentz transformation to some other inertial frame; this restriction is something of which we must be aware in our development.

When we consider isolated atoms, we have no preferred direction, so that it is natural to assume that $\rho_{\text{ext}}(x)$ has spherical symmetry, and is a function of $r = |x|$ only. In this case $\psi_E(x)$ can be put into the form (Grant, 1970)

$$\psi_E(\mathbf{x}) = \frac{1}{r}\begin{pmatrix} P(r) & \chi_{\kappa,m}(\theta,\varphi) \\ iQ(r) & \chi_{-\kappa,m}(\theta,\varphi) \end{pmatrix}. \tag{9}$$

The dependence on the angular coordinates θ, φ in a spherical polar system is contained in the 2-spinors $\chi_{\pm\kappa,m}(\theta,\varphi)$, which are simultaneous eigenfunctions of the total angular momentum j^2, j_3 (eigenvalues $j(j+1)$ and m respectively) and of parity: the explicit definitions are

$$\kappa = -(j+\tfrac{1}{2}) \quad \chi_{\kappa,m} = \begin{bmatrix} \left[\dfrac{j+m}{2j}\right]^{1/2} Y_{j-1/2}^{m-1/2}(\theta,\varphi) \\ \left[\dfrac{j-m}{2j}\right]^{1/2} Y_{j-1/2}^{m+1/2}(\theta,\varphi) \end{bmatrix} \tag{10}$$

$$\kappa = +(j+\tfrac{1}{2}) \quad \chi_{\kappa,m} = \begin{bmatrix} -\left[\dfrac{j+1-m}{2j+2}\right]^{1/2} Y_{j+1/2}^{m-1/2}(\theta,\varphi) \\ \left[\dfrac{j+1+m}{2j+2}\right]^{1/2} Y_{j+1/2}^{m+1/2}(\theta,\varphi) \end{bmatrix}; \tag{11}$$

as usual, j takes values $\tfrac{1}{2}, \tfrac{3}{2}, \ldots$. The parity of $\chi_{\kappa,m}$ is that of the spherical harmonic Y_l^m appearing in its definition. For states representing electrons, the components $P(r)\chi_{\kappa,m}$ in (9) have the Schrödinger–Pauli wave function for a spin-$\tfrac{1}{2}$ particle as their nonrelativistic limit; the mathematical limit $c \to \infty$ corresponds to the physical assumption that light propagates instantaneously. The associated value of the orbital angular momentum $l = j - \tfrac{1}{2}a$, $a = \text{sign}(\kappa)$ can be used to label the state nlj instead of $n\kappa$.

The radial functions $P(r)$ and $Q(r)$ satisfy the reduced equations

$$u(r) = \begin{bmatrix} P(r) \\ Q(r) \end{bmatrix}, \quad \varepsilon = E - mc^2$$

$$h_D u(r) = \varepsilon u(r) \tag{12}$$

where

$$h_D = \begin{bmatrix} V(r) & c\left[-\dfrac{d}{dr} + \dfrac{\kappa}{r}\right] \\ c\left[\dfrac{d}{dr} + \dfrac{\kappa}{r}\right] & -2mc^2 + V(r) \end{bmatrix} \tag{13}$$

in which we have replaced $-e\Phi(r)$ by $V(r)$. The effect of the change of energy zero in (12) is to make a just bound electron have zero energy, so that we are normalizing to the conventional nonrelativistic zero of energy.

Solutions of equations (12) and (13) are investigated in detail in the standard texts (Messiah, 1961; Schiff, 1968). Thus, in the simplest case in which there is no external field, $V(r) = 0$, there are solutions of scattering type when $\varepsilon > 0$ ($E > mc^2$) and when $\varepsilon < -2mc^2$ ($E < -mc^2$). There are no solutions when ε is in the interval $(-2mc^2, 0)$. The solutions when $V(r) = -Z/r$, the pure Coulomb potential, have also been well studied. There are two continua of scattering type solutions as in the free particle case, $\varepsilon > 0$, $\varepsilon < -2mc^2$, and a set of discrete levels with a limit point at $\varepsilon = 0$ in $(-2mc^2, 0)$ when $Z < c$. These discrete energy levels, which are characterized by the quantum numbers n, κ, and are independent of the j_3 quantum number m, have energies given by Sommerfield's fine structure formula. The corresponding eigenfunctions are square integrable; that is to say, the total radial density

$$\int_0^\infty [|P(r)|^2 + |Q(r)|^2] dr \tag{14}$$

is finite. In the scattering case, the density is only locally square integrable; that is to say (14) diverges, although the integral will be finite if it is restricted to a finite interval of r. The quantum number n, the principal quantum number, takes integer values, $n = -\kappa, -\kappa + 1, \ldots$ when $\kappa < 0$, $n = \kappa + 1$, $\kappa + 2, \ldots$ when $\kappa > 0$; for fixed κ, the states n, κ are nondegenerate. As we shall see, these properties are shared by eigenfunctions of the mean field potentials $V(r)$ we are likely to encounter in atomic structure calculations, which smoothly interpolate between the limits $-1/r$ as $r \to \infty$ – seen by an electron at large distances from a neutral atom ($-Z/r$ for a positive ion) – and $-Z/r + \text{constant}$ as $r \to 0$ (when the nucleus is modelled as a point charge of strength $+Ze$). The behaviour of $V(r)$ as $r \to 0$ is modified if a more realistic model of the nuclear charge density is to be used. For example, ignoring any screening by the other electrons, a uniformly charged nucleus of radius a will give a potential $V(r) = (-3Z/2a)(1 - r^2/3a^2)$ for $r \leqslant a$, $-Z/r$ for $r > a$. Other model charge distributions give slightly different expressions for $r < a$, although the potentials they generate all have a finite value for $V(r)$ at the origin

B. Boundary Conditions as $r \to \infty$

In an isolated atom or ion, a single electron detached from the system will see an attractive potential tending to zero at least as fast as $-1/r$. It is then elementary to show that the two-component function $u(r)$ is asymptotically of the form $\exp(\pm \lambda r) \cdot [U_\lambda + O(1/r)]$ where U_λ is a two-component constant, and where

$$\lambda^2 = -2m\varepsilon\left(1 + \frac{\varepsilon}{2mc^2}\right). \tag{15}$$

Clearly, λ^2 is positive and λ is real when ε lies in the interval $(-2mc^2, 0)$; if we denote the *positive* root by λ, then only the solution with the negative exponent, $\exp(-\lambda r)$, will give a square integrable function. Thus requiring that the integral (14) be finite automatically selects functions which go to zero exponentially as $r \to \infty$. No further boundary condition is necessary.

However, when ε lies outside the interval $(-2mc^2, 0)$ on the real line, either $\varepsilon > 0$ (the "positive" continuum) or $\varepsilon < -2mc^2$ (the "negative" continuum), λ^2 is negative so that we can write

$$\lambda = -ip \tag{16}$$

where p can be identified with the momentum of the electron. This is most easily seen by rewriting (15) in terms of the unshifted energy E, (12); with a little rearrangement we obtain the familiar relation

$$E^2 - c^2 p^2 = m^2 c^4. \tag{17}$$

The negative sign in (16) is conventional, and permits analytic continuation of the eigenfunctions in the energy parameter ε from the discrete to the continuous spectrum in the Coulomb field case (Johnson and Cheng 1979, Lee and Johnson 1980). Solutions of Dirac's equation which go like $\exp(\pm ipr)$ for large r are only locally square integrable, that is

$$\int_a^b [|P(r)|^2 + |Q(r)|^2] dr \tag{18}$$

is finite for any finite interval (a, b), but diverges on $(0, \infty)$. Such solutions are of scattering type, a fact which will prove to have some importance later.

C. BOUNDARY CONDITIONS AT $r = 0$

The boundary conditions in the neighbourhood of $r = 0$ are more complex and depend on the model chosen to represent the nuclear charge distribution. We shall, for the sake of generality, assume that the potential $V(r)$ has a series expansion in ascending powers of r

$$V(r) = \frac{-Z_0}{r} - Z_1 - Z_2 r - \ldots \quad r \to 0. \tag{19}$$

For a point nucleus model, $Z_0 = Z$, the nuclear charge, and the remaining coefficients, if non-zero, can represent the effect of screening by orbital electrons. For a distributed nuclear charge model, for example, a uniform charge in $r < a$, with a potential

$$V_{\text{nuc}}(r) = -\frac{3Z}{2a} \cdot \left(1 - \frac{r^2}{3a^2}\right) \quad 0 \leqslant r < a,$$

$$= \frac{-Z}{r} \quad r > a$$

$-Z_1$ will represent $V_{\text{nuc}}(0)$, and higher terms of (19) will represent a combination of the higher terms of $V_{\text{nuc}}(r)$ with orbital screening terms.

Any solution of (12) and (13) can be expanded in a series of the form

$$u(r) = r^\gamma [u_0 + u_1 r + u_2 r^2 + \ldots] \tag{20a}$$

where the coefficients u_i have the form

$$u_i = \begin{bmatrix} p_i \\ q_i \end{bmatrix}. \tag{20b}$$

Inserting the expansions (19) and (20a) into (12) and (13) gives a set of equations for the coefficients; the lowest order term, $r^{\gamma-1}$, gives

$$Z_0 p_0 + c(\kappa - \gamma)q_0 = 0 \tag{21a}$$

$$c(\kappa + \gamma)p_0 + Z_0 q_0 = 0; \tag{21b}$$

the requirement that these have a consistent nontrivial solution leads to the indicial equation

$$\gamma^2 = \kappa^2 - \frac{Z_0^2}{c^2} \tag{21c}$$

for the exponent γ. There are now two cases to consider:

1. Coulomb Potential

Here we take $Z_0 = Z$, $Z_i = 0$, $i > 0$. Either of equations (21a, b) give the ratio

$$\frac{q_0}{p_0} = \frac{Z}{c(\gamma - \kappa)} = \frac{-c(\gamma + \kappa)}{Z} \tag{22}$$

where γ is one of the roots of (21c). Inserting (21a) into (18) with $a = 0$, we see that $u(r)$ is square integrable over $(0, b)$ if $\frac{1}{2} < \gamma \leq |\kappa|$. When $|\kappa| = 1$, this restricts Z to be less than 118. The solution with γ negative is then not square integrable over $(0, b)$ and must be rejected. Note that this is quite independent of the energy parameter ε and of the orbital electron distribution; it is purely a consequence of relativistic dynamics. If $Z > 118$, then solutions with $|\kappa| = 1$ behaving like $r^{-\gamma}$ become square integrable on $(0, b)$, and we have what is called the limit circle case (Richtmyer 1978) in which any linear combination of this solution with that behaving like $r^{+\gamma}$ satisfies the required conditions. We can eliminate the solution proportional to $r^{-\gamma}$ by imposing the physical requirement that the expectation of $V_{\text{nuc}}(r)$ be finite; this extends the range of Z for which we can obtain a square integrable solution near $r = 0$ up to $Z = c \simeq 137$ beyond which no square integrable solutions with $|\kappa| = 1$ exist. This argument extends to higher values of $|\kappa|$ in an obvious way.

2. Extended Nuclear Charge Models

The situation when $Z_0 = 0$ is a little more complicated and depends on $a = \text{sign}(\kappa)$, and so on the value of $l = j - \frac{1}{2}a$. The indicial equation (21c) reduces to $\gamma = \pm|\kappa|$, and it follows that only the upper sign will give a square integrable function. We find $q_0 = 0$, whilst p_0 can be chosen at will. In the next order, r^γ, $p_1 = 0$, $q_1/p_0 = (\varepsilon + Z_1)/c(2l + 3)$. Continuing in this way, we find

$$P(r) = p_0 r^{l+1}[1 + O(r^2)]$$
$$Q(r) = q_1 r^{l+2}[1 + O(r^2)]. \tag{23a}$$

(b) κ positive; $l = \kappa = j + \frac{1}{2}$; $l = 1, 2, \ldots$. This time, $p_0 = 0$, whilst q_0 can be chosen at will. We find $p_1/q_0 = (2mc^2 - \varepsilon - Z_1)/c(2l + 1)$ with $q_1 = 0$, and

$$P(r) = p_1 r^{l+1}\{1 + O(r^2)\}$$
$$Q(r) = q_0 r^l \{1 + O(r^2)\} \tag{23b}$$

For both signs of κ, the power series solution contains either all even or all odd powers of r and one of the two components, P for κ negative, Q for κ

positive, dominates as $r \to 0$. The relative amplitude of the other component depends on $\varepsilon + Z_1$, and so on both the energy parameter and on the strength of the potential at the origin.

III. From QED to Atomic Structure Theory

The spectra of the states of an electron in a typical atomic mean field potential in nonrelativistic and relativistic quantum mechanics are shown schematically in Figures 1(a) and 1(b) respectively. The most noticeable difference is the relativistic continuum for $\varepsilon < -2mc^2$. Clearly, the familiar *aufbau* principle, in which, with the aid of the Pauli (1925) exclusion principle, we build up the ground state of the neutral atom by inserting successive electrons into the states with lowest available energy until electrical neutrality is achieved, cannot work unmodified in the relativistic case. For we have first to fill the lower continuum, which requires an infinite number of particles, having infinite mass and charge. Dirac (1930) showed the way out of this

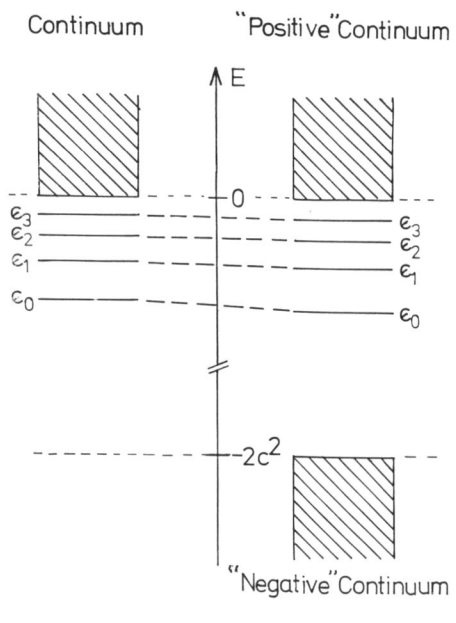

FIG. 1. Schematic spectrum of electron states in a typical mean atomic potential.

difficulty by postulating that these levels are always filled in the reference (vacuum) state and that they can be observed only by exciting an electron out of one into an unoccupied higher level. The vacancy so produced can be related to a positron state (with charge of the opposite sign to that of the electron) by charge conjugation (Messiah, 1961). The excitation energy is of order $2mc^2 \simeq 1$ MeV. Thus in low energy processes, one can proceed by ignoring the lower continuum, as the motion of a test electron will depend only on the field generated by the nuclear potential and the electrons in levels with $\varepsilon > -2mc^2$. Because the theory only conserves total charge, but not the individual numbers of electrons and positrons, excitations producing electron-positron pairs will appear only as intermediate states in perturbation expansions in low energy processes. At higher energies, real pairs may be produced.

This description assumes implicitly that there is a clear separation between the physical electron states in the mean potential field and the lower continuum. It has been established that, when realistic nuclear model charge distributions are employed, this distinction becomes blurred when $Z \geqslant 173$; for higher Z, the 1s level is absorbed into the lower continuum, as are the $2p_{1/2}, 2s, \ldots$ and other levels at still higher Z (Rafelski et al., 1978; Reinhardt and Greiner, 1977). This phenomenology is important for understanding the role of the lower continuum, but plays little part in the problems which concern us here.

A. The Furry Bound Interaction Picture of QED

The one-electron states of a suitable atomic mean field potential are usually a good first approximation to start from in atomic physics. The formalism that conveniently embodies this insight along with the machinery to give effect to Dirac's positron theory is Furry's (1951) bound interaction picture of quantum electrodynamics (QED). We sketch this formalism, and refer the reader to texts on QED for the technical details (Schweber, 1964; Jauch and Rohrlich, 1976). Let $a^\mu(x)$ be the classical four-potential due to the nucleus, $A^\mu(x)$ be the quantized Maxwell field produced by the electrons, and $\psi(x)$ be the Dirac spinor field for the electron-positron system. The Fock space field equations for the coupled system to be solved can be shown to be

$$\{\gamma^\mu[i\partial_\mu + ea_\mu(x)] - mc\}\psi(x) = -e\gamma^\mu A_\mu(x)\psi(x) \qquad (24a)$$

$$\Box\, A_\mu(x) = -\mu_0 j_\mu(x) \qquad (24b)$$

where $j_\mu(x)$ is the electron-positron current operator to be defined presently.

We expand $\psi(x)$ in terms of a complete set of solutions of the Dirac equation with a suitable atomic mean field potential $V(r)$:

$$\psi(x) = \sum_m^{(+)} a_m \varphi_m(x) + \sum_n^{(-)} b_n^\dagger \varphi_n(x) \qquad (25)$$

where $(-)$ denotes the summation is taken over states with $\varepsilon < -2mc^2$ and $(+)$ over all higher states. Equation (25) is written as a summation for simplicity. It will include integrations over continuum energies in general, though the problem can be discretized in this way by confining the system to a finite box whose dimensions are later made infinite. The orbital amplitudes $\varphi_n(x)$ are of stationary state form (7), and we shall assume them to be orthonormal. The symbols a_m and $b_n^\dagger$ are respectively destruction and creation operators for the electron states φ_m and the positron states φ_n respectively. These operators anticommute, that is to say, $\{x, y\} = xy + yx = 0$, where x, y are any pair of operators $a_m, a_m^\dagger, b_n, b_n^\dagger$, except for

$$\{a_m, a_{m'}^\dagger\} = \delta_{m,m'}, \qquad \{b_n, b_{n'}^\dagger\} = \delta_{n,n'} \qquad (26)$$

The number of electrons in the level m is the eigenvalue of $N_m = a_m^\dagger a_m$, and it is an elementary exercise to show that it can only take the values 0 or 1. The total energy of the electron-positron system in the absence of any interaction is then given by the operator

$$H_0 = \sum_m^{(+)} N_m \varepsilon_m - \sum_n^{(-)} N_n \varepsilon_n \qquad (27)$$

where ε_m and ε_n are the Dirac eigenvalues and we have subtracted the energy of the filled vacuum. The current density is given by

$$j^\mu(x) = -\tfrac{1}{2} ec [\tilde{\psi}(x) \gamma^\mu, \psi(x)] \qquad (28)$$

where

$$\tilde{\psi}(x) = \psi^\dagger(x) \gamma^0.$$

The net charge is obtained from

$$Q = \frac{1}{c} \int j^0(x) d^3x = -e \left\{ \sum_m^{(+)} N_m - \sum_n^{(-)} N_n \right\} + Q_{\text{vac}} \qquad (29)$$

where Q_{vac} is the net charge of the vacuum. This vanishes for free electrons, but is finite in the presence of an external field. The charge Q_{vac} may be interpreted as the charge induced in the vacuum by the external field—the vacuum polarization charge. More generally, if we indicate normal ordering of creation and destruction operators—rearranging them so that destruction operators all stand to the right of creation operators as if all anticommutators vanish—by sandwiching them between colons, we see that

$$j^\mu(x) = :j^\mu(x): + e \operatorname{Tr}[\gamma^\mu S_F(x, x)], \qquad (30)$$

where $S_F(x, y)$ is the Feynman causal propagator

$$S_F(x, y) = \langle 0| T(\tilde{\psi}(x)\psi(y))|0\rangle. \tag{31}$$

The vacuum state $|0\rangle$ is defined as the state for which $a_m|0\rangle = b_n|0\rangle = 0$ for all electron-positron states, and T denotes Wick's time-ordering operator. This definition ensures that $\langle 0|:j^\mu(x):|0\rangle = 0$, so that the last term in (30) gives the vacuum polarization current.

We can now treat the interaction between the electron-positron system and the electromagnetic field in terms of an interaction Hamiltonian density

$$H_I(x) = j^\mu(x)A_\mu(x) - \rho(x)V'(r) - \delta M(x); \tag{32}$$

The first term is from the right hand side of (24a); the second is a counter term allowing for the effect of the mean field $V(r)$ after subtracting the nuclear potential energy density, $\rho(x)$ being the electron density $\gamma^0 j^0(x)/c$; and $\delta M(x)$ is the mass renormalization counter term.

B. Perturbation Theory

The simplest way of obtaining bound state energies from this formalism is to apply the usual perturbation theory to states of N electrons, with no positrons or free photons. This is most conveniently done by using a formula due to Gell-Mann and Low (1951) as modified by Sucher (1957) which expresses the energy shift due to the electromagnetic interaction in terms of the adiabatic S-matrix operator defined by

$$S_{\eta,\lambda} = 1 + \sum_{k=1}^{\infty} S^{(k)}_{\eta,\lambda} \tag{33}$$

where

$$S^{(k)}_{\eta,\lambda} = \frac{(-i\lambda)^k}{k!} \int d^4x_k \cdots \int d^4x_1 e^{-\eta|t_k|} \cdots e^{-\eta|t_1|} T(H_I(x_k) \cdots H_I(x_1)).$$

Sucher's formula is

$$\Delta E = \lim_{\eta \to 0} \lim_{\lambda \to 1} \tfrac{1}{2} i\eta \frac{\frac{\partial}{\partial \lambda}\langle S_{\eta,\lambda}\rangle_c}{\langle S_{\eta,\lambda}\rangle_c} \tag{34}$$

where the subscript c indicates that only terms corresponding to connected Feynman diagrams are to appear in the matrix element, and the expectation is to be taken relative to the unperturbed reference state. The leading terms of the perturbation expansion derived from (34) are

$$\Delta E = \lim_{\eta \to 0} \tfrac{1}{2} i\eta \{\langle S^{[1]}_\eta\rangle_c + 2\langle S^{[2]}_\eta\rangle_c - \langle S^{[1]}_\eta\rangle_c^2 + \ldots\}. \tag{35}$$

The result of this process, which we outline below, is a perturbation expansion to which the conventional renormalization procedure can, in principle, be applied order by order. The expansion is most conveniently summarized by exhibiting the Feynman diagrams corresponding to successive terms $k = 1, 2, \ldots$ These are essentially constructed by linking primary elements of the form shown in Figure 2(a), which can represent either the emission or absorption of a photon (the wiggly line) and change of state of the electron (the straight lines) at the space-time point x. There is also a contribution, Figure 2(b) from the mean field counter-term in this order. The rules for relating diagrams to matrix elements may be found in any of the texts on QED listed in the bibliography.

The lowest order contribution to the energy from the interaction between the fields appears at $k = 2$, Figure 2(c). This diagram is constructed by linking two of the primary elements, Figure 2(a), by their photon lines. It generates the full covariant two-body interaction, embodying both the instantaneous

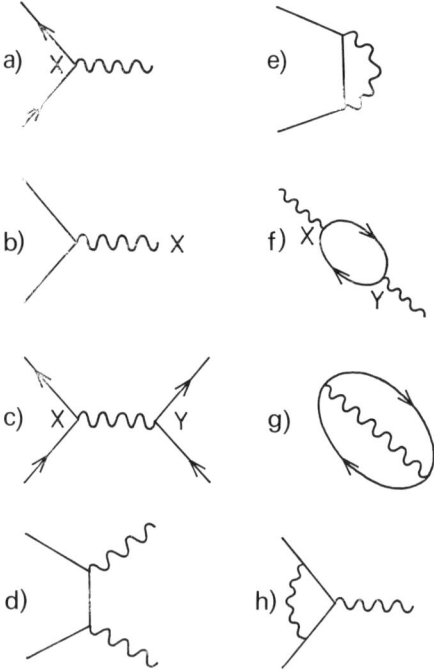

FIG. 2. Feynman diagrams for the lowest order terms of the perturbation series in QED. (a) Emission/absorption of a photon at the space-time point x. (b) One body counter-term, equation (32). (c) Lowest order electron-electron interaction. (d) Two photon scattering process. (e) Electron self-energy diagram. (f) Vacuum polarization diagram. (g) Divergent vacuum diagram making no contribution. (h) Lowest order vertex correction diagram.

Coulomb interaction and a correction for magnetic and retardation effects, which we shall examine in more detail in the next section. The kernel of the matrix element is

$$-\frac{e^2}{4}\mu_0 c : \tilde{\psi}(x)\gamma^\mu \psi(x) g_{\mu\nu} D_F(x-y) \tilde{\psi}(y)\gamma^\nu \psi(y) :$$

where $D_F(x-y)$ is the causal photon propagator, represented by the wiggly line joining to the two vertices. For free electrons, we can think of this as a scattering process in which one particle emits a photon which is subsequently absorbed by the other particle. In like manner, Figure 2(d) represents a process in which two photons interact with the same particle; its representation is

$$\frac{e^2}{4}:\tilde{\psi}(x)\gamma^\mu A_\mu(x) S_F(x,y)\gamma^\nu A_\nu(y)\psi(y):$$

where $S_F(x, y)$ is the electron propagator defined in (31).

Other diagrams appearing in this order of perturbation theory are infinite before applying the renormalization procedure; thus the diagram of Figure 2(e) represents the lowest order contribution to the electron self-energy; its representation has the kernel

$$-\frac{e^2}{8}\mu_0 c : \tilde{\psi}(x)\gamma^\mu S_F(x,y)\gamma^\nu g_{\mu\nu} D_F(x-y)\psi(y):$$

Similarly, Figure 2(f) represents the lowest order vacuum polarization correction generated by the vacuum charge-current distribution of equation (30), and has the matrix element kernel

$$\frac{e^2}{8}:A_\mu(x)A_\nu(y): \text{Tr}[\gamma^\nu S_F(x,y)\gamma^\mu S_F(y,x)].$$

These two processes contribute to the Lamb shift of electronic bound levels; whilst this is experimentally measurable, it is the contribution of Figure 2(c) that dominates in this order, and is most important for subsequent developments.

We can carry this process to higher orders. The most important terms are those based on the covariant Coulomb interaction, and if we neglect diagrams incorporating self-energy insertions of the form of 2(e) and 2(f), the resulting series will be exactly analogous to the many-body theory of nonrelativistic atomic structure as described in texts such as Lindgren and Morrison (1982). We refer to this as the "standard" model of relativistic atomic structure theory, based on an effective Hamiltonian

$$H = H_0 + H_1 \tag{36}$$

where H_0 is given by equation (27) and H_1 has the form

$$H_1 = \frac{1}{2} \sum_{pqrs}^{(+)} :a_p^\dagger a_q^\dagger a_r a_s: \langle pq|v|sr\rangle + \sum_{pq}^{(+)} :a_p^\dagger a_q: \langle p| - V(r) - \frac{Z(r)}{r}|q\rangle. \quad (37)$$

The leading summation comes from the diagram 2(c), whilst the second is just the counter-term 2(b). This model neglects all pair excitation processes and all radiative corrections in a manner consistent with the Dirac positron theory outlined at the beginning of this section. We shall briefly describe what can be done about the corrections in a later section.

C. The Two-Body Interaction Kernel

In this section we explain briefly how the form of the two-body potential of (37) can be obtained from (35). The integrand of the second order S-matrix element 2(c) can be expressed in the form

$$S^{[2]} = \frac{(ie)^2}{2!} T[j^\mu(x) A_\mu(x) \cdot j^\nu(y) A_\nu(y)] \quad (38)$$

This operator will be placed between two state vectors representing N electrons with no positrons or photons, so that we require the matrix element (photons only)

$$\langle 0| T[A_\mu(x) A_\nu(y)]|0\rangle = -\tfrac{1}{2} c\mu_0 g_{\mu\nu} D_F(x - y) \quad (39)$$

where μ_0 is the permeability of the vacuum, $g_{\mu\nu}$ is the metric and $D_F(x - y)$ is the causal photon propagator in configuration space which we shall rewrite as an electron-electron interaction kernel of more conventional form. By using the definitions (25) and (28), and retaining only those terms in the expansion (25) which give electron-electron interactions, we arrive at an expression of the form given on the first line of (37) with

$$\langle pq|v|rs\rangle = \int d^3y\; \varphi_p^\dagger(x)\varphi_q^\dagger(y) v_{sq}^F(x, y) \varphi_r(x)\varphi_s(y) \quad (40)$$

where

$$v_{sq}^F(x, y) = \frac{e^2}{4\pi\varepsilon_0 R} e^{i\omega_{sq} R}(1 - \boldsymbol{\alpha}_x \cdot \boldsymbol{\alpha}_y).$$

with $R = |x - y|$ and

$$\omega_{sq} = \frac{(\varepsilon_s - \varepsilon_q)}{c}.$$

Subscripts x and y have been appended to the α matrices to indicate that they refer to the electron whose line passes through the relevant vertex. The real part of $v_F(x, y)$ gives an energy shift; there is a small nonvanishing imaginary part, usually neglected, which will also give a small width to the level. In addition, because p, q, r, and s are dummy summation variables, it is convenient to symmetrize the interaction kernel and write

$$\bar{v}^F(x, y) = \tfrac{1}{2}[v^F_{sq}(x, y) + v^F_{rp}(x, y)]. \tag{41}$$

The unsymmetric appearance of equation (40) may cause cause a little surprise at first glance. It can be explained by rewriting the integral as the interaction of the current

$$j^\mu_{rp}(x) = \varphi^\dagger_r(x)\gamma^0\gamma^\mu\varphi_p(x)$$

with the retarded four-potential $A_{sq}(x)$ due to the current $j^\mu_{sq}(x)$.

The same interaction appears for all the components of the diagram 2(c) including electron-positron and positron-positron interactions. The electron-positron interactions will involve large energy differences, and the dependence on ω can never be neglected. However, for low Z, electron-electron orbital energy differences are small, and (40) reduces to the standard nonrelativistic Coulomb interaction integral in the limit $c \to \infty$. It is usually convenient for practical calculation to separate off the Coulomb part, leaving an expression which can be regarded as a relativistic correction. This separation is most explicit when we express the photon propagator in the traditional nonrelativistic transverse photon gauge (Berestetskii et al., 1971), in which case v^F is replaced by a kernel v^T, where

$$v^T_{sq} = \frac{e^2}{4\pi\varepsilon_0 R} - \frac{e^2}{4\pi\varepsilon_0}\left\{\alpha_x \cdot \alpha_x \frac{e^{i\omega_{sq}R}}{R} + (\alpha_x \cdot \nabla)(\alpha_y \cdot \nabla)\frac{e^{i\omega_{rp}R} - 1}{\omega^2_{rp}R}\right\}. \tag{42}$$

The final result should not depend on our choice of gauge if gauge independence is to be maintained, although we can expect to see differences if we truncate our perturbation expansion. The equivalence of v^F and v^T when the orbitals are defined by a *local* mean potential has been demonstrated in several publications, for example Hata and Grant (1984). The form (42) of the interaction is convenient for demonstrating the relation to the familiar interaction kernel of Breit (1929, 1930), which appears as the $\omega \to 0$ limit of v^T. (We can then drop the orbital subscript labels.)

$$\lim_{\omega \to 0} v^T = \frac{e^2}{4\pi\varepsilon_0 R} + v^B$$

where

$$v^B = -\frac{e^2}{8\pi\varepsilon_0 R}\left[\alpha_x \cdot \alpha_y + \frac{(\alpha_x \cdot \mathbf{R})(\alpha_y \cdot \mathbf{R})}{R^2}\right]. \tag{43}$$

This can in turn be decomposed into the dominant magnetic interaction kernel introduced by Gaunt (1929)

$$v^G = -\boldsymbol{\alpha}_x \cdot \boldsymbol{\alpha}_y \frac{e^2}{4\pi\varepsilon_0 R} \tag{44}$$

and a residual part attributed to retardation of the interaction. The reasons for the often repeated rule (Bethe and Salpeter, 1957, §38) that v^B should only be used in first order perturbation theory are a direct consequence of its derivation as a low-Z, low energy limit.

D. The "Standard" Model of Relativistic Atomic Structure Theory and The Hartree–Fock Approximation

There has been frequent criticism of relativistic self-consistent field theory on two grounds. The first is that the existence of the negative energy states means that it is impossible to use variational methods (Swirles, 1935; Grant, 1961, 1970) to derive the SCF equations in a consistent manner. The second is that the theory is not renormalizable and that it is not obvious how to use it as the starting point of a systematic improvement of the solution using many-body perturbation theory. It should be clear by now that the last objection is not applicable to the machinery we have developed and that there are no difficulties of principle to be overcome. It thus remains to put SCF methods into context.

The problem here is that by tradition (Swirles, 1935, Grant 1961, 1970) the relativistic Hartree–Fock (Dirac–Fock) equations have been derived from a variational principle, using what has become known (Sucher 1980) as the Dirac–Coulomb Hamiltonian

$$H_{DC} = \sum_{i=1}^{N} H_D(i) + \frac{1}{2} \sum_{i,j=1}^{N} \frac{1}{r_{ij}}. \tag{45}$$

This gives the same energy as the standard model Hamiltonian (37) when referred to a determinantal wave function, Φ say, for a closed shell state constructed from the same orbital basis. (The one-body potential drops out of the energy expression. The magnetic and retarded parts of the two-body interaction were discarded by Swirles and Grant because of their relatively small contribution and because of their mathematical complexity.) The derivation of SCF equations follows the usual lines:

Make $\langle \Phi | H_{DC} | \Phi \rangle$ stationary,

subject to $\langle \Phi | \Phi \rangle = 1$

and $\langle \varphi_p | \varphi_q \rangle = \delta_{pq} \forall p, q.$

The orthonormality constraints lead to a trial function Φ built from bound-type one-electron orbitals φ_p, satisfying boundary conditions discussed in §II.C. The resulting variational differential equations determine these orbitals φ_p self-consistently; the choice of orbital quantum numbers n_p, κ_p and the appropriate *aufbau* construction are needed to make the problem well-defined. This gives the lowest energy solution Φ within the N electron space of N-products of trial functions $\{\varphi_p\}$ defined by the construction. If we place further constraints on Φ, requiring it to be orthogonal to states already constructed having the same symmetry, we shall obtain excited states. Because the boundary conditions explicitly exclude any continuum-type orbitals there is no danger whatsoever of trying to solve an insoluble problem suffering from the so-called "Brown–Ravenhall disease" (Sucher, 1980, 1985). Even though the formulation appears superficially, by its use of H_{DC}, to be vulnerable to the "disease," the boundary conditions which must be imposed to make the problem well-defined provide an effective "vaccine."

There is, however, another way of looking at the Hartree–Fock model within the context of many-body perturbation theory that avoids the need to formulate the equations variationally. In effect, the potential V used to define the orbital basis in the standard model is identified with the potential that would be obtained from the Hartree–Fock equations so that the expectation of H_1, (37), vanishes for the Hartree–Fock wavefunction. The book "Atomic Many-Body Theory" by I. Lindgren and J. Morrison gives a very clear presentation of the particle-hole formalism, and their development can be taken over essentially unchanged.

The first step is to redefine the "Fermi level" below which all sub-shells are occupied by the N electrons of the system. The relevant orbitals are designated as "core" orbitals, and those unoccupied orbitals of higher energy as "virtual" orbitals.

Core orbitals: $\varepsilon < F$. Labels $a, b, c, \ldots$

Virtual orbitals: $\varepsilon > F$. Labels $r, s, t, \ldots$

General orbitals: Any ε. Labels $i, j, k, \ldots$

We implement the shift from $-2mc^2$ to F of the Fermi level by defining a new normal ordering of creation/destruction operators so that an operator $a^\dagger$ appears to the right for a core orbital, whilst a continues to be placed on the right for a virtual orbital. In this new ordering

$$H_0 = E_{\text{core}} + \sum_i \{a_i^\dagger a_i\} \varepsilon_i, \qquad E_{\text{core}} = \sum_a \varepsilon_a, \tag{46}$$

whilst H_1 splits into three terms: a constant energy shift

$$H_1^0 = \sum_a \langle a| - V|a \rangle - \frac{1}{2} \sum_{a,b} [\langle ab|v|ab \rangle - \langle ba|v|ab \rangle]; \tag{47a}$$

a one-body operator

$$H_1^1 = \sum_{i,j} \{a_i^\dagger a_j\} \langle i|W|j\rangle; \tag{47b}$$

and a two-body operator

$$H_1^2 = \frac{1}{2} \sum_{i,j,k,l} \{a_i^\dagger a_j^\dagger a_l a_k\} \langle ij|v|kl\rangle. \tag{47c}$$

The new effective non-local two-body potential W is defined by matrix elements of the primary two-electron interaction v

$$\langle i|W|j\rangle = \sum_a [\langle ia|v|ja\rangle - \langle ai|v|ja\rangle] - \langle i|V|j\rangle. \tag{47d}$$

The choice $W \equiv 0$ defines the Hartree–Fock potential; the first order shift H_1^0 vanishes identically, as does the one body correction H_1^1. Only H_1^2 will contribute in higher orders of perturbation theory. In this sense, all we are doing in a Hartree–Fock calculation is to choose a one-body potential V satisfying $W \equiv 0$.

This point of view carries a big bonus: it indicates how we can improve on the Dirac–Fock approximation perturbatively just as in the nonrelativistic case and deals with a large part of the second objection mentioned at the beginning of this subsection.

E. THE OPEN SHELL PROBLEM. MCDF AND OTHER APPROACHES

Open shell problems bring greater technical complexity, even though the principles of calculation remain unaltered. A considerable quantity of computational machinery has been developed for this purpose in general purpose packages such as those of Desclaux (1975), Klapisch et al. (1977), and Grant et al. (1980) and McKenzie et al. (1980). Here we shall consider how these schemes can be accommodated by this machinery.

1. CI Methods Using Parametric Potentials

From the point of view of perturbation theory based on the standard model, the simplest approach to understand is that of Klapisch et al. (1977); they employ a parameterized model potential V to construct the required orbitals and so generate a (finite) set of CSF (configurational state functions—linear combinations of N-fold product functions, $\Phi(\gamma J\Pi)$ having prescribed total angular momentum, J, parity, Π, and other symmetry properties) to span a subset of the N-electron space of interest. A general ASF (atomic state function) of the same symmetry is a linear combination (wavefunction of CI type)

$$\Psi(\Gamma J\Pi) = \sum_\gamma c_{\Gamma\gamma}(J\Pi)\Phi(\gamma J\Pi) \equiv c_\Gamma^\dagger(J\Pi)\cdot\Phi(J\Pi).$$

The atomic energy levels and wavefunctions may then be approximated by the eigenvalues $E_\Gamma(J\Pi)$ and eigenvectors $c_\Gamma(J\Pi)$ of the matrix of the effective Hamiltonian (36) over the set of CSF. These estimates are what one would get as the sum of a perturbation expansion to all orders within the finite subspace spanned by the CSF. The parameters of the generating potential, V, can be varied to optimize the value of some observable: for example, to fit a prescribed set of term or level energies, or to minimize the ground state energy. (This optimization is permissible for exactly the same reasons as in the Hartree–Fock problem.) It is a cheap and effective method.

In principle, we should be able to get exact results within the standard model provided we use a "sufficiently large" basis of CSF. The main difficulty is to know what is meant by "sufficiently large." In most cases, the choice is made either by subjectively accepting the results as "good enough" for the purpose, or by recognizing that the cost of further calculation is prohibitive. Whilst it is sometimes reasonably straight-forward to improve a CSF basis set in a systematic way, best results are usually obtained by people with "green fingers." Nevertheless, the relativistic parametric potential method has proved invaluable for treating highly ionized spectra of the heavy metals of interest in plasma spectroscopy and diagnostics; the paper of Klapisch *et al.* (1977) is typical of this kind of application.

2. MCDF Methods

MCDF (multi-configuration Dirac–Fock) packages have been published by Desclaux (1975) and by Grant *et al.* (1980) and are in widespread use for calculation of atomic energy levels, transition probabilities and other properties for atoms of all rows of the periodic table. There is widespread misunderstanding about the meaning and validity of such calculations, mostly concerning the little used MCDF-OL or -EOL method. The former corresponds to the classical MCHF method of Hartree *et al.* (1939), Sabelli and Hinze (1969) and Froese Fischer (1972). The self-consistent field equations are derived as in the closed shell case, but with an energy expression based on a trial ASF of CI type. The derivation leads to a set of coupled integro-differential orbital equations similar to those of the closed shell problem; however, the potential determining the orbitals now depends on a specific eigenvector c_Γ of the Hamiltonian matrix. The solution algorithm therefore has a two-step iterative cycle: a solution of the integro-differential equations with an estimated c_Γ is followed by a CI step to determine a revised estimate of c_Γ and the level energy $E_\Gamma(J\Pi)$. The MCDF–EOL method is basically the same, save that the SCF equations, and the equivalent potential V are derived from a weighted mean energy of several ASF which are redetermined in each CI step.

The attraction of SCF methods of this type is that one gets a solution which is optimal in the sense that the energy functional defined by a specified ASF (–OL case) or set of ASF (–EOL case) is made stationary within the space of trial functions. If this is all that is needed, the method is not too expensive, although experience shows that MCDF–(E)OL calculations are often hard to converge. However, the potential V will in general give poor results for ASF other than those used to generate it, and it is necessary to do more MCDF–(E)OL calculations if we want acceptable energies and wavefunctions for other states. Another drawback is that orthogonality integrals $\langle (n\kappa)_1 | (n'\kappa)_2 \rangle$ between functions computed in different –(E)OL calculations will not be equal to the Kronecker delta, $\delta_{n,n'}$ because the orbitals are generated by different potentials. This complicates the calculation of transition probabilities. A further, and more serious, problem is that the potentials acting on orbitals nlj with $j = l \pm \frac{1}{2}$ do not, in general, converge to the potential for the same nonrelativistic nl orbital in the limit $c \to \infty$. This causes a nonphysical spin-orbit splitting in the nonrelativistic limit (see for example, Cheng et al., 1978; Hata and Grant, 1983b). It is obvious that spin-orbit splittings determined (for the physical value of c) from independent MCDF–OL calculations are unlikely to be very reliable, so that –EAL methods are often preferable. Despite these difficulties, there remain cases where the orbital relaxation between states is a genuine physical feature of the problem, and then there is a real incentive to use –OL methods.

These considerations influenced Grant et al. (1976) to introduce the MCDF–(E)AL schemes. The idea was to use a mean potential V determined from a weighted sum of energies of all the participating CSF. Ideally, the CSF should span the totality of states of all symmetries which can be constructed within the core and valence shells. Such a scheme automatically ensures that the same orbitals are used for all ASF; it is computationally cheap, convergence usually being no more difficult than for single configuration Hartree–Fock problems; and it is possible to choose the weights so that the nlj orbitals, $j = l \pm \frac{1}{2}$, converge to the same nl orbital in the nonrelativistic limit. Of course, once the potential has been determined, the rest of the calculation is of CI type, and the scheme has similar advantages and disadvantages. Indeed, for highly ionized systems, there seems to be little to choose between them. The MCDF–EAL method has been the most popular of all those described here in the last few years.

It remains to place these methods in the context of many-body theory based on the standard model. The particle-hole formalism outlined in the last subsection has to be extended for open shell systems (Lindgren and Morrison, 1982 Chapter 13) to distinguish the orbitals belonging to open shells (valence orbitals) from the unoccupied (virtual) orbitals. The diagrams are more complicated than in the closed shell case, but the perturbation operator has essentially the form of (47a-d). The effective potential W only vanishes if it is chosen to be the potential generated by the closed shell atomic core. If we use an MCDF calculation to define V, whether it be of -(E)OL or -(E)AL type, the potential W will not cancel all the diagrams in second order, and will have to be included in all orders of perturbation Lindgren and Morrison, 1982, §13.1). Nobody has attempted such calculations at the time of writing.

3. The GRASP Code; Radiative and Other Corrections

The programs described by Grant et al. (1980) and McKenzie et al. (1980) provide for very general atomic structure calculations based on the principles just outlined. The technical construction of Dirac–Fock equations and their numerical solution using finite difference methods has been described in detail by Grant (1970) and Lindgren and Rosén (1974); most of this can be applied to the open shell problem. However, techniques for reducing matrix elements of the one- and two-electron operators of the effective Hamiltonian have only been sketched in the literature (Grant 1981). (Related techniques used by workers in USSR have been described (in English) by Kaniauskas and Rudzikas (1980) and Šimonis et al. (1984), and in other Russian publications.) A new version of the system of Grant et al. (1980) and McKenzie et al. (1980), renamed GRASP (for General-purpose Relativistic Atomic Structure Program), is being field tested, and will soon be made available to a wide range of users. It is hoped that technical improvements will both speed up the code and make *ab initio* studies of more complex problems such as the spectra of transition metals a practical possibility. GRASP and its predecessors therefore make provision for a variety of corrections to remedy the deficiencies of the standard model. These will be described in the next section.

F. Corrections to The Standard Model

1. Nuclear Motion and Properties

The most important corrections to the standard model are due to the fact that the nucleus is not a static point charge but has a finite mass and extension. Hughes and Eckart (1930) were the first to discuss the effect of a finite nuclear mass for the many-electron atom in nonrelativistic quantum mechanics, and Stone (1961) extended this analysis to the relativistic case by treating the nucleus as a (non-interacting) collection of Dirac particles. This has limited use for our purposes as he immediately expanded his results in a formal expansion in powers of α^2. Some attempts have been made to treat a two-particle system using the covariant Bethe–Salpeter equation (Bethe and Salpeter, 1957). Here we follow the treatment of Grotch and Yennie (1969) who derived an effective potential model for treating the nuclear motion in hydrogen. Consider a set of uncoupled electrons along with a nucleus of rest mass Mc^2. The Hamiltonian is

$$H_{\text{free}} = \sum h_i + \mathbf{P}^2/2M, \qquad h_i = c\boldsymbol{\alpha}_i \cdot \mathbf{p}_i + (\beta_i - 1)mc^2 \qquad (48)$$

For the centre of mass to be at rest,

$$\mathbf{P} + \sum \mathbf{p}_i = 0;$$

this allows us to eliminate $\mathbf{P}$ from H_{free}. Using $\mathbf{p}^2 = (\boldsymbol{\alpha} \cdot \mathbf{p})^2$, we find

$$H_{\text{free}} = \sum_i \left\{ 1 + \frac{h_i + 2mc^2}{2Mc^2} \right\} h_i + \sum_{i<j} \frac{\mathbf{p}_i \cdot \mathbf{p}_j}{M} \qquad (49)$$

For the orbitals of interest in most problems, the value of $|h_i| \ll mc^2$, so that the expression in braces $\{...\}$ reduces to $(1 + m/M)$. In other words, in the nonrelativistic limit, we recover the usual prescription for replacing the free electron mass by the reduced mass

$$m_{\text{red}} = mH/(m + M). \qquad (50)$$

The equations of motion are thereby unaltered except for a rescaling of the atomic units in terms of natural units through the usual replacement $m \to m_{\text{red}}$.

The second summation in (49) is the familiar Hughes–Eckart correction; it is relatively small, contributing to the atomic isotope shift of spectral lines as the so-called specific mass term, especially in light elements. It has not, as yet, been implemented in GRASP. Higher order effects, mainly of importance in high precision work on very light atoms, are described by Erickson and Yennie (1969).

The distribution of electric charge and the electric and magnetic multipole moments of the nucleus can also produce observable effects. We have already discussed the boundary conditions in the presence of a spherically symmetric distributed nuclear charge in §II.C.2. It is particularly important to use a suitable nuclear model potential for elements with Z greater than about 50;

this directly affects the electron density near the nucleus for orbitals with $j = \frac{1}{2}$, and the consequent adjustment has an indirect effect on other orbitals (Shun, 1977, Grant, 1980). Provided care has been taken to ensure that the different numerical processes in the calculation have truncation errors of the same order in the radial step size, and the computer has a word length long enough to make the noise due to rounding errors unimportant, analysis shows (Grant, 1980) that a Dirac–Fock SCF calculation can even yield sufficient accuracy to make reasonable estimates of the "nuclear volume effect" resulting from isotopic variations in nuclear size. This direct method has recently been used by Bouazza et al. (1986) for Pb II. Such isotope shifts are roughly proportional to the nuclear mean square radius and it would be desirable to make the determination of the constant of proportionality a routine product of GRASP calculations as an aid to spectrum analysis. This has not yet been done.

Hyperfine interactions arise from the interaction of nuclear electromagnetic moments with the orbital electron distribution. Relativistic effects are important for the heavy elements and and are often modelled using an effective operator in which nonrelativistic radial moments are replaced by weighted sums of relativistic ones (Sandars and Beck, 1965). Direct calculations using Dirac–Fock–Slater orbitals have been made by Rosén (1969) and Rosén and Lindgren (1972), whilst Desclaux (1971) used restricted and unrestricted Dirac–Fock orbitals. Whilst there have been a number of nonrelativistic many-body calculations of hyperfine interactions, for example Lindgren et al. (1976), we are not aware of any analogous work using relativistic wavefunctions.

2. Corrections to the Coulomb Potential

We have seen that the full covariant two-body interaction is a fairly complicated expression; for this reason, and because of folklore regarding use of the correction terms, the latter have been omitted in Dirac–Fock and MCDF calculations. GRASP allows for the transverse interaction v^T, (42), to be computed after completion of the SCF or CI stage. The matrix of v^T is first transformed from the CSF basis to a basis of ASF; the eigenvalues from the SCF stage are then added to the diagonal elements and the resulting matrix is then rediagonalized to give the final level energies and ASF. Ideally, the full electron–electron interaction should be kept intact, particularly if one wishes to take the MBPT expansion beyond the lowest order. It would be possible, though very expensive in codes such as GRASP, to do this iteratively by re-entering the SCF step after including v^T corrections in the orbital equations. We shall see in §IV that v^T can be incorporated without difficulty in SCF calculations using basis set expansion methods.

Relativistic field theory will generate many-body potentials as well as the two-body potentials of our effective Hamiltonian (Primakoff and Holstein, 1939), and Mittleman (1971) has suggested that these may be large enough to account for some discrepancies between theoretical and experimental inner-shell binding energies. A detailed investigation of these contributions has just been carried out by Zygelman and Mittleman (1986); this shows that the contribution of three-body potentials to heavy element (and *a fortiori* to light element) binding energies is negligible.

3. Radiative Corrections

The contributions due to insertion of terms described by the diagrams, Figures 2(e) and 2(f), into the perturbation series give rise, in general, to corrections comparable in size to those of v^T. They should therefore not be ignored if the physical problem demands predictions of sufficient precision. The self-energy term, Figure 2(e), is a one-body operator which, unfortunately, is difficult to calculate in a routine many-electron calculation, and there are currently no satisfactory methods available to do this. The best that can be done is to make use of the hydrogenic calculations of Mohr (1974, 1982), Desiderio and Johnson (1971), Cheng and Johnson (1976) and Soff *et al.* (1982). For a recent review and tabulation of the data for the whole Lamb shift in hydrogenic atoms (which includes both the vacuum polarization, nuclear size and mass effects) see Johnson and Soff (1985). Vacuum polarization is somewhat easier to handle, since the results of calculation of Figure 2(f) and other diagrams of this type can be expressed as an effective short-range modification of the photon propagator (39) and therefore of the effective interaction between charged particles. The dominant contribution is to the electron-nucleus interaction; the contribution to the electron-electron interaction is relatively tiny and has been universally ignored.

Most calculations of radiative corrections for many-electron atoms have therefore been of the type used in GRASP, described in McKenzie *et al.* (1980). The self-energy contribution is calculated by assuming that the shift is hydrogenic with an effective nuclear charge Z' determined by the orbital mean radius, and then interpolating for this value in the hydrogenic tables of Mohr (1974, 1982) for $1s$, $2s$, $2p_{1/2}$ and $2p_{3/2}$ states. The choice of mean radius to give an effective charge is quite arbitrary; it will however, give a reasonable order of magnitude for inner shells where the orbital shape is likely to be not too far from hydrogenic in the region where it matters for determination of the self-energy. Work on helium-like systems (Hata and Grant 1983a) suggests that the electron density near the nucleus might be a more reliable guide, but this has never been tested on many-electron systems. The self-energy of orbitals with principal quantum numbers $n > 2$ has been estimated by assuming the hydrogenic dependence of n^{-3}. Vacuum polarization potentials have been treated in GRASP using the finite nucleus modification of the Uehling (1935) potential given by Fullerton and Rinker (1976) and higher order terms given by Wichmann and Kroll (1956) have been incorporated in the current code revision.

There is one further radiative correction of importance in light elements which may also be needed for accurate work on fine structure in heavier systems. The third order diagram Figure 2(h) (which requires renormalizing like the self-energy and vacuum polarization corrections) adds a term to the electron current (28) which can be interpreted as a correction to the intrinsic magnetic moment of the electron predicted by Dirac one electron theory. If g_0 is the uncorrected electron g-factor, $g_0 = 2$, then

$$g = g_0(1 + a) \quad (51)$$

where a is known as the electron anomaly. Diagram 2(h) contributes $\alpha/2\pi$ to the value of a, which has the value

$$a = \left(\frac{\alpha}{2\pi}\right) - 0.328478966\left(\frac{\alpha}{\pi}\right)^2 + 1.1835(61)\left(\frac{\alpha}{\pi}\right)^3 \ldots$$

when higher order contributions are taken into account (Wichmann and Kroll, 1956). Whilst the effect of the electron anomaly has not yet been considered in many-electron relativistic calculations in general, it may be of some interest to record additions to the effective Hamiltonian of the standard model to take it into account (Hata, Cooper and Grant, 1985). Their (unquantized) potential is

$$H_2 = \frac{-i\alpha^2}{4\pi} \sum_i \frac{Z}{r_i^3} (\mathbf{r}_i \cdot \boldsymbol{\gamma}_i)$$
$$+ \frac{i\alpha^2}{4\pi} \sum_{i<j} \left[\frac{1}{r_{ij}^3} (\mathbf{r}_{ij} \cdot \boldsymbol{\gamma}_i - \mathbf{r}_{ij} \cdot \boldsymbol{\gamma}_j) + \beta_i \boldsymbol{\Sigma}_i \cdot (\mathbf{r}_{ij} \times \boldsymbol{\alpha}_j) - \beta_j \boldsymbol{\Sigma}_j \cdot (\mathbf{r}_{ij} \times \boldsymbol{\alpha}_i) \right]$$

(52)

where

$$\boldsymbol{\Sigma} = \begin{bmatrix} \boldsymbol{\sigma} & 0 \\ 0 & \boldsymbol{\sigma} \end{bmatrix}, \quad \boldsymbol{\gamma}_i = \beta_i \boldsymbol{\alpha}_i.$$

This agrees with the lowest order phenomenological anomalous magnetic moment correction given by Barut and Xu Bo-wei (1982). The effect on the Breit–Pauli Hamiltonian is also considered in the paper of Hata, Cooper and Grant (1985).

IV. New Developments—Approximation by Finite Basis Sets

The majority of truly relativistic atomic structure programs use finite difference algorithms to solve the radial equations and obtain energies, wavefunctions and predictions of observable quantities. However, a few attempts have been made to use the method of expansion in a finite set of basis functions as first proposed for nonrelativistic quantum mechanics by Hall (1951) and Roothaan (1951). Synek (1964) first formulated relativistic equations for this approach, but the first successful matrix Dirac–Fock calculations were done by Kim (1967) using expansions in Slater type functions with cusp exponents of index γ as in equation (21c). Kagawa (1975) later used Slater type functions with integer cusp exponents. Multiconfiguration open shell equations were written down by Leclercq (1970) but the first real open shell calculations using the matrix method were done by Kagawa (1980). These calculations achieved reasonable results, though they were not really competitive with the finite difference methods, and only relatively few investigations have been made along these lines.

These calculations suggested that the matrix DF solutions were not always true upper bounds to the exact solution, but it was not until attempts were made by quantum chemists to apply the matrix DF method to molecules that serious problems were encountered. These were dubbed "variational collapse" (Schwarz and Wallmeier 1982) or "finite basis set disease" (Schwarz and Wechsel-Trakowski 1982); their characteristics were largely mapped out in a series of papers (Mark *et al.*, 1980; Mark and Rosicky, 1980; Wallmeier and Kutzelnigg, 1981; Malli and Oreg, 1979; Matsuoka *et al.*, 1980; Aoyama *et al.*, 1980; Datta and Ewig, 1982; Lee and McClean, 1982). Considerable ingenuity has been devoted to finding ways round the problems as documented by Kutzelnigg (1984), who characterized the pathologies described in these papers as follows:

(a) The prediction of a spectrum in simple hydrogenic problems which bore little resemblance to what one expected on physical grounds (Schwarz and Wallmeier, 1982; Wallmeier and Kutzelnigg, 1981). Spurious intruder solutions appeared with low kinetic energy that were either close to the physical solutions or else had an unphysically low energy. There was no criterion available for judging which of these solutions were of any use.

(b) In some cases (Mark and Rosicky, 1980; Wallmeier and Kutzelnigg, 1981; Mark and Schwarz, 1982) the results were very sensitive to details of the basis set, and the results showed no systematic convergence with increasing basis set size.

Kutzelnigg (1984) assessed the remedies then available, focussing attention on the fact that the matrix representations of the Dirac operator failed to give the right formal nonrelativistic limit as $c \to \infty$. He characterized some fifteen recipes belonging to four general classes: A. Accept the inevitability of variational collapse but try to extract information about "relativistic corrections" from the results. B. Transform the Dirac operator in some way to a form which gives the correct nonrelativistic limit. C. Transform the matrix representation of the Dirac operator to a form which gives the correct nonrelativistic limit. D. Replace the Dirac operator by some other operator which is essentially positive, such as its square.

There have been a number of papers since then developing one or other of these recipes. In our view, none of them deal with the real causes of the pathologies that have been described. We shall present here a theoretical framework for carrying out "disease-free" calculations in which the considerations of §II play a central part. We shall then demonstrate how relaxation of the conditions we impose on the basis sets leads to trouble.

A. Principles of Basis Set Calculations in Atomic and Molecular Structure

We shall start by considering the purely radial equation (13); the principles involved are easily extended to the Dirac–Fock case. We shall therefore be approximating the radial functions $P(r)$ and $Q(r)$ in

$$u(r) = \begin{bmatrix} P(r) \\ Q(r) \end{bmatrix}, \qquad \varepsilon = E - mc^2 \qquad (12)$$

$$h_D u(r) = \varepsilon u(r)$$

where

$$h_D = \begin{bmatrix} V(r) & c\left[-\dfrac{d}{dr} + \dfrac{\kappa}{r}\right] \\ c\left[\dfrac{d}{dr} + \dfrac{\kappa}{r}\right] & -2mc^2 + V(r) \end{bmatrix} \qquad (13)$$

by linear combinations of the form

$$P(r) = P_N(r) = \pi_1(r)p_1 + \cdots + \pi_N(r)p_N \equiv \boldsymbol{\pi}^\dagger(r) \cdot \mathbf{p}$$
$$Q(r) = Q_{N'}(r) = \rho_1(r)q_1 + \cdots + \rho_{N'}(r)q_{N'} \equiv \boldsymbol{\rho}'^\dagger(r) \cdot \mathbf{q}, \qquad (53)$$

the expressions in bold type on the right being vectors of the appropriate dimension N or N'. We saw in §II B,C that to represent bound states of the system, the functions required must have the following properties:

(i) The radial density $D(r) = |P(r)|^2 + |Q(r)|^2$ must be integrable on the positive real line $0 < r < \infty$:

$$\int D(r)dr \quad \text{is finite.}$$

(ii) The potential $V(r)$ must have a finite expectation:

$$\int D(r)V(r)dr \quad \text{is finite.}$$

(We have seen that this condition is fulfilled automatically for Coulomb boundary conditions when Z is sufficiently small.)

(iii) The functions $P(r)$ and $Q(r)$ must be related as prescribed by §IIC in the neighbourhood of $r = 0$.

The finiteness of the total radial density means that the solutions P and Q must vanish at least as fast as $\exp(-\lambda r)$, for some $\lambda > 0$, as $r \to \infty$, and this suggests that functions of Slater type (STF) $r^\nu e^{-\lambda r}$ or Gaussian type (GTF) $r^\nu \exp(-\lambda r^2)$ are worth considering. Although GTF do not have the correct asymptotic behaviour as $r \to \infty$, the extreme tail contributes little to the physics of the problem, and the relative ease with which multi-centre integrals can be evaluated in the many-electron molecular problem makes GTF attractive as trial functions. Clearly, it will be to our advantage if we can exploit the highly efficient algorithms available for nonrelativistic calculations as far as possible. There are, of course, other possibilities: for example there has recently been some interest in the use of piecewise polynomial basis sets (Johnson, 1987) of a type extensively studied by Silverstone et al. (1978), and these show some promise for atomic calculations of high accuracy. We shall not consider them here.

However, it is the form of the functions at the singular point of the Dirac operator at the origin that is most important in gaining control over the wild behaviour of basis sets. We need trial functions which satisfy the differential equations to as much precision as possible near $r = 0$, in particular by eliminating any singular terms; we can only assure this for expansions of the form (53) if we can match the *individual basis functions $\pi_i(r)$ and $\rho_i(r)$ in pairs*. This means that only expansions with $N \equiv N'$ can be entertained; we cannot have unmatched functions in general in either the expansion of P or of Q and expect to satisfy the requirements. The easiest way to pair the functions is to construct the functions $\rho_i(r)$ from the relation

$$\rho_i(r) = \left[\frac{d}{dr} + \frac{\kappa}{r}\right]\pi_i(r), \quad i = 1, 2, \ldots, N; \quad (54)$$

this relation was named "kinetic balance" by Stanton and Havriliak (1984). The analysis of §IIC. shows that (54) gives the correct relation of the leading coefficients, p_0 and q_0, of the power series expansion of the analytic solution near $r = 0$. The prescription works particularly well with GTF having integer exponent ν for the case of a nuclear model with distributed charge, and with STF having $\nu = \gamma$, (21c), for the case of a point nuclear charge. The use of

STF with *integer* exponents cannot completely satisfy equations (21) and (22) in the latter case, so that the error has to be accomodated by the coefficients p and q in the expansion (53). We shall see that this causes no serious failures for low Z, but it becomes progressively harder to get high accuracy and to avoid some minor form of variational collapse for $Z \geqslant 20$. Obviously larger basis sets will be needed to get a given precision when the boundary conditions are not properly satisfied by each pair of basis functions. The influence of singularities on the rate of convergence of basis set expansions in (nonrelativistic) variational calculations has been discussed by Klahn and Morgan (1984).

It is possible to give a rigorous characterization of the properties of the matrix problem resulting from the use of expansions (53) with the bais set construction (54) (Grant 1982, 1986a). To understand this characterization, consider what happens when we use (53) as a variational trial solution to obtain stationary values of the Rayleigh quotient

$$W[u] = \frac{(u_N|h_D|u_N)}{(u_N|u_N)} \quad (55)$$

where

$$u_N(r) = \begin{bmatrix} P_N(r) \\ Q_N(r) \end{bmatrix}.$$

The coefficient vectors p and q satisfy the eigenvalue equation

$$\begin{bmatrix} \mathbf{V} & c\mathbf{\Pi} \\ c\mathbf{\Pi}' & \mathbf{V}' - 2c^2\mathbf{S}' \end{bmatrix} \begin{bmatrix} \mathbf{p} \\ \mathbf{q} \end{bmatrix} = \varepsilon \begin{bmatrix} \mathbf{Sp} \\ \mathbf{S'q} \end{bmatrix} \quad (56)$$

where $\mathbf{V}, \mathbf{V}', \mathbf{S}, \mathbf{S}', \mathbf{\Pi}, \mathbf{\Pi}'$ are $N \times N$ matrices with elements

$$V_{ij} = \langle \pi_i|V|\pi_j \rangle \qquad V'_{ij} = \langle \rho_i|V|\rho_j \rangle$$

$$S_{ij} = \langle \pi_i|\pi_j \rangle \qquad S'_{ij} = \langle \rho_i|\rho_j \rangle$$

$$\Pi_{ij} = \left\langle \pi_i \left| \left(\frac{-d}{dr} + \frac{\kappa}{r} \right) \right| \rho_j \right\rangle \qquad \Pi'_{ij} = \left\langle \rho_i \left| \left(\frac{d}{dr} + \frac{\kappa}{r} \right) \right| \pi_j \right\rangle.$$

Notice that $\Pi_{ij} = \Pi_{ji}^\dagger$ when the basis functions satisfy bound state boundary conditions, so that Π and $\Pi^\dagger$ are then matrix adjoints as the superscript † implies. In this form, one can prove rigorously (see Grant 1982, 1986a for the details) that if $V_{\min}$ is a lower bound to the lowest eigenvalues of the matrices $\mathbf{V}$ and $\mathbf{V}'$, and if we have chosen the basis sets so that $V_{\min} \geqslant -2mc^2$, than there will be N eigenvalues below $-2mc^2$ and N eigenvalues above $V_{\min}$. The N eigenvalues in the lower continuum are square integrable approximations to continuum solutions of the Dirac operator, and one does not expect—nor does one find in practice—that these will converge to any identifiable solution as the dimension $N \to \infty$. The same is true of the uppermost (say) $N - M$ in the upper continuum. The remaining M eigenvalues in the gap $(V_{\min}, 0)$ converge from above to bound energy levels as $N \to \infty$. Finally, the highest N eigenvalues and eigenvectors converge to the matrix nonrelativistic Schrödinger solutions in the limit $c \to \infty$.

It follows that we can expect those solutions corresponding to the bound eigenstates to converge with N in much the same way as in nonrelativistic calculations provided *all* the conditions set out in the last paragraph are met. We have already seen that there are likely to be problems if the boundary conditions of §II B,C are not satisfied by matched pairs of basis functions. The boundary conditions can only be satisfied in a least square sense when this is not the case, and we shall need larger values of N to achieve the required accuracy. This proves to be particularly catastrophic when we use integer exponent STF (or GTF) for cases with $\kappa > 0$ with a point nucleus. (We discuss this pathology below.) If, as do many authors, we construct our small component basis $\{\rho_j(r)\}$ by *adding* additional functions to the set $\{\pi_i(r)\}$ to allow for linear combinations which *can* satisfy the $r = 0$ boundary conditions so that $N' > N$, the equations (56) are formally unchanged as long as we adjust the ranges of the indices. However, there will be at least $N' - N$ spurious eigenvalues which can have no physical counterpart—the corresponding eigenvectors cannot behave correctly as $r \to 0$—and we have no idea where these will occur in the spectrum. In fact, *all* the eigenvectors are likely to be affected by the mismatch, and they are quite likely to be poor approximations to the physical states. Basis sets with $N' \neq N$ or sets with $N' = N$ which are not properly matched are therefore useless in problems in which one does not know the correct answer in advance. Basis sets of properly matched pairs of radial functions will henceforth be referred to as (STF or GTF) spinor basis functions.

Much of the literature of the subject has concerned the need to ensure that equations (56) have the correct Schrödinger limit, and the fact that the prescription (54) guarantees the correct nonrelativistic kinetic energy as $c \to \infty$ is an integral component of the proof of the basic theorem. The reader must appreciate that finite matrix representations of unbounded quantum mechanical operators such as **r** and **p** do not behave in quite the same way as their counterparts in the usual Hilbert space $L^2(\mathbf{R}_+)$. For example, it is well known that the canonical operator commutator relation $[x_i, p_j] = i\hbar \delta_{ij}$ cannot be satisfied in a finite matrix representation. (To prove this, just take the trace of both sides.) One consequence of this is that the matrix representation of an operator product AB, when this has a meaning, is not in general equal to the product of the matrix representations of A and B. A particular case of this is the matrix product that appears when one tries to reduce (56) to the nonrelativistic limit; we want the matrix $\frac{1}{2}\Pi(S')^{-1}\Pi^\dagger$ to approximate the matrix of $\frac{1}{2}[-d^2/dr^2 + l(l+1)/r^2]$ with respect to the non-orthogonal basis $\{\pi_i(r)\}$. The problem was studied in detail by Dyall, Grant and Wilson (1984) who showed that by choosing the basis sets according to the prescription (54) the matrix $\frac{1}{2}\Pi(S')^{-i}\Pi^\dagger$ gives exactly the required result (up to arithmetic rounding error). It is easy to show that any other choice of basis $\{\rho_j(r)\}$ will lead to estimates of the nonrelativistic kinetic energy which are smaller than they should be, so that it is not surprising that naive use of basis set methods generally produces eigensolutions which exhibit variational collapse.

The arguments of this section are not restricted to radially reduced equations. The boundary condition at the centre of the nuclear charge distribution will generate precise relations between the four components of a basis spinor in Cartesian coordinates, and the trial function will once more be expressible as a linear combination of four component basis spinor functions. This is likely to prove an important element in developing a successful procedure for the relativistic calculation of the electronic structure of molecules.

B. The Matrix Dirac–Fock Equations Including The Breit Interaction

We present in this section the matrix form of the Dirac–Fock equations for a closed shell atom when we use the low frequency (Breit) form (43) of the transverse interaction. Whilst this is not strictly correct when the orbital energy difference ω is not small, it is relatively easy to program; the finite ω corrections are very small and are best accomodated as a perturbation, or, more accurately, as an iterative correction at a later stage.

The equations, which are very similar to those derived earlier by Kim (1967) and Kagawa (1975, 1980), have been given by Quiney *et al.* (1987). We suppose, for convenience, that we are using the same size, N, of basis set for all orbital symmetry types, κ. Then, if there are K symmetry types, the Dirac–Fock matrix will be of size $2NK \times 2NK$. Examination of the block structure shows that this can be rewritten as a set of equations

$$\mathbf{F}_\kappa \mathbf{X}_\kappa = \mathbf{E}_\kappa \mathbf{S}_\kappa \mathbf{X}_\kappa, \tag{57}$$

one for each symmetry type, coupled through the Fock operators $\mathbf{F}_\kappa$. $\mathbf{E}_\kappa$ is a diagonal matrix whose entries are the $2N$ orbital eigenvalues for symmetry κ. The $2N \times 2N$ matrix $\mathbf{S}_\kappa$ can be partitioned into $N \times N$ blocks in the form

$$\mathbf{S}_\kappa = \begin{bmatrix} \mathbf{S}_\kappa^{LL} & 0 \\ 0 & \mathbf{S}_\kappa^{SS} \end{bmatrix} \tag{58}$$

where the superscripts TT' (here given the values LL or SS) indicate which basis sets are used in the construction in the manner of (56). The Fock matrix $\mathbf{F}_\kappa$ takes the form

$$\mathbf{F}_\kappa = \mathbf{h}_\kappa + \mathbf{g}_\kappa + \mathbf{b}_\kappa \qquad (59)$$

where, in a notation corresponding to (56),

$$\mathbf{h}_\kappa = \begin{bmatrix} \mathbf{V}_\kappa^{LL} & c\Pi_\kappa^{LS} \\ c\Pi_\kappa^{SL} & \mathbf{V}_\kappa^{SS} - 2c^2 \mathbf{S}_\kappa^{SS} \end{bmatrix};$$

$\mathbf{h}_\kappa$ is the matrix representation of the Dirac operator in the basis corresponding to symmetry κ; $\mathbf{g}_\kappa$ is the matrix of the electron–electron Coulomb repulsion,

$$\mathbf{g}_\kappa = \begin{bmatrix} \mathbf{J}_\kappa^{LL} - \mathbf{K}_\kappa^{LL} & -\mathbf{K}_\kappa^{LS} \\ -\mathbf{K}_\kappa^{SL} & \mathbf{J}_\kappa^{SS} - \mathbf{K}_\kappa^{SS} \end{bmatrix};$$

and $\mathbf{b}_\kappa$ is the matrix of the Breit interaction operator,

$$\mathbf{b}_\kappa = \begin{bmatrix} \mathbf{B}_\kappa^{LL} & \mathbf{B}_\kappa^{LS} \\ \mathbf{B}_\kappa^{SL} & \mathbf{B}_\kappa^{SS} \end{bmatrix}.$$

Each of the submatrices has dimension $N \times N$; they can, as usual, be expressed in terms of the density matrices $D_\kappa^{TT'}$, whose elements are

$$D_{\kappa ij}^{TT'} = \chi_{\kappa i}^T \cdot \chi_{\kappa j}^{T'}, \qquad i,j = 1,\ldots,N \qquad (60)$$

where T and T' represent either of the letters L or S and, in terms of the notation used for the one electron problem, $\chi_\kappa^L = \mathbf{p}_\kappa$ and $\chi_\kappa^S = \mathbf{q}_\kappa$ are the coefficients of (53), along with integral over basis functions. Thus J_κ^{TT} and K_κ^{TT} have the matrix elements

$$J_{\kappa ij}^{TT} = \sum_{\kappa'l}(2j'+1)\,[D_{\kappa'kl}^{TT} J_{\kappa ij,\kappa'kl}^{0,TT,TT} + D_{\kappa'kl}^{\overline{T}\overline{T}} J_{\kappa ij,\kappa'kl}^{0,TT,\overline{T}\overline{T}}]$$

$$K_{\kappa ij}^{TT'} = \sum_v \sum_{\kappa'kl}(2j'+1)b_v(jj')D_{\kappa'kl}^{TT'} K_{\kappa ij,\kappa'kl}^{v,TT'T'T} \qquad (61)$$

whilst the Breit matrix elements have the form

$$B_{\kappa ij}^{TT} = \sum_v \sum_{\kappa'kl}(2j'+1)e_v(jj')D_{\kappa'kl}^{\overline{T}\overline{T}} K_{\kappa ij,\kappa'kl}^{v,TT,\overline{T}\overline{T}}$$

and

$$B_{\kappa ij}^{T\overline{T}} = \sum_v \sum_{\kappa'kl}(2j'+1)[d_v(\kappa\kappa')D_{\kappa'kl}^{\overline{T}T} K_{\kappa ij,\kappa'kl}^{v,TT,TT} \qquad (62)$$
$$+ g_v(\kappa\kappa')D_{\kappa'kl}^{TT} M_{\kappa ij,\kappa'kl}^{v,TT,TT}]$$

In these formulas, based on the paper of Grant and Pyper (1976), the superscripts T, T' can take the values L and S, and are associated with the basis set indices appearing in the subscripts below. Also the notation $T\bar{T}$ implies a pair LS or a pair SL. The radial integrals are defined by

$$J^{v,TT,T'T'}_{\kappa ij,\kappa'kl} = \int_0^\infty \int_0^\infty f^T_{\kappa i}(r) f^T_{\kappa j}(r) U_v(r,s) f^{T'}_{\kappa'k}(s) f^{T'}_{\kappa'l}(s)\,ds\,dr \tag{63}$$

and

$$K^{v,TT',TT'}_{\kappa ij,\kappa'kl} = J^{v,TT,T'T'}_{\kappa i,\kappa'k,\kappa j,\kappa'l}$$

whilst

$$M^{v,T\bar{T},\bar{T}T}_{\kappa ij,\kappa'kl} = \int_0^\infty \int_r^\infty f^T_{\kappa i}(r) f^{\bar{T}}_{\kappa'k}(r) U_v(r,s) f^T_{\kappa j}(s) f^{\bar{T}}_{\kappa'l}(s)\,ds\,dr \tag{64}$$

where $f^T_{\kappa i}(r)$ denotes a basis function, either π_i or ρ_i, and

$$U_v(r,s) = \frac{r^v}{s^{v+1}} \quad \text{if} \quad r < s,$$

$$= \frac{s^v}{r^{v+1}} \quad \text{if} \quad s > r.$$

The coefficients are most simply generated using the MCBP module of the GRASP package from McKenzie et al. (1980), whilst the two-electron integrals for the exponential type functions used in our calculations can be found in or deduced from the paper of Roothaan and Bagus (1962).

C. Choice of Basis Sets

The choice of a set of basis functions in which to parametrize the set of four component spinor solutions inevitably represents a compromise between the accuracy and economy of any calculation. Ideally, we wish to carry over into the Dirac problem as much of the highly developed nonrelativistic algorithms for integral evaluation as possible. The spinor description should be as economical as possible, the basis set should tend to completeness as it is enlarged systematically and should be free of computational linear dependence. An efficient parametrization will be essential if we are to make worthwhile calculations for molecules.

A certain amount of "skill" is usually required to select basis set parameters which will result in an economic and accurate calculation. This knowledge is acquired from experience of previous calculations of similar systems, but often does not allow any means of systematic improvement as the basis is extended. Methods of optimization which involve the nonlinear variation of exponential parameters do not necessarily lead to a set which tends towards completeness. Such sets may suffer from computational linear dependence and make their use highly inefficient. We have argued, §IV A, that such schemes of "improvement" cannot converge in the Dirac case unless they satisfy certain conditions. We must use efficient algorithms with spinor calculations since the matrices have at least double the dimension of the corresponding nonrelativistic ones. This will exacerbate the inherent weaknesses and expense of non-linear optimization procedures and make them unattractive for further study.

We have therefore turned to even-tempered basis sets which avoid the failings of nonlinear optimizing methods and possess many desirable properties. First suggested by Reeves (1963), and later revived and developed by Ruedenberg and co-workers (see Wilson 1984 for an extensive list of references) even-tempered basis sets are defined as sets of functions of the form

$$N_k \exp(-\zeta_k r^p) r^{l+1} \qquad k = 1, 2, \ldots \qquad (65)$$

where $p = 1$ for STF and $p = 2$ for GTF. The exponents are chosen to form a geometric sequence

$$\zeta_k = \alpha_N \beta_N^{k-1}, \qquad k = 1, 2, \ldots, N \qquad (66)$$

Schmidt and Ruedenberg (1979) and Feller and Ruedenberg (1979) have devised schemes for systematically extending such exponent sets to completeness. This requires the selection of a sequence of values α_N and β_N such that $\alpha_N \to 0+$, $\beta_N \to 1+$ from above (Klahn and Bingel, 1977a, 1977b). We have used the sequences

$$\alpha_N = \left(\frac{\beta_N - 1}{\beta_{N-1} - 1}\right)^a \alpha_{N-1}, \qquad \ln \beta_N = \left(\frac{N}{N-1}\right)^b \ln \beta_{N-1} \qquad (67)$$

where $a > 0$ and $b < 0$ are constants. Linear dependence is not a problem as long as β_N is greater than unity, and it is easy to control ill-conditioning. We have also exploited the use of efficient recursive methods to generate integrals over the basis functions.

It is still necessary to decide what basis functions should be used for the large component $\{\pi_i(r)\}$ and small component $\{\rho_i(r)\}$ basis sets. Early workers with matrix DF schemes (Kim, 1967; Kagawa, 1975, 1980), used the same functions for both sets. Kim, however, chose STF with the correct Coulomb cusp exponents

$$N_k \exp(-\zeta_k r) \cdot r^\gamma \qquad (68)$$

whereas Kagawa employed integer exponents as in (65). Drake and Goldman (1981) used expansions in functions of the form

$$r^{\gamma+i} \exp(-\zeta r) \qquad i = 0, 1, \ldots, (N-1) \qquad (69)$$

for both sets. This worked satifactorily for $\kappa < 0$, but gave a spurious eigenvalue for the lowest positive energy state with $\kappa > 0$ degenerate with the corresponding eigenvalue for $\kappa < 0$. Goldman (1985) attempted to overcome the difficulty by choosing (in our notation)

$$\pi_i(r) = r^{\gamma+i} \exp(-\zeta r)$$

$$\rho_i(r) = \left[\frac{(2\gamma + i)}{r} + \frac{Z}{\kappa} - \zeta\right] r^{\gamma+i} \exp(-\zeta r)$$

for $i = 1, 2, \ldots, N$, augmented with an additional function proportional to $r^\gamma \exp(-\zeta r)$ when $\kappa > 0$ with either the same variational exponent ζ or the specific exponent $Z/|\kappa|$. Several authors have assessed kinetic balance (Stanton and Havriliak 1984) as a partial means of achieving "variational safety," though they have interpreted the idea in different ways. Some (Lee and McLean, 1982; Aerts and Nieuwpoort, 1986) have expanded the set $\{\rho_j\}$ to include both those in the set $\{\pi_i\}$ and additional functions so that the dimension N' of the former is greater than N. We have argued above that this is likely to be unsatisfactory, and indeed the results obtained by these authors are consistent with our point of view. Others, especially Ishikawa, Binning and Sando (1984), who did minimal basis set calculations with a point nucleus, and Ishikawa, Baretty and Binning (1985), who considered the use of GTF with finite nuclear charge models, have operated much more in the spirit of our investigation.

It is not easy to compare the results of different authors as they have studied different systems and adopted optimization procedures that are not fully defined by the printed descriptions. We have therefore studied a range of representative test problems whose results are reported in the next section.

D. COMPARISON OF METHODS

The first of our illustrative calculations for hydrogenic atoms is represented by Table I, in which the basis sets were of the classic nonrelativistic STF type with strict kinetic balance

$$\pi_i(r) = r^{l+1} \exp(-\zeta_i r),$$

$$\rho_i(r) = [(l + 1 + \kappa)r^l - \zeta_i r^{l+1}] \exp(-\zeta_i r) \tag{70}$$

as defined by (54). The choice $Z = 92$ was made to ensure a genuinely relativistic problem. The basis set parameters, determined in accordance with (67), were chosen more or less arbitrarily so as not to favour any one method

TABLE I

(i) Hydrogenic Atom, $Z = 92$.[a]

κ		α	β	E_{MATRIX}	E_{DIRAC}
-1	$1s_{1/2}$	110.273	1.50000	-4861.1817858398	-4861.1817858398
$+1$	$2p_{1/2}$	27.7273	1.89453	-1257.4065338220	-1257.3906767178
-2	$2p_{3/2}$	21.0036	1.50000	-1089.6107790759	-1089.6107790759
$+2$	$3d_{3/2}$	19.8173	1.60004	-489.03686731398	-489.03670400437
-3	$3d_{5/2}$	20.6622	1.50000	-476.26147664590	-476.26147664590
$+3$	$4f_{5/2}$	23.3027	1.44015	-268.96581997541	-268.96597005073
-4	$4f_{7/2}$	15.4468	1.50000	-266.38941053008	-266.38941053008
$+4$	$5g_{7/2}$	18.5363	1.37100	-170.82890869533	-170.82890697718
-5	$5g_{9/2}$	12.3339	1.50000	-170.04991942815	-170.04991942815

(ii) Alternative calculation for states of negative κ

κ		α	β	E_{MATRIX}	E_{DIRAC}
-1	$1s_{1/2}$	121.958	1.40000	-4861.1817858398	-4861.1817858398
-2	$2p_{3/2}$	46.6146	1.40000	-1089.6107790759	-1089.6107790759
-3	$3d_{5/2}$	15.8102	1.40000	-476.26147664590	-476.26147664590
-4	$4f_{7/2}$	16.5307	1.40000	-266.38941053008	-266.38941053008
-5	$5d_{9/2}$	13.2043	1.40000	-170.04991942815	-170.04991942815

[a] Comparison of computed eigenvalue E_{MATRIX} for basis set, (70), having 9 even-tempered exponents, with the eigenvalue E_{DIRAC} from Sommerfeld's formula.

Note: The value $\alpha^{-1} = 137.0373$ has been used for all the calculations reported in this paper for historical reasons.

unduly. On the other hand, we have tried to ensure that they do give fair, though not necessarily optimal, representations of the states of interest. With $N = 9$, it is apparent that the eigenvalue of the lowest state of each symmetry agrees with Sommerfeld's formula to within rounding errors for negative κ values ($1s_{1/2}, 2p_{3/2}, 3d_{5/2}, \ldots$). In the case of positive κ ($2p_{1/2}, 3d_{3/2}, 4f_{5/2}, \ldots$) the matrix eigenvalue is below the exact one, the most serious failure to give an upper bound being for the $2p_{1/2}$ case. The lower part of the table shows that the results for the negative κ states are insensitive to the basis set parameters; in all cases, the eigenvalues agree within rounding error. This kind of behaviour is typical of that seen in other calculations with kinetically balanced sets (Stanton and Havriliak, 1984). We remark that bounds failures are less severe with higher states so that it is legitimate to pay most attention to the lowest states of each symmetry. Clearly, the $2p_{1/2}$ state is the critical one.

The next, Table II, compares the behaviour of two different types of even-tempered basis set for the $1s$ state with $Z = 1$; columns 4 and 5 give,

TABLE II

Convergence of Hydrogenic $1s_{1/2}$ States Using Systematic Sequences of Even-Tempered Basis Sets.[a]

(a) $Z = 1$

N	α	β				
2	0.835	2.534	−0.49462	0.4969(−02)	−0.49462	0.4969(−02)
3	0.719	2.136	−0.49649	0.2794(−02)	−0.49649	0.2795(−02)
4	0.650	1.930	−0.49933	0.4819(−03)	−0.49933	0.4819(−03)
5	0.603	1.800	−0.49997	0.2762(−04)	−0.49997	0.2762(−04)
6	0.568	1.710	−0.50000	0.1229(−05)	−0.50000	0.1230(−05)
7	0.541	1.644	−0.50000	0.5922(−05)	−0.50000	0.5921(−05)
8	0.519	1.592	−0.50000	0.3881(−05)	−0.50000	0.3882(−05)
9	0.500	1.550	−0.50000	0.1340(−05)	−0.50000	0.1340(−05)
10	0.484	1.536	−0.50001	0.5828(−06)	−0.50001	0.5829(−06)
(EXACT)			−0.50001		−0.50001	

(b) $Z = 92$

N	α	β				
2	28.000	1.750	−4391.34355	0.5424(−01)	−3709.75116	0.1340
3	24.606	1.579	−4782.72612	0.6749(−02)	−4315.92258	0.4017(−01)
4	22.526	1.485	−4856.00826	0.3899(−03)	−4603.62583	0.1229(−01)
5	21.069	1.424	−4861.13733	0.3218(−05)	−4721.42645	0.5060(−02)
6	19.967	1.381	−4861.17714	0.3283(−06)	−4771.04173	0.2882(−02)
7	19.091	1.349	−4861.18151	0.1881(−07)	−4796.94893	0.1931(−02)
8	18.372	1.322	−4861.18177	0.1214(−08)	−4812.96123	0.1387(−02)
9	17.764	1.301	−4861.18178	0.1236(−09)	−4824.37128	0.1020(−02)
10	17.241	1.284	−4861.18179	0.1660(−10)	−4832.61622	0.7684(−03)
11	16.784	1.269	−4861.18179	0.5274(−12)	−4838.50752	0.5961(−03)
12	16.379	1.256	−4861.18179	0.3355(−12)	−4842.91463	0.4714(−03)
13	16.018	1.245	−4861.18179	0.4718(−15)	−4846.35243	0.3768(−03)
14	15.395	1.226	−4861.18179	0.2054(−14)	−4847.10836	0.3546(−03)
(EXACT)			−4861.18179		−4861.18179	

[a] Eigenvalues (columns 4, 6) and overlap deficiencies (columns 5, 7). Basis sets: equation (71), columns 4, 5; equation (70), columns 6, 7.

respectively, the eigenvalue and the difference of the overlap of the trial solution with the exact solution from unity in a calculation with even-tempered cusp functions

$$\pi_i(r) = r^\gamma \{C_1 F(a+1, b; 2\zeta_i r) - C_2 F(a, b; 2\zeta_i r)\}$$
$$\rho_i(r) = r^\gamma \{C_1 F(a+1, b; 2\zeta_i r) + C_2 F(a, b; 2\zeta_i r)\}$$

(71)

TABLE III

CONVERGENCE OF HYDROGENIC $2p_{1/2}$ STATES USING SYSTEMATIC SEQUENCES OF EVEN-TEMPERED BASIS SETS.[a]

(a) $Z = 1$

N	α	β						
2	0.835	2.534	−0.08355	0.1303	−0.08355	0.1303	−0.08355	0.1303
3	0.719	2.136	−0.11241	0.5185(−01)	−0.11241	0.5184(−01)	−0.11241	0.5184(−01)
4	0.650	1.930	−0.12093	0.1900(−01)	−0.12093	0.1900(−01)	−0.12093	0.1900(−01)
5	0.603	1.800	−0.12373	0.6168(−02)	−0.12373	0.6167(−02)	−0.12373	0.6167(−02)
6	0.568	1.710	−0.12465	0.1684(−02)	−0.12465	0.1684(−02)	−0.12465	0.1684(−02)
7	0.541	1.644	−0.12493	0.3487(−03)	−0.12493	0.3486(−03)	−0.12493	0.3486(−03)
8	0.519	1.592	−0.12499	0.3960(−04)	−0.12499	0.3959(−04)	−0.12499	0.3959(−04)
9	0.500	1.550	−0.12500	0.6632(−12)	−0.12500	0.1794(−11)	−0.12500	0.1794(−11)
10	0.484	1.516	−0.12500	0.7650(−05)	−0.12500	0.7651(−05)	−0.12500	0.7651(−05)
(EXACT)			−0.12500		−0.12500		−0.12500	

(b) $Z = 92$

N	α	β						
2	58.231	2.875	−1234.39567	0.1090(−01)	−1249.51049	0.2688(−02)	−1246.80712	0.2490(−02)
3	57.629	2.617	−1240.70206	0.8157(−02)	−1255.15578	0.1250(−02)	−1252.16015	0.1338(−02)
4	56.062	2.448	−1247.83250	0.4716(−02)	−1256.16827	0.6968(−03)	−1254.89237	0.6205(−03)
5	54.902	2.332	−1251.61563	0.2872(−02)	−1257.17604	0.1974(−03)	−1256.41967	0.2149(−03)
6	53.990	2.246	−1253.77986	0.1794(−02)	−1257.47805	0.8238(−04)	−1257.01470	0.7157(−04)
7	53.242	2.178	−1255.10262	0.1134(−02)	−1257.52538	0.1922(−04)	−1257.23863	0.1707(−04)
8	52.610	2.123	−1255.93451	0.7184(−03)	−1257.58651	0.1031(−04)	−1257.30335	0.9160(−05)
9	52.064	2.077	−1256.47049	0.4513(−03)	−1257.51153	0.2620(−04)	−1257.29222	0.2229(−04)
10	51.585	2.038	−1256.81834	0.2788(−03)	−1257.49511	0.4347(−04)	−1257.25591	0.4320(−04)
11	51.158	2.004	−1257.04540	0.1669(−03)	−1257.42100	0.6913(−04)	−1257.20895	0.6644(−04)
12	50.775	1.975	−1257.19113	0.9575(−04)	−1257.38569	0.8909(−04)	−1257.16209	0.8881(−04)
13	50.427	1.948	−1257.28396	0.5079(−04)	−1257.33078	0.1097(−03)	−1257.12040	0.1082(−03)
14	50.109	1.925	−1257.34002	0.2392(−04)	−1257.29742	0.1256(−03)	−1257.08258	0.1253(−03)
15	49.816	1.903	−1257.37174	0.8868(−05)	−1257.26009	0.1395(−03)	−1257.05279	0.1386(−03)
16	49.545	1.883	−1257.38676	0.1820(−05)	−1257.23675	0.1492(−03)	−1257.02894	0.1490(−03)
(EXACT)			−1257.39067		−1257.39067		−1257.39067	

[a] As Table II, with additional columns 8, 9: integer STF, uncontracted kinetic balance.

where $F(a, b; z)$ is a confluent hypergeometric function, $-a = n_r = n - |\kappa|$ is the number of radial nodes in $\pi_i(r)$, $b = 2\gamma + 1$, $C_1 = l - |\kappa| + 1$ and $C_2 = H - |\kappa|$ with

$$H = +\{(l + 1)^2 - 2C_1(|k| - \gamma)\}^{1/2}.$$

These functions satisfy the full boundary conditions as $r \to 0$ and recover kinetic balance, (54), as $c \to \infty$. Columns 6 and 7 give similar information for the classic integer STF set (70). It is clear that there is little to choose between the functions for such a low value of Z. However, for $Z = 92$, it becomes obvious that the set (71) converges faster and more smoothly than the others; there are no cases in which any set fails to give an upper bound. The superior performance of cusp functions is not surprising, because $\gamma \simeq 0.74$ for $Z = 92$, so that the solution has infinite slope at the origin, and this is properly represented in (71). The larger exponents needed when using the functions (70) are to compensate for the singular derivative of the analytic solution; obviously this can never be more than a partial remedy for incorrect analytic behaviour. Such large exponents may lead to values of the lowest eigenvalue of the potential matrix V_{min} which are below the critical value of $-2mc^2$, and then there is no admissible lower bound to the positive spectrum.

Table III gives similar information for the more critical $2p_{1/2}$ state. This table includes two additional columns, 8 and 9, which refer to the case in which the set $\{\rho_i\}$ has been constructed not according to (54) but as the union of the large component set $\{\pi_i\}$ with additional functions of the form $r^l \exp(-\zeta_i r)$, so that $N' = 2N$. Once again, there is little to choose between the three types of basis set for $Z = 1$. The main feature of interest is the remarkable accuracy for $N = 9$, which arises because an exponent in one of the basis functions accidentally coincides with the exponent of the analytic solution. Thus the cusp function is virtually exact; for the kinetic balance sets, the poor fit of the integer exponent expansions near $r = 0$ has to be compensated by other functions with large ζ_i values. However, it is a different story for $Z = 92$. Columns 4 and 5 show smooth monotone convergence in both eigenvalue and overlap deficiency as N increases, whereas the strict kinetic balance eigenvalues for integer l, column 6, fall below the exact eigenvalue for $N = 6 - 12$ after which the energy begins to rise again. Such behaviour will make it difficult to use this basis when we do not know the correct answer in advance. Moreover, there is no systematic improvement in the overlap deficiency in column 7 as N increases. Similar behaviour is shown in columns 8 and 9, where the solutions approach those of the two previous columns as N increases. The doubled size of the small component basis leads to severe computational linear dependence problems and loss of accuracy.

Table IV illustrates one of two specimen sets of matrix DF calculations carried out for the present investigation. The $1s^2$ ground state of the helium

TABLE IVA

HELIUM ($1s_{1/2}^2$)($^1S_0$)

(i) Even-tempered basis of integer Slater-type functions

N	α	β	ε_{1s}	E_{SCF}
2	1.750	1.850	−0.87452670	−2.8438816
3	1.533	1.653	−0.91407246	−2.8611277
4	1.401	1.545	−0.91807281	−2.8618106
5	1.309	1.476	−0.91805190	−2.8617996
6	1.239	1.426	−0.91799521	−2.8618078
7	1.184	1.389	−0.91798887	−2.8618121
8	1.139	1.360	−0.91798997	−2.8618132
9	1.100	1.336	−0.91799055	−2.8618133
10	1.068	1.317	−0.91799068	−2.8618133
(MCDF)			−0.91799074	−2.8618134

(ii) Even-tempered basis of cusp Slater-type functions

N	α	β	ε_{1s}	E_{SCF}
2	1.750	1.850	−0.87453247	−2.8438879
3	1.533	1.653	−0.91406531	−2.8611251
4	1.401	1.545	−0.91807405	−2.8618104
5	1.309	1.476	−0.91805161	−2.8617997
6	1.239	1.426	−0.91799524	−2.8618077
7	1.184	1.389	−0.91798889	−2.8618121
8	1.139	1.360	−0.91798995	−2.8618132
9	1.100	1.336	−0.91799055	−2.8618133
10	1.068	1.317	−0.91799068	−2.8618133
(MCDF)			−0.91799074	−2.8618134

TABLE IVB

HELIUM-LIKE URANIUM ($1s_{1/2}^2$)($^1S_0$)

(i) Kinetically balanced basis of even-tempered integer Slater-type functions

N	α	β	ε_{1s}	E_{SCF}
2	140.000	2.800	−4471.0086	−9029.2638
3	119.795	2.318	−4776.4223	−9627.0584
4	107.992	2.071	−4774.8318	−9622.4121
5	99.970	1.917	−4789.6706	−9649.8237
6	94.032	1.812	−4789.6061	−9650.1950
7	89.393	1.734	−4788.7447	−9648.0997
8	85.625	1.673	−4790.2263	−9651.2851
9	82.451	1.625	−4789.4190	−9649.6128
10	79.790	1.585	−4790.2496	−9651.3178
(MCDF)			−4790.2724	−9651.3622

(ii) Even-tempered basis of cusp Slater-type functions

N	α	β	ε_{1s}	E_{SCF}
2	28.000	1.750	−4343.3545	−8734.6953
3	24.606	1.579	−4718.6229	−9501.3149
4	22.526	1.485	−4785.8319	−9641.7602
5	21.069	1.424	−4790.2373	−9651.2826
6	19.967	1.381	−4790.2703	−9651.3554
7	19.091	1.349	−4790.2724	−9651.3620
8	18.372	1.322	−4790.2723	−9651.3621
9	17.764	1.301	−4790.2724	−9651.3622
10	17.241	1.284	−4790.2724	−9651.3622
(MCDF)			−4790.2724	−9651.3622

TABLE VA
HELIUM $(2p_{1/2}^2)(^1S_0)$

(i) Kinetically balanced basis of even-tempered integer Slater-type functions

N	α	β	$\varepsilon_{2p_{1/2}}$	E_{SCF}
2	0.434	2.042	−0.19199494	−0.67268584
3	0.378	1.791	−0.19308214	−0.67415834
4	0.346	1.656	−0.19310687	−0.67414082
5	0.323	1.569	−0.19306287	−0.67414272
6	0.305	1.509	−0.19306303	−0.67415885
7	0.291	1.464	−0.19306925	−0.67415931
8	0.280	1.428	−0.19306787	−0.67415945
9	0.270	1.400	−0.19306754	−0.67415956
(MCDF)			−0.19306763	−0.67415961

(ii) Even-tempered basis of cusp Slater-type functions

N	α	β	$\varepsilon_{2p_{1/2}}$	E_{SCF}
2	0.434	2.042	−0.19255044	−0.67355883
3	0.378	1.791	−0.19308308	−0.67415834
4	0.346	1.656	−0.19310649	−0.67414070
5	0.323	1.569	−0.19306287	−0.67414260
6	0.305	1.509	−0.19306307	−0.67415887
7	0.291	1.464	−0.19306922	−0.67415324
8	0.280	1.428	−0.19306787	−0.67415944
9	0.270	1.400	−0.19306755	−0.67415956
(MCDF)			−0.19306763	−0.67415961

TABLE VB
HELIUM-LIKE URANIUM $(2p_{1/2}^2)(^1S_0)$

(i) Kinetically balanced basis of even-tempered integer Slater-type functions

N	α	β	$\varepsilon_{2p_{1/2}}$	E_{SCF}
2	60.000	2.900	−1221.1861	−2466.0085
3	57.629	2.617	−1232.8193	−2487.9673
4	56.062	2.448	−1233.9793	−2490.1830
5	54.902	2.332	−1235.3526	−2492.5162
6	53.990	2.246	−1235.7306	−2493.1938
7	53.242	2.178	−1235.9495	−2493.4584
8	52.610	2.123	−1236.0608	−2493.6290
(MCDF)			−1235.8786	−2493.2369

(ii) Even-tempered basis of cusp Slater-type functions

N	α	β	$\varepsilon_{2p_{1/2}}$	E_{SCF}
2	31.331	1.406	−1234.0633	−2489.1619
3	30.415	1.361	−1235.8204	−2493.0974
4	29.789	1.332	−1235.8763	−2493.2317
5	29.317	1.311	−1235.8787	−2493.2367
6	28.939	1.296	−1235.8789	−2493.2369
7	28.626	1.283	−1235.8790	−2493.2369
8	28.357	1.237	−1235.8790	−2493.2369
(MCDF)			−1235.8786	−2493.2369

isoelectronic sequence is a familiar prototype. The table shows the same basis sety dependence as the corresponding hydrogenic problems studied above. There is little to choose between the various basis sets for the neutral atom, but at $Z = 92$ cusped spinors give more rapid convergence to the benchmark solutions of the finite difference MCDF code (Grant et al., 1980) as N increases. There are no bounds failures, but the kinetic balance set (70) again needs larger exponents ζ_i.

Table V deals with the more sensitive $2p_{1/2}^2$ two-electron state. As before, the variation between basis sets is what would be expected from the hydrogenic calculations. For $Z = 2$, there is little to choose between cusp spinors and kinetically balanced integer exponent functions, both of which give good total energies, but cusped spinors show their marked superiority for $Z = 92$. The total energy for the kinetically balanced integer exponent basis falls below the MCDF value for $N \geqslant 7$, and the exponents are again much larger than those for cusp functions.

V. Outlook and Conclusions

We have argued that the traditional methods of relativistic atomic structure theory based on the papers of Swirles (1935), Grant (1961, 1970, 1981), Desclaux (1971), Lindgren and Rosén (1974), and others can be placed on a firm foundation in a manner consistent with the Furry bound interaction picture of quantum electrodynamics. In the past, the major obstacle to acceptance of this statement has been the presence of negative energy states, which seemed to demand special treatment to exclude their effects from bound state calculations and to prevent the so-called "continuum dissolution" (Sucher, 1980, 1985). We have shown that, as has been known for many years to those using finite difference numerical methods, it is only necessary to ensure that the one-body spinors on which the theory is built should satisfy the correct bound state boundary conditions as given in §II. Negative energy states can appear in higher orders of perturbation theory within the "standard" model (on the same footing as bound core orbitals), or in QED perturbation theory, through excitation of virtual or real electron-positron pairs. There is no difficulty in handling such excitations properly as long as the problem is formulated in the appropriate Fock space. By placing self-consistent field and model potentials in the context of standard QED in the Furry picture as in §III, it becomes possible to carry out the renormalization of divergent terms in the perturbation series according to well understood prescriptions. Similar conclusions have been reached independently in recent conference presentations by, for example, Brown (1987) and Sapirstein (1987).

Many Russian papers on relativistic atomic structure theory have used the Furry picture and the Gell–Mann and Low formula (34) as their starting point. Thus Labzovskii (1971) has presented formulae for calculating the relativistic and correlation energy of atoms. Although much of the work is only available in Russian, several calculations have appeared in English-language journals. For example, Safronova and Rudzikas (1976) used Dirac hydrogenic orbitals to develop Z expansion formulae for states of up to 10 electrons, whilst Ivanov and Pobodebova (1977) have used a zero-order model potential to construct energies and transition rates in complex atoms. In the main, these papers have emphasised applications to the ionic spectra of astrophysical and laboratory plasmas.

The basis set expansion technique seems likely to be a powerful tool for the calculation of atomic and molecular structure if the technical problems so far encountered can be overcome. We have argued that it is vital to use only matched basis sets in which each pair of functions, $\pi_i(r)$ and $\rho_i(r)$, satisfies the correct Dirac kinematic conditions at the origin, and have carried out comparison calculations, §IV, which support this contention. We have not had space to describe other applications in this paper. Those interested should see Grant (1986a) for the ground state of neon and Grant (1986b) for calculations on the ground state of the beryllium isoelectronic sequence. The Dirac–Fock equations including the Breit interaction in the form of §IV B, has has been studied by Quiney et al. (1987a) for helium-like and beryllium-like systems, and the results compared with what happens when it is treated as a first-order perturbation in the traditional manner (Grant, 1970). The accuracy attained is comparable with that of finite difference calculations. All this work used STF basis sets, but we have done some experiments with GTF which demonstrate their promise, especially if finite size nuclear charge models are to be used. Our conclusions agree with those of Ishikawa, Baretty and Binning (1985) on this point.

The use of finite basis sets in atomic calculations has other potential benefits. Thus, by applying many-body perturbation theory, it appears to offer a way to study the non-additivity of relativistic and correlation effects, particularly for few-electron systems. The main advantage of using basis sets in this context is that sums over a complete set of states can be replaced by matrix multiplications in the algebraic method (Quiney et al., 1985a, 1985b), and so avoid difficult problems of numerical integration over continuum functions (Kelly, 1963, 1964). First steps in this direction have been taken by Quiney et al. (1987b) by studying radial correlation in the ground states of He- and Be-like ions; this has recently been extended to include angular correlations with promising results. These ideas, when applied to excited states, may be of considerable use for studying the spectra of highly ionized few-electron ions, for example He- or Li-like uranium on which experiments are currently under way (Gould, 1985).

It is as yet unclear if basis set methods can be made efficient enough to make it possible to do calculations of adequate precision on complex atoms with more than a few electrons. Yet it is here, and in the application to molecules, that the real potential of basis set methods is to be found. We have currently been carrying out some calculations for the ground state of the HCl molecule (Laaksonen, Grant and Wilson, in preparation); these so far use only kinetically balanced GTF sets of the extended type. Further work is needed to see what can be gained by using properly matched basis functions. It will be interesting to see how the subject develops in the next few years.

ACKNOWLEDGEMENTS

The Association of Commonwealth Universities and the British Council are thanked for the award of a Scholarship to H. M. Quiney.
This review was completed in June 1986.

REFERENCES

Aerts, P. J. C., and Nieuwpoort, W. C. (1986). *Chem. Phys. Lett.* **125** 83.
Aoyama, T., Yamakawa, H. and Matsuoka, O. (1980). *J. Chem. Phys.* **73**, 1329.
Barut, A. O., and Xu Bo-Wei (1982). *Physica Scripta* **26**, 129.
Berestetskii, V. B., Lifshitz, E. M., and Pitaevskii, L. P. (1971). "Relativistic Quantum Theory," Pergamon, Oxford.
Bethe, H. A., and Salpeter, E. E. (1957). "Quantum Mechanics of One- and Two-Electron Systems", §42. Springer-Verlag, Berlin, Göttingen, Heidelberg.
Bouazza, S., Guern, Y., and Bauche, J. (1986). *J. Phys. B* **19**, 1881.
Breit, G. (1929). *Phys. Rev.* **34**, 553.
Breit, G. (1930). *Phys. Rev.* **36**, 383.
Brown, G. E. (1987). *Phys. Scr.*, **36**, 76.
Cheng, K. T., Desclaux, J. P., and Kim, Y.-K. (1978). *J. Phys. B* **11**, L359.
Cheng, K. T., and Johnson, W. R. (1976). *Phys. Rev. A* **14**, 1943.
Datta, S. N., and Ewig, C. S. (1982). *Chem. Phys. Lett.* **85**, 443.
Desclaux, J. P. (1971). Thèse, University of Paris. Unpublished.
Desclaux, J. P. (1975). *Computer Phys. Commun.*, **9**, 31.
Desiderio, A. M., and Johnson, W. R. (1971). *Phys. Rev. A* **3**, 1267.
Dirac, P. A. M. (1930). *Proc. R. Soc. A* **126**, 360.
Drake, G. W. F., and Goldman, S. P. (1981). *Phys. Rev. A* **23**, 2093.
Dyall, K. G., Grant, I. P., and Wilson, S. (1984). *J. Phys. B* **17**, 493.
Erickson, G. W., and Yennie, D. R. (1965). *Ann. Phys. (NY)* **35**, 271.

Feller, D. F., and Ruedenberg, K. (1979). *Theor. Chim. Acta.* **52**, 231.
Foldy, L. L., (1956). *Phys. Rev.* **102**, 568.
Froese Fischer, C. (1972). *Computer Phys. Commun.* **4**, 107.
Fullerton, L. W. and Rinker, G. A. (1976). *Phys. Rev. A* **13**, 1283.
Furry, W. H. (1951). *Phys. Rev.* **81**, 115.
Gaunt, J. A. (1929). *Proc. R. Soc. A* **122**, 513.
Gell-Mann, M., and Low, F. E. (1951). *Phys. Rev.* **81**, 115.
Goldman, S. P. (1985). *Phys. Rev. A* **31**, 3541.
Gould, H. E. (1985). *AIP Conf. Proc. (USA)* no. 136, p. 66.
Grant, I. P. (1961), *Proc. R. Soc. A* **262**, 555.
Grant, I. P. (1970), *Advan. Phys.* **19**, 747.
Grant, I. P. (1980). *Phys. Scr.* **21**, 443.
Grant, I. P. (1981). *In* "Relativistic Effects in Atoms, Molecules and Solids" (G. L. Malli, ed.), p. 89, Plenum. New York.
Grant, I. P. (1982). *Phys. Rev. A* **25**, 1230.
Grant, I. P. (1986a). *J. Phys. B* **19**, In press.
Grant, I. P. (1986b). *Aust. J. Phys.* **39**, 649.
Grant, I. P., McKenzie, B. J., Norrington, P. H., Mayers, D. F., and Pyper, N. C. (1980). *Computer Phys. Commun.* **21**, 207.
Grant, I. P., and Pyper, N. C. (1976). *J. Phys. B* **9**, 761.
Grotch, H., and Yennie, D. R. (1969). *Revs. Mod. Phys.* **41**, 350.
Hall, G. G. (1951). *Proc. R. Soc. A* **205**, 541.
Hartree, D. R., Hartree, W., and Swirles, B. (1939). *Phil. Trans. R. Soc. A*, **238**, 229.
Hata, J., Cooper, D. L., and Grant, I. P. (1985). *J. Phys. B* **18**, 1907.
Hata, J., and Grant, I. P. (1983a). *J. Phys. B* **16**, 523.
Hata, J., and Grant, I. P. (1983b). *J. Phys. B* **16**, 915.
Hata, J., and Grant, I. P. (1984). *J. Phys. B* **17**, L107.
Hughes, D. S., and Eckart, C. (1930). *Phys. Rev.* **36**, 694.
Ishikawa, Y., Binning, R. C., and Sando, K. M. (1984). *Chem. Phys. Lett.* **105**, 189.
Ishikawa, Y., Baretty, R., and Binning, R. C. (1985). *Chem. Phys. Lett.* **121**, 130.
Ivanov, L. N., and Pobodebova, L. I. (1977). *J. Phys. B* **10**, 1001.
Jauch, J. M., and Rohrlich, F. (1976). "The Theory of Photons and Electrons", 2nd ed., Springer, New York.
Johnson, W. R. (1987). *Phys. Scr.* To be published.
Johnson, W. R., and Cheng, K. T. (1979). *J. Phys. B* **12**, 863.
Johnson, W. R., and Soff, G. (1985). *Atomic Data & Nucl. Dat. Tables* **33**, 405.
Kagawa, T. (1975). *Phys. Rev. A* **12**, 2245.
Kagawa, T. (1980). *Phys. Rev.* **22**, 2340.
Kelly, H. P. (1963). *Phys. Rev.* **131**, 684.
Kelly, H. P. (1964). *Phys. Rev. B* **136**, 896.
Kim, Y. K., (1967a). *Phys. Rev.* **154**, 17.
Kim, Y. K., (1976b). *Phys. Rev.* **159**, 190.
Klahn, B., and Bingel, W. A. (1977a). *Theor. chim. Acta* **44**, 9.
Klahn, B., and Bingel, W. A. (1977b). *Theor. chim. Acta* **44**, 27.
Klahn, B., and Morgan, J. D. (1984). *J. Chem. Phys.* **81**, 410.
Klapisch, M., Schwob, J. L., Fraenkel, B. S., and Oreg, J. (1977). *J. Opt. Soc. Amer.* **67**, 148.
Kutzelnigg, W. (1984). *Int. J. Quant. Chem.* **25**, 107.
Labzovskii, L. N., (1971). *Sov. Phys. JETP* **32**, 94.
Leclercq, J. M. (1970). *Phys. Rev. A* **1**, 1358.
Lee, C. M., and Johnson, W. R. (1980). *Phys. Rev. A* **22**, 979.

Lee, Y. S., and McLean, A. D. (1982). *J. Chem. Phys.* **76**, 735.
Lamb, H. (1960). "The Dynamical Theory of Sound". Dover, New York, (Reprint of 1925 2nd ed., Arnold, London)
Lindgren, I., Lindgren, J., and Mårtensson, A.-M. (1976). *Z. Physik* **A279**, 113.
Lindgren, I., and Rosén, A. (1974). *Case Studies in Atomic Physics* **4**, 93.
Lindgren, I., and Morrison, J. (1982). "Atomic Many-Body Theory". Springer-Verlag, Berlin, Heidelberg, New York.
Malli, G., and Oreg, J. (1979). *Chem. Phys. Lett.* **69**, 313.
Mark, F., Lischka, H., and Rosicky, F. (1980). *Chem. Phys. Lett.* **71**, 507.
Mark, F., and Rosicky, F. (1980). *Chem. Phys. Lett.* **74**, 562.
Mark, F., and Schwarz, W. H. E. (1982). *Phys. Rev. Lett.* **48**, 673.
McKenzie, B. J., Grant, I. P., and Norrington, P. H. (1980). *Computer Phys. Commun.* **21**, 233.
Matsuoka, O., Suzuki, N., Aoyama, T., and Malli G. (1980). *J. Chem. Phys.* **73**, 1320.
Messiah, A. (1961) "Quantum Mechanics" North-Holland, Amsterdam.
Mittleman, M. H. (1971). *Phys. Rev. A* **4**, 893.
Mohr, P. J. (1974). *Ann. Phys. (NY)* **88**, 26, 52.
Mohr, P. J. (1982). *Phys. Rev. A* **26**, 2338.
Pauli, W. (1925). *Z. Phys.* **31**, 765.
Primakoff, H., and Holstein, T. (1939). *Phys. Rev.* **55**, 1218.
Quiney, H., Grant, I. P., and Wilson, S. (1987a). *J. Phys. B* **20**, 1413.
Quiney, H., Grant, I. P., and Wilson, S. (1987b). *Phys. Scr.* **36**, 460.
Rafelski, J., Fulcher, L. P., and Klein, A. (1978). *Phys. Rep. C* **38**, 277.
Reeves, C. M. (1963). *J. Chem. Phys.* **39**, 1.
Reinhardt, J., Greiner, W. (1977). *Rep. Prog. Phys* **40**, 219.
Richtmyer, R. D. (1978). "Principles of Advanced Mathematical Pysics", Vol. 1, §§10.17, 11.6 Springer, New York.
Roothaan, C. C. J., (1951). *Revs. Mod. Phys.* **23**, 69.
Roothaan, C. C. J., and Bagus, P. (1963). *Meth. Comp. Phys.* **2**, 47.
Rosén, A., (1969). *J. Phys. B* **2**, 1257.
Rosén, A., and Lindgren, I. (1972). *Physica Scripta* **6**, 109.
Sabelli, N., and Hinze, J. (1969). *J. Chem. Phys.* **50**, 684.
Safronova, U. I., and Rudzikas, Z. B. (1976). *J. Phys. B* **9**, 1989.
Sapirstein, J. R. (1987). **36**, 801.
Schiff, L. I. (1968). "Quantum Mechanics", 3rd ed. McGraw-Hill, New York.
Schmidt, M. W. and Ruedenberg, K. (1979). *J. Chem. Phys.* **71**, 3951.
Schwarz, W. H. E., and Wallmeier, H. (1982). *Mol. Phys.* **46**, 1045.
Schwarz, W. H. E., and E. Wechsel–Trakowski, (1982). *Chem. Phys. Lett.* **85**, 94.
Schweber, S. S. (1964). "An Introduction to Relativistic Quantum Field Theory", Harper and Row, New York.
Shirokov, Yu.M. (1958). *Sov. Phys. JETP* **6**, 919.
Shun, D. H. (1977). D. Phil. thesis, University of Oxford. Unpublished.
Silverstone, H. J., Carroll, D. P., and Silver, D. M. (1978). *J. Chem. Phys.* **68**, 616.
Soff, G., Schlüter, P., Müller, B., and Greiner, W. (1982). *Phys. Rev. Lett.* **48**, 1465.
Stanton, R. E., and Havriliak, S. (1984). *J. Chem. Phys.* **81**, 1910.
Stone, A. P. (1961). *Proc. Phys. Soc. (London)* **77**, 786.
Sucher, J. (1957). *Phys. Rev.* **107**, 1448.
Sucher, J. (1980). *Phys. Rev A* **22**, 348.
Sucher, J. (1985). *AIP Conf. Proc. (USA)*, no. 136, p. 1.
Swirles, B. (1935). *Proc. R. Soc. A* **152**, 625.
Synek, M. (1964). *Phys. Rev.* **136**, A1552.

Uehling, E. A. (1935). *Phys. Rev.* **48**, 55.
Wallmeier, H., and Kutzelnigg, W. (1981). *Chem. Phys. Lett.* **78**, 341.
Wichmann, E. H., and Kroll, N. M. (1956). *Phys. Rev.* **101**, 843.
Wilson, S. (1984). "Electron Correlation in Molecules". Clarendon Press, Oxford.
Zygelman, B., and Mittleman, M. H. (1986). *J. Phys B* **19**, 1891.

ADVANCES IN ATOMIC AND MOLECULAR PHYSICS, VOL. 23

POINT-CHARGE MODELS FOR MOLECULES DERIVED FROM LEAST-SQUARES FITTING OF THE ELECTRIC POTENTIAL

*D. E. WILLIAMS and JI-MIN YAN**

Department of Chemistry
University of Louisville
Louisville, Kentucky

I. Introduction	87
II. Calculation of the Electric Potential	90
A. Definition and Calculation of the Electric Potential	90
B. About *Ab Initio* Calculation	92
C. About Choice of Basis Set	93
III. Calculation of the PD/LSF Point Charges in Molecules	94
A. Contour Maps of the Electric Potential	94
B. Modelling of the Electric Potential by Atomic Site Charges	96
C. Inclusion of Non-atomic Sites	101
IV. Examples	101
A. Atomic Site Charges for Small Molecules	101
B. Lone-Pair Electron Sites in Azabenzenes	110
C. Bond Sites in Fluorocarbons	117
D. Water Monomer and Dimer	122
V. Conclusion	128
Acknowledgement	129
References	129

I. Introduction

A molecule consists of some nuclei and electrons, with the nuclei vibrating around their equilibrium positions and electrons moving around the nuclei. In quantum mechanics, the motion of electrons around nuclei is called the electron cloud. The electron cloud has an approximate limit of extension, which we will call the van der Waals envelope, where the electron density falls effectively to zero. Inside the van der Waals envelope, close to the nuclei, the electric potential is strongly positive; outside the van der Waals envelope the electric potential is small but not zero.

*Permanent address: Institute of Chemistry, Academia Sinica, Beijing, China.

As molecules approach, their electric potentials will interact even though their van der Waals envelopes do not overlap. When the intermolecular interaction outside the van der Waals envelope is considered, because of the long distance between atoms, the nucleus and the electrons moving around it can be referred to as a uniform point charge fixed in space. This is called the point-charge model of the molecule. It can be seen that the point-charge model of the molecule is only useful for long-range interaction between molecules. If we confuse the interaction inside the molecule with the interaction outside the molecule, then some problems will be hard to understand. For example, why does a molecule composed of identical atoms have a surrounding electric potential? Why does a homonuclear diatomic molecule such as nitrogen have a significant surrounding electric potential?

When the interaction between molecules is considered, the point-charge model of molecule has an immediate intuitive appeal and has been in use for many years. The point-charge model permeates nearly every facet of the study of molecules. Such net atomic point charges have been used as an index of reactivity of atomic sites in molecules, as an aid to understanding molecular complexes, for packing energy calculations in crystals, and for a variety of other purposes.

Mulliken's population analysis (1955) presents one approach to the calculation of atomic point charges in a molecule. But population analysis has some problems, which can be described in four aspects (Xu, Li and Wang, 1985). (1) It is not correct to divide electron cloud overlap $2C_{\mu i}^* C_{\nu i} S_{\mu \nu}$ equally between atomic orbitals φ_μ and φ_ν, particularly for polar bonds and lone-pair electrons, because in those cases the charge obviously tends towards the nucleus with higher charge. (2) Except for s orbitals, p, d, and f orbitals are spread away from the nucleus, so that charge originally belonging to orbitals of atom X may distribute nearly to atom Y. It is not correct to put this part of the charge back to atom X. (3) In some cases, incorrect values may be obtained, with the electron number in some orbitals being negative and electron number in other orbitals being larger than 2. This is contradictory to the Pauli principle. (4) The charge distribution depends on the basis set too much, and if the basis set is not atomic orbitals then the population analysis is not defined.

Since the atomic point-charge concept is deeply kept in the chemists' heart, there are many scientists who have tried to improve the population analysis method. Some examples (Xu, Li and Wang, 1985) are the following: (1) To solve the allocation of the $2C_{\mu i}^* C_{\nu i} S_{\mu \nu}$, Christofferson and Baker (1971) advise to divide it unequally according to orbital coefficient C. Pollak and Rein (1967) advise that overlap charge should be allocated to keep the dipole moment of overlap charge invariant, a technique which is given by Löwdin

(1953). Davidson (1967) advises to orthogonalize the basis set to remove the overlap term. (2) Politzer and Harris (1970) advise to allocate the charge according to the space which is occupied by the atom in the molecule. (3) Yáñez, Stewart and Pople (1978) advise to project the charge of molecule to the density basis set.

However, none of these treatments is completely satisfactory. It is not difficult to see that the reason why the problem cannot be completely overcome is that we cannot use a viewpoint inside the molecule to obtain a satisfactory result outside the molecule. In fact, electric potential is an exactly defined physical property different from net atomic point charge. It is the electric potential (Scrocco and Tomasi, 1978) that determines the interaction outside the molecule, so it is better to define point charge inside the molecule by using the electric potential outside the molecule. We refer to such intramolecular point charges obtained from the extramolecular electric potential as potential-derived (PD) point charges.

Some initial efforts to represent the extramolecular electric potential approximately by intramolecular point charges were made by Kollman (1978). Momany (1978) introduced a least-squares fitting method which allowed systematic derivation of net atomic charges. Cox and Williams (1981) further examined the method by deriving atomic point charges for a number of small molecules using wavefunctions of differing quality. Williams and Weller (1983) extended the method to include lone-pair electron site point charges in azabenzenes; Williams and Houpt (1986) included bond site point charges in perfluorocarbons. Singh and Kollman (1984) optimized extra lone-pair electron sites in the water molecule but found "inverted" lone-pair sites. Williams and Craycroft (1985) extended this treatment to water dimer and made estimates of site-charge polarization in the dimer.

Many other researchers have also obtained satisfactory results with this method. Smit, Derissen and van Duijneveldt (1979) derived PD charges for methanol, formaldehyde and formic acid; Agresti, Bonaccorsi and Tomasi (1979), for benzenic compounds; Singh and Kollman (1984), for a variety of molecules; Ray, Shibata, Bolis and Rein (1984), for DNA base components; Crowder, Alldredge and White (1985), for urea and thiourea; and Ray, Shibata, Bolis and Rein (1985), for hydrogen bonded systems.

Hall (1985) has recently reviewed atomic point-charge models for molecules. In this chapter, we would like to introduce the potential-derived point-charge least-squares fitting model for the electric potential (PD/LSF model) to persons who are interested in it in detail. It should be mentioned that the electric potential around a molecule may also be represented by a distribution of multipoles (Stone, 1981). Price (1985) gives an example of this technique applied to aromatic hydrocarbons.

II. Calculation of the Electric Potential

As mentioned above, to establish the PD/LSF point-charge model of a molecule, the first step is to calculate the electric potential at point **r** in vicinity of the given molecule, and the second step is to calculate point charges from the electric potential data by means of the least-squares fitting method. In the present section, we discuss how to calculate the electric potential accurately from the molecular wavefunction.

A. Definition and Calculation of the Electric Potential

The electric potential at point **r** in the vicinity of the given molecule is the electric force acting on a unit positive charge at that point causes by the nuclei and the electrons of the molecule (Fig. 1). Suppose there are N nuclei and n electrons in the molecule, and then according to Coulomb's law, the electric potential $V(\mathbf{r})$ at point **r** arising from the nuclei will simply be

$$V_n(\mathbf{r}) = \sum_{\alpha=1}^{N} \frac{Z_\alpha}{|\mathbf{r} - \mathbf{R}_\alpha|} \tag{1}$$

The electric potential at point **r** arising from the electrons $V(\mathbf{r})$ is more complicated because of motion of the electrons. Let the μth electron be in the ψ_μ state, and then the electric charge in the volume element $d\tau'$ given by the μth electron will be $-|e|\psi_\mu^*\psi_\mu\, d\tau'$. Thereby, the electric potential $V(\mathbf{r})$ is

$$V_e(\mathbf{r}) = \sum_{\mu=1}^{n} \int \frac{-|e|\psi_\mu^*\psi_\mu\, d\tau'}{|\mathbf{r} - \mathbf{r}'|} \tag{2}$$

The total electric potential at point **r** is

$$V(\mathbf{r}) = V_n(\mathbf{r}) + V_e(\mathbf{r}) \tag{3}$$

In Eqs. (1) and (2), Z_α is the charge of αth nucleus and e is the charge of an electron. The integration goes over all space.

In the molecule, the state function ψ_μ of a single electron is called a molecular orbital (MO), which is usually expressed by a linear combination of the atomic orbitals (AO), namely LCAO–MO. If there are atomic orbitals $\varphi_i (i = 1, 2, \ldots, p)$, ψ_μ will be

$$\psi_\mu = C_{\mu 1}\varphi_1 + C_{\mu 2}\varphi_2 + \cdots + C_{\mu i}\varphi_i + \cdots + C_{\mu p}\varphi_p \tag{4}$$

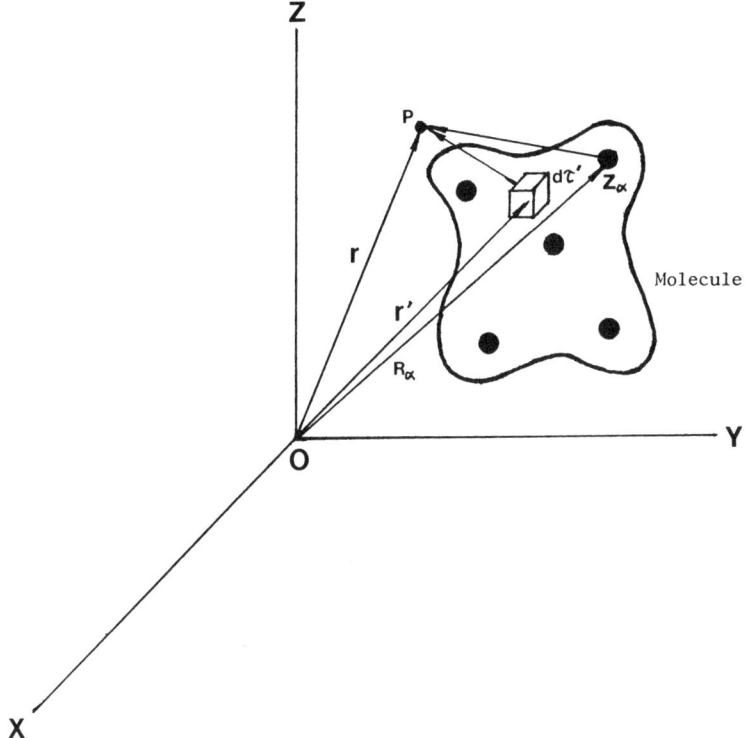

FIG. 1. Definition of the electric potential.

and then we have

$$V_e(\mathbf{r}) = -|e| \sum_{i=1}^{p} \sum_{j=1}^{p} p_{ij} \int \frac{\varphi_i^* \varphi_j}{|\mathbf{r} - \mathbf{r}'|} d\tau' \qquad (5)$$

where

$$p_{ij} = \sum_{\mu=1}^{n} C_{\mu i}^* C_{\mu j} \qquad (6)$$

The p_{ij} are called density matrix elements of LCAO-MO.

From Eqs. (1)–(6) it has been shown that, unlike net atomic charges, the molecular electric potential is a rigorously defined quantum mechanical property, and it also has been shown that if we can accurately calculate the LCAO-MO coefficients $C_{\mu i}(\mu = 1, 2, \ldots, n)$ by solving the Schrödinger equation, the electric potential $V(\mathbf{r})$ can be obtained easily. Furthermore, since the electric potential is the expectation value of a one-electron operator $|\mathbf{r} - \mathbf{r}'|^{-1}$, the calculation of the electric potential is correct to one order higher than the wavefunction employed (Scrocco and Tomasi, 1978).

B. About Ab Initio Calculation

In order to obtain the electric potential accurately, we must calculate the molecular orbital ψ or LCAO–MO coefficient using the *ab initio* method. In this subsection, we give some information about that.

Improvement in computers has led to the availability of *ab initio* calculations for small and medium sized molecules. Examples of computer programs distributed for this purpose are POLYATOM (Neumann et al., 1971), IBMOL (Veillard, 1972), HONDO (Dupuis, Rys and King, 1977) and GAUSSIAN (Binkley et al., 1984). These programs calculate a Hartree–Fock molecular orbital expressed as a linear combination of atomic orbitals. The atomic orbital is almost always modelled by gaussian-type functions. In some cases, the programs have capability for obtaining an improved wavefunction including the effects of electron correlation. This is accomplished through use of configuration interaction or by the Møller–Plesset method. A recent text by Hehre, Radom, Schleyer and Pople (1986) explains the procedures used to obtain *ab initio* wavefunctions.

The accuracy of an *ab initio* calculation usually is limited by the computing capacity available. For small molecules, it is generally possible to compute physical properties that are as accurate as those obtained by experiment. In a few cases, such calculations have turned out to be more reliable than experiment and have caused earlier experimental results to be remeasured and revised. An example of this is the methylene radical, which was predicted to be bent by theorists and linear by experimentalists. The discrepancy was resolved in favor of theory when new experiments showed a bent structure. In a review of this situation, Schaeffer (1986) concluded that theoretical chemistry has reached a new stage where the goal was to be a "full partner with experiment." Also, Wasserman and Schaefer (1986) state that in the future "we expect to find an increasing number of situations in which theory will be the preferred source of information for aspects of complex chemical systems." However, more powerful computers are needed to make possible accurate treatment of larger molecules.

For rather large molecules, it seems impossible to calculate with the *ab initio* method now. In this situation, if we only want to obtain relative values of electric potential at different parts of the molecule, semi-empirical methods of quantum chemistry, such as CNDO and EHMO, also can be used

(Scrocco and Tomasi, 1978). Also, point charges derived from smaller molecules may be approximately transferable to larger molecules (see examples section).

C. About Choice of Basis Set

It is also important to make a correct choice of basis set for φ in order to obtain an accurate electric potential. For not very simple molecules, atomic orbitals are almost invariably modelled with gaussian-type basis sets because of the ease of evaluating integrals involving these functions. The approach of a Slater-type atomic orbital modelled with K gaussian-type basis sets is referred to as STO–KG. The quality of such models increases as K increases. Many such combinations of gaussians have been proposed. To illustrate the essentials involved, we give some illustrations of gaussian basis functions proposed by Professor J. A. Pople's research group. Hehre, Stewart and Pople (1969) investigated gaussian basis functions for various values of K. They found that three primitive gaussian basis functions was a good compromise to model Stater-type atomic orbitals. Their STO-3G basis set has been widely used even through it must be characterized as a minimal basis set.

It is known that Slater-type atomic orbitals can be dramatically improved in quality if two Slater exponents (called ζ) are used for each function (Tang, Yang and Li, 1982; Xu, Li and Wang, 1985; Lowe, 1978). These so-called double zeta STO's are widely used in atomic calculations and give excellent agreement with experimental data. An analogous procedure which is used with GTO's is called split valence, where the outer shell gaussian primitives are divided into two groups. Bonding between atoms, which is a relatively long-distance phenomenon, is most sensitive to the gaussian primitives with the smallest exponents. Hehre, Ditchfield and Pople (1971) proposed split-valence gaussian basis sets where the inner shell orbitals are modelled with one STO–KG and the outer shell orbitals are modelled with two STO–KG's. These basis sets were christened n-31G, and showed marked improvement over the STO-3G sets. The 4-31G and 6-31G basis sets are popular, in which inner shell orbitals are modelled with one STO-4G or one STO-6G, and outer shell orbitals are modelled with one STO-3G and one STO-1G. Recently, Binkley, Pople and Hehre (1980) defined 3-21G basis sets which have the advantage of split-valence with a smaller number of primitives.

In order to allow polarization of the atoms it was necessary to add d-type functions to second-row atoms Li through Ne, and p-functions for hydrogen and helium. The d-type functions are added first, indicated by a single asterisk, followed by inclusion of extra *p*-type functions on hydrogen, indicated by a double asterisk. Thus, the *n*-31G basis sets with polarization functions for the heavy atoms are designated *n*-31G* and those with polarization functions on hydrogen as well *n*-31G** (Hariharan and Pople, 1972). A further improvement in quality results of the outer gaussians are triply split (3, 1, 1) to form the 6-311G** basis sets (Krishnan, Frisch and Pople, 1980). Further improvement in the wavefunction can be obtained by consideration of electron correlation.

The more complicated the basis sets, the more computer time is needed. It has been found that the average relative central processor time for electric potential calculation in STO-3G, 6-31G and 6-31G** basis sets was about 1 to 3 to 8 (Cox and Williams, 1981). It has also been found that the electric potential values obtained from 6-31G and 6-31G** differ by a scale factor (Cox and Williams, 1981), so that we can obtain approximate electric potential values for 6-31G** by scaling 6-31G values.

III. Calculation of PD/LSF Point Charges in Molecules

A. CONTOUR MAPS OF THE ELECTRIC POTENTIAL

The electric potential may be evaluated from the wavefunction at points in space around the molecule. As molecules approach to form a dimer (such as a van der Waals molecule or a protein-substrate complex) or an aggregate (such as a crystal), their electric potentials will interact. A contour map of the molecular electric potential will show regions in space around the molecule which can interact favorably to the approach of positive or negative groups. To study this interaction the electric potential can be evaluated on a grid defined outside of the nominal van der Waals radii of the atoms in the molecule plus the radii of atoms in the approaching group.

Fig. 2 shows a contour plot of the electric potential of ethylene as calculated from the 6-31G** wavefunction (Cox and Williams, 1981). The two views are in the molecular plane and perpendicular to this plane. It is apparent that the electric potential is always positive in the plane. However, strong negative regions exist perpendicular to the plane, especially near the carbon atoms. The magnitude of this potential is significant, reaching 25 kJ/mol at van der Waals contact distance. An approaching electrophilic moiety would be directed by this potential perpendicular to the molecular plane, and nucleophilic moieties would be directed in plane, especially toward the hydrogen regions.

Fig. 3 shows a contour plot of the electric potential of water, also from the 6-31G** wavefunction. In the plane of the molecule the potential is strongly negative on the oxygen side, and strongly positive on the hydrogen side. Perpendicular to the plane, it can be seen that the negative region is displaced toward the hydrogens.

The potential energy of interaction of approaching molecules may be obtained by integration over the electric potentials of both molecules, but in practice this a cumbersome procedure. To simplify the calculation of the electric interaction between molecules, a PD/LSF point-charge model is sought. The intermolecular energy can then simply be obtained according to Coulomb's law.

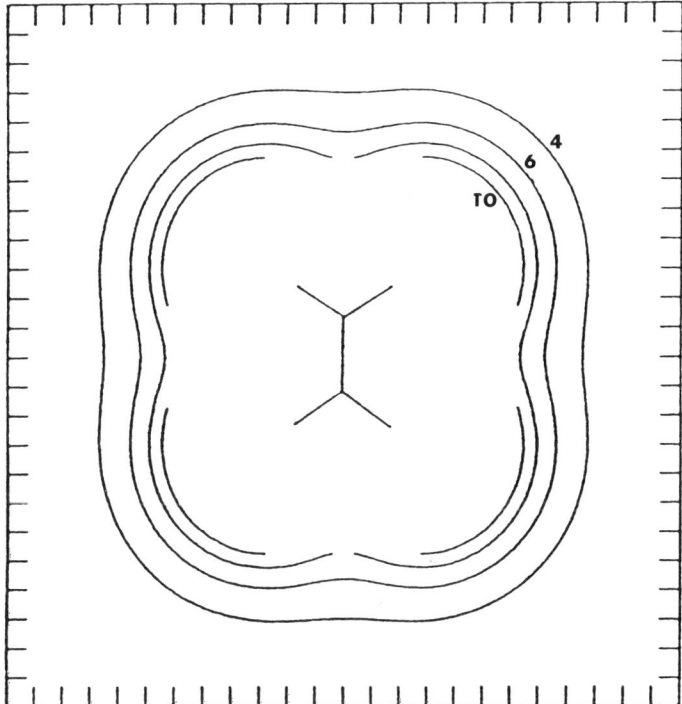

FIG. 2. Electric potential of ethylene calculated from the 6-31G** wavefunction. Contours are marked in KJ/mol and distances in atomic units ($0.529 \cdot 10^{-10}$ m). (a) In the molecular plane.

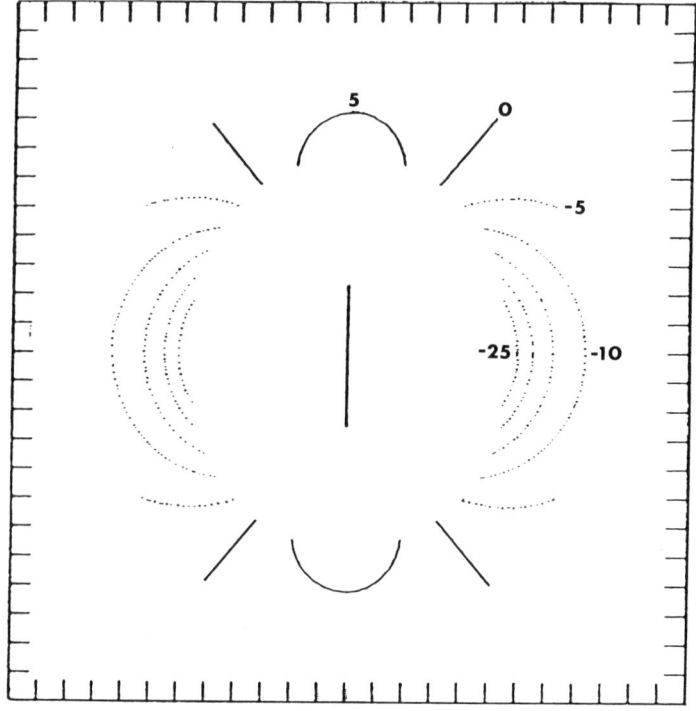

FIG. 2. (b) Perpendicular to the molecular plane.

B. Modelling of the Electric Potential by Atomic Site Charges

In the model we suppose that the QM (quantum-mechanical) electric potential surrounding the molecule can be regarded as originating from point charges q_j placed at atomic locations $\mathbf{r}_j$. The sum of the atomic point charges must be zero if a neutral molecule is considered. If an ion is considered, the sum of the net atomic point charges must be an integer Z representing the net charge on the ion in electron units. For n atoms we have

$$\sum_j^n q_j = Z \qquad (7)$$

where Z is the charge on the ion, or zero for a neutral molecule. If we take $(n-1)$ net charges as independent, then the dependent charge is

$$q_n = Z - \sum_j^{n-1} q_j \qquad (8)$$

At m points in space, $\mathbf{r}_i$, around the molecule or ion we have available the QM electric potential V_i^0. The model potential V_i^c at those points is given by Coulomb's law:

$$V_i^c = \sum_j^n \frac{q_j}{r_{ij}} \qquad (9)$$

where r_{ij} is the magnitude of the distance between point charge j and grid point i. Inserting the value for the dependent charge q_n, we have

$$V_i^c = \sum_j^{n-1} \frac{q_j}{r_{ij}} + \frac{Z - \sum_j^{n-1} q_j}{r_{in}} \qquad (10)$$

We need the derivative of the potential with respect to the kth charge:

$$\frac{\partial V_i^c}{\partial q_k} = \frac{1}{r_{ik}} - \frac{1}{r_{in}} \qquad (11)$$

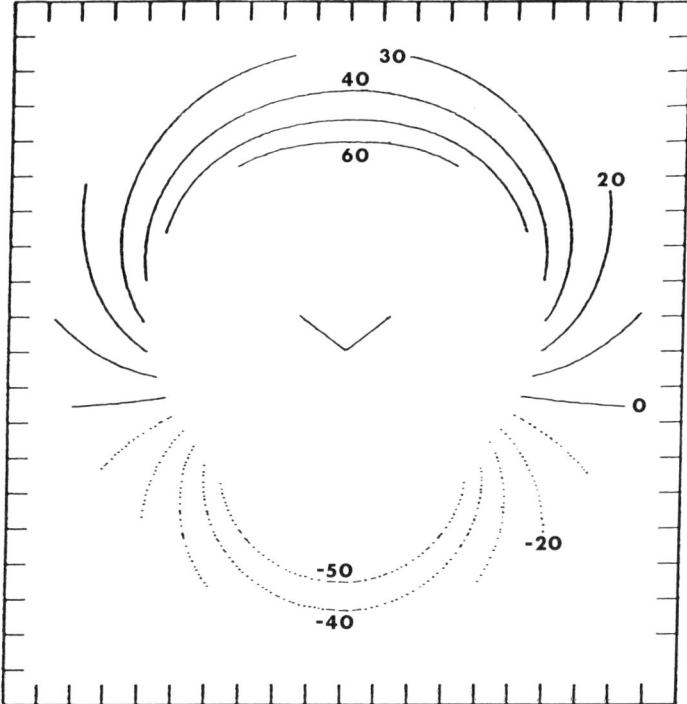

FIG. 3. Same as Fig. 2, but for the water molecule. (a) In the molecular plane.

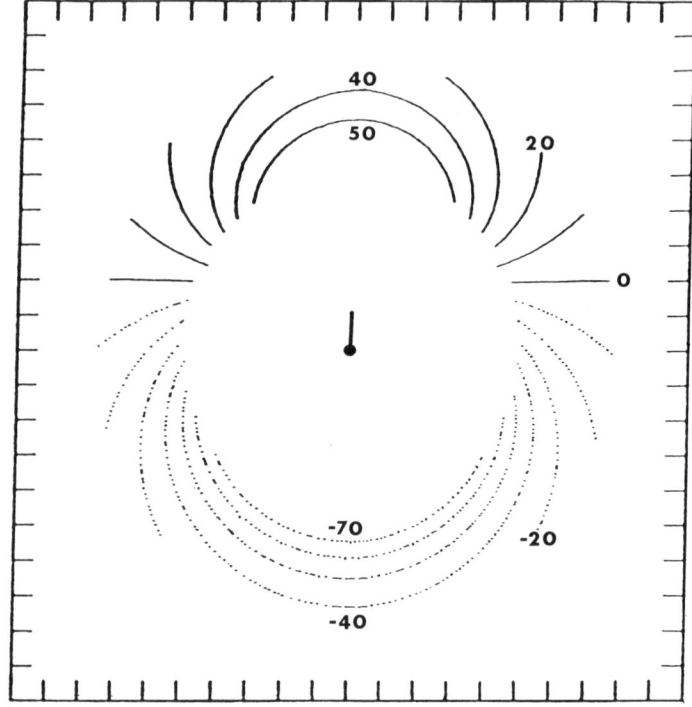

FIG. 3. (b) Perpendicular to the molecular plane.

Let V_i^0 be the QM electric potential calculated from Eq. (3); then the least-squares fitting problem is to minimize the sum

$$R = \sum_i^m w_i(V_i^0 - V_i^c)^2 \qquad (12)$$

by setting the first derivatives to zero:

$$\frac{\partial k}{\partial q_k} = \sum_i^m w_i(V_i^0 - V_i^c)\frac{\partial V_i^c}{\partial q_k} = 0 \qquad (13)$$

where w_i is a statistical weight factor. After substitution and rearrangement, Eq. (13) can be written as

$$\sum_i^m w_i\left(V_i^0 - \frac{Z}{r_{in}}\right)\left(\frac{1}{r_{ik}} - \frac{1}{r_{in}}\right) = \sum_j^{n-1}\left[\sum_i^m w_i\left(\frac{1}{r_{ik}} - \frac{1}{r_{in}}\right)\left(\frac{1}{r_{ij}} - \frac{1}{r_{in}}\right)\right]q_j \qquad (14)$$

In matrix form, it is

$$\mathbf{AQ} = \mathbf{a} \qquad (15)$$

where $\mathbf{A}$ is a $(n-1) \times (n-1)$ dimension matrix, and its elements A_{ij} are

$$A_{hj} = \sum_i^m w_i \left(\frac{1}{r_{ik}} - \frac{1}{r_{in}}\right)\left(\frac{1}{r_{ij}} - \frac{1}{r_{in}}\right) \qquad j, k = 1, 2, \ldots, n-1 \qquad (16)$$

$\mathbf{Q}$ and $\mathbf{a}$ are column matrices, in which the elements of $\mathbf{a}$ are

$$a_k = \sum_i^m w_i \left(V_i^0 - \frac{Z}{r_{in}}\right)\left(\frac{1}{r_{ik}} - \frac{1}{r_{in}}\right) \qquad k = 1, 2, \ldots, n-1 \qquad (17)$$

The matrix $\mathbf{A}$ is symmetric. The solution is

$$\mathbf{Q} = \mathbf{A}^{-1}\mathbf{a} \qquad (18)$$

which can be found numerically using standard matrix inversion methods. Appropriate row and column operations may be performed to reduce the dimension of the problem when certain charges are assumed to tbe related by symmetry. If s is the number of symmetry conditions, the resulting matrix equations are of an order $(n - s - 1)$. Cox and Williams (1981) found that the statistical weight, w_i, for each point could reasonably be taken as unity. The standard deviations of the q_j and correlation coefficients between them can be found from A_{jk}^{-1} by standard methods.

It was necessary to establish criteria for the selection of points where the molecular electric potential was to be evaluated and subsequently fitted by a PD/LSF point-charge model. The region of interest is a shell that is just outside the van der Waals envelope of the molecule.

It is necessary to avoid getting too close to nuclei, where the electric potential is always positive. Well established values of van der Waals radii are available (Pauling, 1960, for example), which define empirically the van der Waals envelope. As molecules approach, the electric potential of the second molecule interacts with the electric potential of the first molecule. Since we wish to allow an arbitrary orientation of the second molecule, which contains arbitrary atoms, we take the van der Waals radius of hydrogen, the smallest atom considered here, as a possible atom in an interacting molecule.

Thus the molecular boundary may be defined as the van der Waals envelope, and this boundary may be extended 1.2 Å farther to allow for the possibility of an interaction with hydrogen, the smallest atom normally considered. This defines the electric potential envelope. Tests were made of the effect of choosing an electric potential envelope smaller or larger by 1Å. The main effect observed was that the fit to the thc QM potential, both in the root-man-square (rms) fit and in the relative root-mean-square (rrms) fit (defined below), became worse at the smaller boundary, and became better at the larger boundary. The PD/LSF charges themselves changed by a negligible amount. Thus the choice of the electric potential envelope, within reasonable limits, primarily affects only the goodness of fit indices.

The grid points for evaluation of the electric potential are located between the electric potential envelope and a thickness, t, beyond this envelope. It is convenient to use an evenly spaced grid of points in this electric shell, with spacing d. The axial system for generation of the grid should coincide with any molecular symmetry elements so as to retain the required symmetry of atomic charges. In practice, the molecular inertial axis system is a good choice.

The effect of varying the shell thickness, t, and the cubic grid spacing used, d, was investigated. As t increases and d decreases the molecular electric potential must be calculated at a larger number of points. Since the electric potential is largest in magnitude and most detailed closer to the molecule, if we fit points inside the potential shell the region outside the shell is well represented also. Only an evenly spaced grid inside the potential shell is needed, and additional grid points farther away from the molecule do not contribute much additional information toward deriving the atomic point charges.

A number of test calculations were performed using various values of d and t with an eye toward reducing the number of points to a minimum commensurate with good definition of the net atomic point-charges. It was found that d values of 1.0–1.2 Å and a t value of 1 Å were satisfactory. For small molecules, these choices of d and t gave around 100 to 200 sampling points for the molecular electric potential. Of course, for symmetrical molecules only the asymmetrical portion of the grid needs consideration. Singh and Kollman (1984) recommend a different choice for the location of the grid points based on the Connolly (1982) surface algorithm.

To evaluate the suitability of the net atomic charge model the root-mean-square fit may be used:

$$\text{rms} \equiv \sqrt{\frac{\sum_i^m (V_i^0 - V_i^c)^2}{n}} \qquad (19)$$

expressed in some convenient energy unit such as kJ/mol. In some cases the relative rms fit is a more convenient index to use:

$$\text{rrms} \equiv \frac{\text{rms}}{\sqrt{\frac{\sum_i^m (V_i^0)^2}{n}}} \qquad (20)$$

Section IV.A below gives examples where the point charges are only at the atomic sites.

C. Inclusion of Non-atomic Sites

In the process of bonding the electron cloud is slightly polarized so that it is no longer spherically symmetric around atomic sites. If this polarization is relatively small, the PD/LSF atomic site point charge will be adequate. Larger polarization can occur, in which case it may be desirable to include in the PD/LSF model additional charge sites. These sites may be located at expected bonding or lone-pair electron positions, or at completely arbitrary, but optimized, positions. Optimized positions are those which lead to the lowest values of rms and rrms. The calculation method for the nonatomic site point charge is the same as that described in the above section, except that non-atomic site point charges are added.

Section IV.B below gives examples where additional lone-pair electron sites were added to the nitrogen atoms of azabenzene molecules. In the case of s-tetrazine an optimum lone-pair distance was obtained by fitting the electric potential. In the case of pyridine the optimum lone-pair distance was found to be short, in which case the electrical effect at the nitrogen atom could be described by an atomic dipole.

Section IV.C below gives examples where additional bonding pair electron sites were added to the C—F bonds of perfluorocarbons. The bond sites were taken at the center of the respective bonds.

Section IV.D below gives an example where a full optimization of the location of additional charge sites was made for the water molecule. Surprisingly, these extra charge sites were not located near tetrahedral locations where lone-pair electrons might be expected. It was found that these extra site charges also improved the modeling of the electric potential about various configurations of water dimer.

From Eq. (7) to Eq. (20) for deriving PD-LSF point charges, it can be seen that the PD-LSF point charges are not exactly defined physical properties. The values obtained will depend on both the choice of grid points for electric potential evaluation and the choice of charge sites. The choice of grid points has been discussed above, and the choice of charge sites will be discussed in the following section.

IV. Examples

A. Atomic Site Charges for Small Molecules

To examine the utility of the PD-LSF point-charge model it is desirable to carry out tests on several classes of molecules using a variety of calculation conditions, making comparisons with experimental data where possible. Cox and Williams (1981) have systematically determined PD-LSF atomic site charges (PD/AC) for a variety of small molecules. They obtained results with three different atomic basis sets: the minimal STO-3G basis, the split-valence 6-31G basis, and the extended 6-31G** basis. Table I shows a summary of their results.

Since computing time increases rapidly with larger basis sets, it is of interest to compare the PD-AC values obtained with the smaller basis sets with those obtained from the larger 6-31G** basis set. Relative to the 6-31G** values, Table I shows that the charges obtained from the STO-3G basis functions are generally smaller, and those obtained from the 6-31G basis functions are generally larger. This situation shows the sensitivity of the PD/AC values to the choice of basis set, and also that the 6-31G values differ from the 6-31G** values mainly by a scaling constant. The least-squares scale factor relating the 6-31G values to the 6-31G** values was found to be 0.83. The scaling relationship is not as good for the STO-3G values, but application of a scale factor of 1.15 to the STO-3G values approximates the 6-31G** values.

TABLE I

PD–AC FOR VARIOUS MOLECULES[a]

Molecule	Atom	Basis set		
		STO-3G	6-31G	6-31G**
HF	F	−294(1)	−524(1)	−449(3)
	H	294(1)	524(1)	449(3)
	rms	0.5	0.6	1.6
	rrms	2.2	1.6	5.0
H_2O	O	−614(1)	−940(3)	−786(6)
	H	307(1)	470(2)	393(3)
	rms	0.5	1.2	2.1
	rrms	1.6	3.0	6.1
NH_3	N	−972(4)	−1278(8)	−1047(9)
	H	324(1)	426(3)	349(3)
	rms	0.8	1.6	2.0
	rrms	3.0	5.0	7.3
CH_4	C	−484(4)	−528(6)	−556(6)
	H	121(1)	132(2)	139(1)
	rms	0.2	0.4	0.3
	rrms	7.8	9.5	9.4
C_2H_2	C	−180(1)	−286(1)	−297(1)
	H	180(1)	286(1)	297(1)
	rms	0.1	0.1	0.2
	rrms	1.1	0.6	1.0

continued on next page

TABLE I—Continued

		Basis set		
Molecule	Atom	STO-3G	6-31G	6-31G**
C_2H_4	C	−164(4)	−356(5)	−352(4)
	H	82(2)	178(3)	176(2)
	rms	0.7	0.9	0.8
	rrms	18.1	10.3	9.5
CO_2	O	−405(3)	−569(3)	−434(2)
	C	810(5)	1138(6)	868(4)
	rms	0.1	0.4	0.5
	rrms	0.6	2.2	3.8
CH_2O	O	−293(7)	−522(6)	−459(4)
	C	359(2)	530(18)	423(13)
	H	−33(7)	−4(6)	18(4)
	rms	1.6	1.3	0.9
	rrms	8.5	3.7	2.9
CH_3OH	O	−501(13)	−827(15)	−680(13)
	C	188(13)	416(60)	244(52)
	eclipsed H	14(4)	14(16)	48(14)
	staggered H	−9(14)	−44(16)	−18(14)
	hydroxyl H	317(5)	485(6)	424(5)
	rms	1.6	1.9	1.6
	rrms	8.1	6.0	6.2
$HCONH_2$	O	−435(9)	−666(9)	−556(9)
	N	−841(2)	−1114(26)	−883(24)
	C	676(3)	890(29)	637(27)
	carbon H	−62(9)	−15(10)	30(9)
	trans H	323(8)	435(9)	374(8)
	cis H	337(8)	470(8)	398(8)
	rms	1.6	1.7	2.2
	rrms	5.7	3.9	3.8
HCOOH	O	−554(8)	−817(8)	−628(6)
	carbonyl O	−439(7)	−676(7)	−568(5)
	C	699(17)	948(17)	674(14)
	H	−61(6)	4(7)	59(5)
	hydroxyl H	355(5)	541(4)	462(4)
	rms	1.4	1.4	1.2
	rrms	8.7	5.3	4.6
CH_3CN	N	−457(3)	−555(2)	−514(18)
	C	−574(20)	−602(22)	−577(17)
	nitrile C	515(10)	557(7)	488(6)
	H	172(7)	200(5)	201(4)
	rms	0.7	0.5	0.4
	rrms	2.2	1.9	1.4
CO_3^{-2}	O	−938(1)	−1152(2)	−1070(4)
	C	814(3)	1456(7)	1210(11)
	rms	0.3	0.7	1.1
	rrms	0.1	0.1	0.1

[a] Atomic site charges $\times 10^3$ in electron units. Standard deviations are given in parentheses; these reflect only the goodness of fit to the QM electric potential, and do not include any errors in the calculated values of the potential.

The effect of the choice of basis set is further illustrated with the double zeta (DZ) basis set of Snyder and Basch (1972). This basis set is split-valence, similar in concept to 6-31G, but with the inner shells also split into two contractions. Using formaldehyde as an example, the DZ QM potential was very similar to the 6-31G QM potential; the DZ PD-AC values are: oxygen, -0.530; carbon, 0.520; and hydrogen, 0.005 $|e|$. These values are almost the same as 6-31G PD–AC (Table I) and therefore it is a good approximation to not split the inner shells.

Quantum mechanical operator dipole or quadrupole moments of the molecules may be calculated directly from the wavefunctions. Also, distributed dipole or quadrupole moments[1] may be calculated from the PD/AC. Table II gives dipole or quadrupole moments calculated by both methods, as well as the observed moments. Note that there is excellent agreement between moments calculated from the same wavefunction using either the rigorous operator definition or distributed moments obtained from the PD/AC. This agreement is still maintained when additional non-atomic sites are used. However, dipole moments calculated from population analysis charges (PA) can be quite different from the QM or PD/AC values (Cox and Williams, 1981).

It is known that the 6-31G** basis set, and especially the 6-31G basis set, generally yields too large dipole and quadrupole moments, and this is apparent from an examination of Table II. When even larger basis sets are used or electron correlation is included the observed dipole moments are usually approached from the top side.

The calculated dipole moments are valuable in evaluation of the models because reference observed values are available. Table II shows that the PD-AC distributed dipole moments are very similar to the QM calculated dipole moments. The scale factors for the three basis sets suggest that the reasons for over- or underestimation of the net atomic charges and dipole moments are similar, and lie in the SCF calculation and choice of basis set.

The observed quadrupole moments are not as accurately known as are the dipole moments, so reference values are taken as the 6-31G** QM quadrupole moments, instead of the observed values. Table II shows that quadrupole moments calculated from the 6-31G** PD/AC for acetylene and carbon dioxide are rather close to the QM values. For ethylene the PD–AC yields a quadrupole moment larger than the QM value.

---

[1] A point dipole as considered here is an electrically neutral site in space which gives rise to a surrounding anisotropic electric potential due to its dipole moment. A distributed dipole is a set of two sites in space each having an electrical charge but each site itself having no dipole moment. The distributed dipole moment of the two sites is in the direction of the site-site vector and has a magnitude equal to one-half of the differences between the site charges multiplied by the intersite distance. As a special case the sum of the site charges may be zero.

TABLE II

DIPOLE OR QUADRUPOLE MOMENTS CALCULATED FROM THE PD-AC[a]

Molecule		Basis set			Observed[b]
		STO-3G	6-31G	6-31G**	
HF	u_{emp}	1.296	2.308	1.981	1.82
	u_c	1.288	2.296	1.973	
H_2O	u_{emp}	1.727	2.647	2.211	1.846
	u_c	1.726	2.630	2.184	
NH_3	u_{emp}	1.780	2.340	1.917	1.468
	u_c	1.783	2.319	1.887	
CH_2O	u_{emp}	1.510	3.006	2.765	2.339
	u_c	1.525	3.010	2.759	
CH_3OH	u_{emp}	1.491	2.357	1.899	1.7
	u_c	1.500	2.367	1.915	
$HCONH_2$	u_{emp}	2.596	4.205	3.929	3.72
	u_c	2.616	4.200	3.940	
HCOOH	u_{emp}	0.787	1.402	1.585	1.415
	u_c	0.789	1.399	1.583	
CH_3CN	u_{emp}	3.046	4.131	4.092	3.915
	u_c	3.063	4.132	4.092	
C_2H_2	Q_{emp}	4.148	6.590	6.844	3.0
	Q_c	4.160	6.590	6.829	
CO_2	Q_{emp}	−5.245	−7.371	−5.620	4.3
	Q_c	−5.258	−7.307	−5.503	
C_2H_4	Q_{emp}	1.002	2.175	2.155	1.5[c]
	Q_c	0.652	1.740	1.751	

[a] u_{emp} and Q_{emp} are calculated from the PD-AC; quantum u_c and Q_c are calculated directly from the SCF-MO wavefunctions.
[b] References for the observed values are given by Cox and Williams (1981).
[c] Experimental Q_{zz} value.

The agreement between the calculated and observed dipole moments may be improved by applying a scale factor characteristic of the basis set used. As expected, the best fit to the observed dipole moments is given by the largest basis set 6-31G**. If the 6-31G** PD-AC values are scaled by a factor of 0.91, the best least-squares agreement between the observed and PD-AC dipole moments is obtained. The 6-31G PD-AC also overestimates dipole moments, with the scale factor being 0.83. The STO-3G PD-AC underestimate the dipole moments, with the scale factor being 1.27. The use of these scale factors for the PD/AC helps correct for deficiencies in the quality of the basis set used for the wavefunction.

A comparison of the PD–ACs can be made with charges optimized to give best fit to crystal structure data, which can be viewed as experimentally derived charges. It should be noted that atomic site charges derived from crystal structure data may show polarization effects from the surrounding molecules. Such polarization effects are not included in the simple PD–AC. Williams & Starr (1977) obtained a charge of 0.153 $|e|$ on hydrogen attached to aromatic carbon, from a fit to observed crystal structures. If one makes the reasonable assumption that the charge for an aromatic hydrogen should be intermediate between hydrogens of single- and double-bonded carbons, then the average of the PD/AC for methane and ethylene should be close to 0.153 $|e|$. In fact, this average for the 6-31G** basis is 0.158 $|e|$, which represents excellent (although perhaps fortuitous) agreement. The STO-3G PD–AC values average to 0.102 $|e|$ which is less than this experimental value.

Yuen, Lister and Nyburg (1978) determined the atomic site charges on the carbonate anion by fitting the crystal structure of calcium carbonate (calcite). They obtained $q(C) = 0.95(26)$ and $q(O) = -0.98(9)$. The 6-31G** PD–AC values (Table I) are $q(C) = 1.210(11)$ and $q(O) = -1.070(4)$, which are narrowly within the confidence limits of the crystal structure derived charges.

Hagler, Huler and Lifson (1974) derived a set of atomic site charges from the crystal structures of amides and from dipole moment data. They assumed as an approximation that the CO and NH_2 groups are electrically neutral, and that hydrogen bonded to carbon has the same charge as an alkane hydrogen. In the case of formamide, the HCO group was considered to be neutral, with an assumed charge of 0.101 $|e|$ on hydrogen. The agreement of their charges with PD–AC values is fairly good for the amino group, but the PD–AC values show that the carbonyl group is considerably more polar than the Hagler, Huler and Lifson model.

A similar approach was used by Lifson, Hagler and Dauber (1979) to obtain atomic site charges from the crystal structures and dipole moments of carboxylic acids. They found that site charges on both the carbonyl oxygen and the hydroxyl oxygen could be taken equal and set to the same value found for the carbonyl oxygen in amides. This assumption left only one atomic site charge to be determined, that of the hydroxyl hydrogen. The carbonyl carbon charge was determined from the molecular neutrality condition. Again, PD–AC values show the carbonyl group to be considerably more polar than this model.

A further comparison can be made with the charges derived by Hirshfeld and Mirsky (1979), which were obtained by fitting the calculated molecular quadrupole moments of acetylene and carbon dioxide. For acetylene, the quadrupole charges found were $q(C) = -0.312$ and $q(H) = 0.312$; the 6-31G** PD–AC are $q(C) = -0.297$ and $q(H) = 0.297$, in good agreement. The STO-3G PD–AC values are smaller. For carbon dioxide, the quadrupole charges were $q(O) = -0.410$ and $q(C) = 0.820$, while the 6-31G** PD–AC values are $q(O) = -0.433$ and $q(C) = 0.867$, also in good agreement. In this case the STO-3G PD–AC values do not change much from the 6-31G** PD–AC values, and are also in good agreement with the cited values. There seems to be a pattern that STO-3G PD–AC values for hydrogen are too small, while the heavy atom STO-3G PA–AC values agree better with the 6-31G** PD–AC values.

Fig. 2 shows electric potential maps for the doubly bonded molecule ethylene, using the 6-31G** basis set. Fig. 4 shows difference maps obtained by subtraction of the PD–AC electric potential. The difference maps show that the PD–AC give a reasonable representation of the QM potential. The region of poorest fit is not perpendicular to the double bond; it is in the plane of the molecule. Wasiutynski, van der Avoird and Berns (1978) found that atomic site charges corresponding to a molecule with shrunken bond lengths gave better agreement to the calculated interaction energy between the

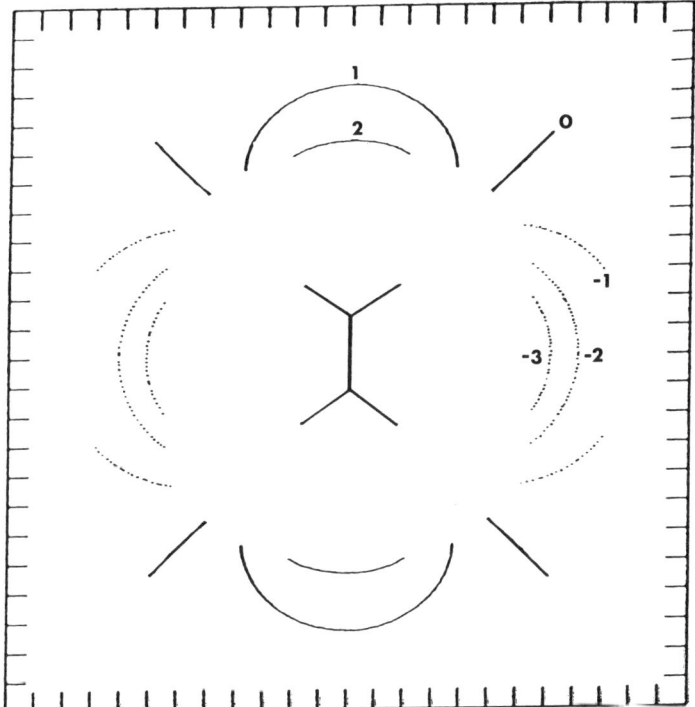

FIG. 4. Difference contour map showing the electric potential of the ethylene molecule calculated from PD–AC minus the QM potential. (a) In the molecular plane.

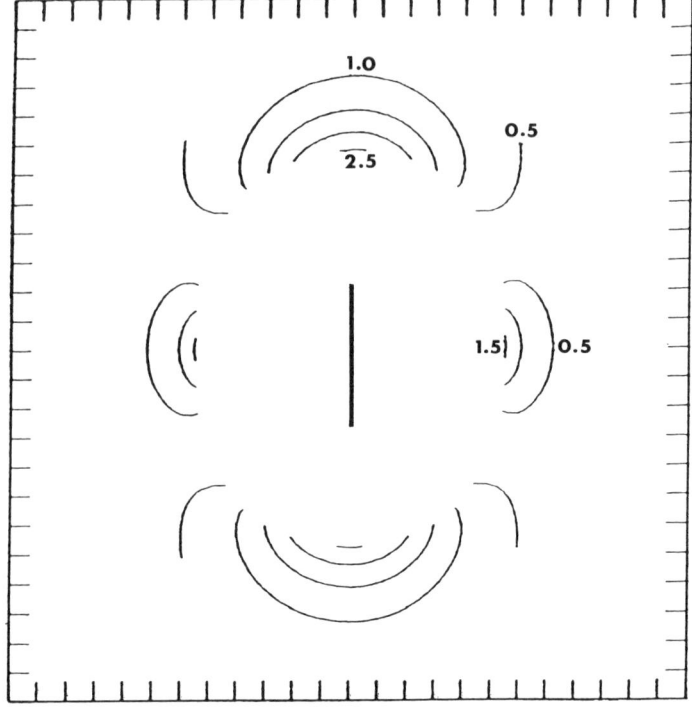

FIG. 4. (b) Perpendicular to the molecular plane.

molecules of ethylene dimer. Cox and Williams (1981) also found that rms improves if the molecule is shrunken by a similar amount.

Fig. 5 shows difference maps for water obtained by subtracting the PD-AC electric potential from the QM potential. It can be seen that a poorer fit is obtained for water than for ethylene. In the plane of the molecule there is excess negative potential near the hydrogens; out of plane there is excess positive potential above and below the molecular center. In Section IV.D below it is shown that the fit may be improved by adding additional out-of-plane negative site charges, positioned on the hydrogen side of the oxygen.

Momany (1978) obtained PD-AC values and plotted electric potentials for methanol, formic acid and formamide. The Cox and Williams results for the PD-AC values are in good agreement with those for methanol and formic acid. However, there are some differences in the case of formamide. Momany used sampling points for the electric potentials closer to the atoms than used by Cox and Williams. The Momany charges for formamide show more positive oxygen and nitrogen atoms, while the carbon atom becomes more negative to preserve electrical neutrality.

Smit, Derissen and van Duijneveldt (1979) obtained PD-AC values for formaldehyde, methanol, and formic acid, using a split-valence basis wavefunction. Their results are similar to the Cox and Williams values using the 6-31G basis wavefunction.

Cox, Hsu and Williams (1981) obtained PD-AC values using the STO-3G basis set for a group of oxohydrocarbon molecules which included carbon dioxide, trioxane, tetroxocane, pentoxecane, succinic anhydride, diglycolic anhydride, 1,4-cyclohexanedione, 1,4-benzoquinone, and furan. These charges were used in connection with derivation of nonbonded potential parameters for oxohydrocarbons.

Singh and Kollman (1984) obtained PD-AC values for a number of small molecules using several basis sets. Their results are generally in good agreement with those of Cox and Williams (1981). In addition, they found PD-AC values for dimethyl phosphate ion, deoxyribose, adenine, 9-methyladenine, thymine, 9-methylthymine, guanine, 9-methylguanine, cytosine, 1-methylcytosine, uracile and 1-methyluracil.

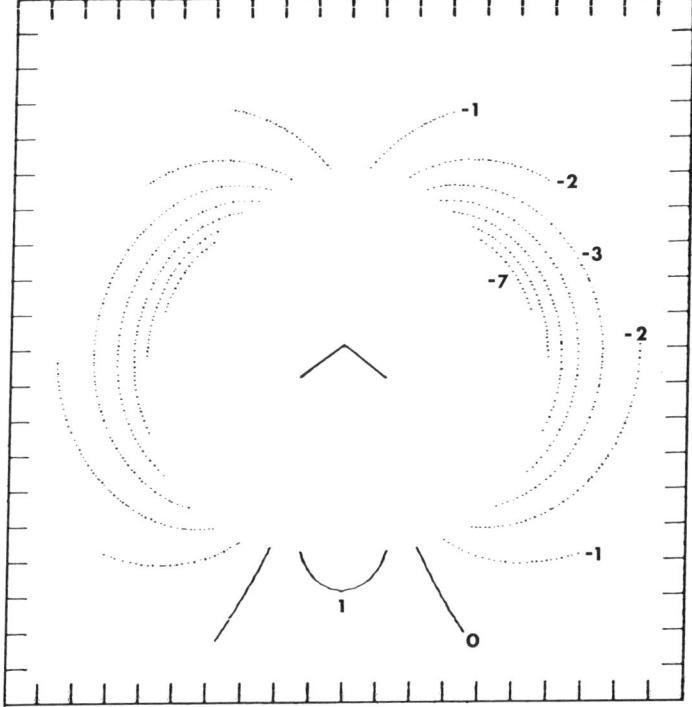

FIG. 5. Same as FIG. 4, but for the water molecule. (a) In the molecular plane.

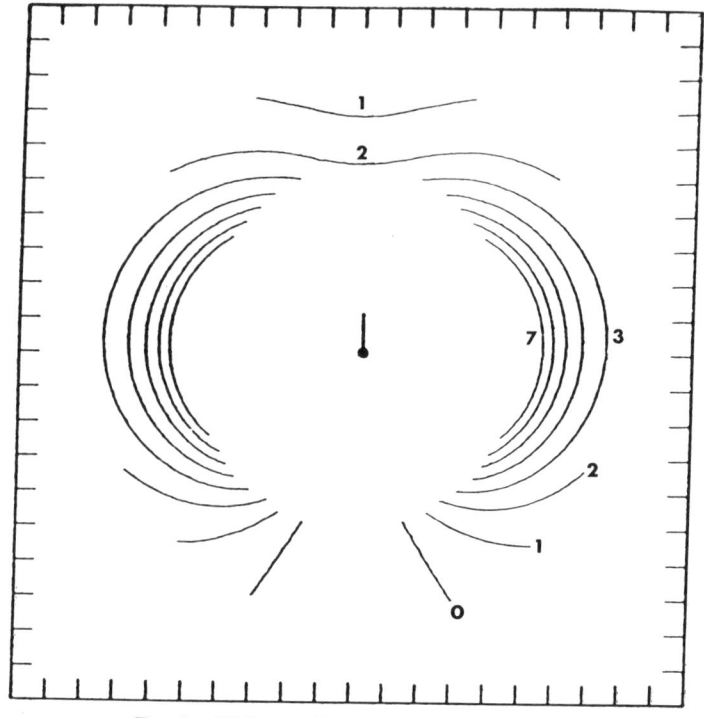

Fig. 5. (b) Perpendicular to the molecular plane.

B. Lone-Pair Electron Sites in Azabenzenes

In studying PD–AC for azahydrocarbons, Cox (1980) found that a good fit to the electric potential of s-tetrazine, for example, could not be obtained with atomic site charges alone. He found that placement of lone-pair electron sites in the ring plane several tenths of an Angstrom out from the nitrogen nucleus greatly improved the fit of the PD–LSF model for the electric potential. Coincidentally, it was noticed that only a poor fit could be obtained for the crystal structure of s-tetrazine by molecular packing analysis (MPA) if only atomic site charges (and no lone-pair charges) were used.

Williams and Weller (1983) made a systematic study of atomic and lone-pair electron site charges in azabenzenes, aiming toward successful application of MPA for prediction of the crystal structures of these molecules. Previous work on the modelling of the crystal structures of azabenzenes had not utilized lone-pair electron sites.

Fig. 6 shows the molecular structures of the azabenzenes which were considered. Wavefunctions were obtained for these molecules using the

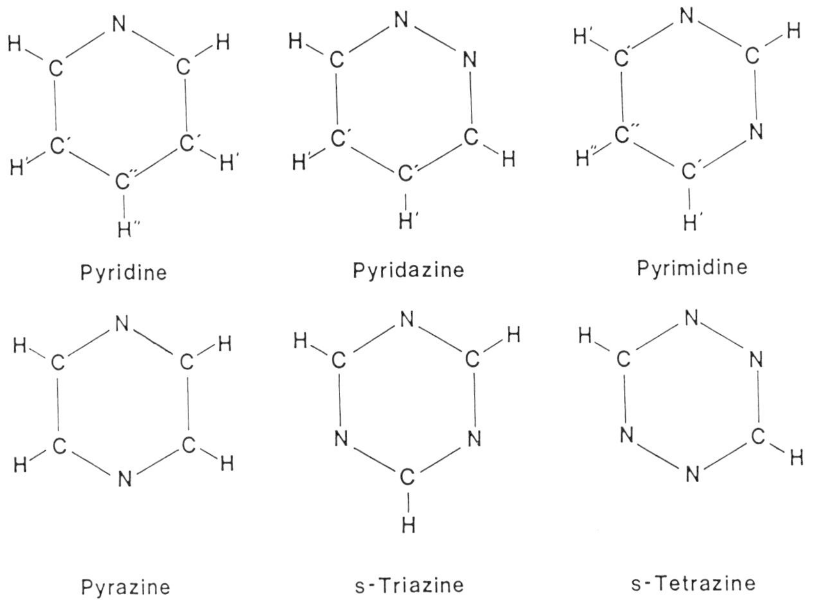

[Structures of Pyridine, Pyridazine, Pyrimidine, Pyrazine, s-Triazine, s-Tetrazine]

FIG. 6. Molecular structures of azabenzenes.

6-31G basis set. Three kinds of site charges were obtained for each molecule: population analysis charges (PA), atomic site charges (PD–AC) and with further inclusion of lone-pair electron charge sites (PD–LP) on nitrogen. The lone-pair electron site of nitrogen was assumed to be in the ring plane at a variable radial distance from the nitrogen.

Figure 7 shows how the rms fit index and the value of the lone-pair electron site charge vary with the lone pair distance for each molecule. The figure shows that the best fit to the calculated electric potential was obtained with the lone pair being located from < 0.02 Å (pyridine) to 0.23 Å (tetrazine) from the nitrogen atom. The optimum distances (Å) and charges (electrons) are: pyridine, < 0.02, < -25.76; pyridazine, 0.03, -12.65; pyrimidine, 0.06, -6.52; pyrazine, 0.06, -7.68; triazine, 0.04, -8.90; tetrazine, 0.23, -1.25.

It could be argued that since a filled lone-pair electron orbital cannot contain more than two electrons, the lone-pair electronic charge should not be more negative than -2 electrons. Fig. 7 shows that most of the benefit of the PD–LP model is retained for all molecules considered for a lone-pair distance of up to about 0.25 Å. At this distance all lone-pair electron site charges are less than 2.0 in magnitude; they range from -1.62 for pyridine up to -1.25 for s-tetrazine. In MPA it is desirable to avoid large site charges because the crystal lattice coulombic energy converges better with small site

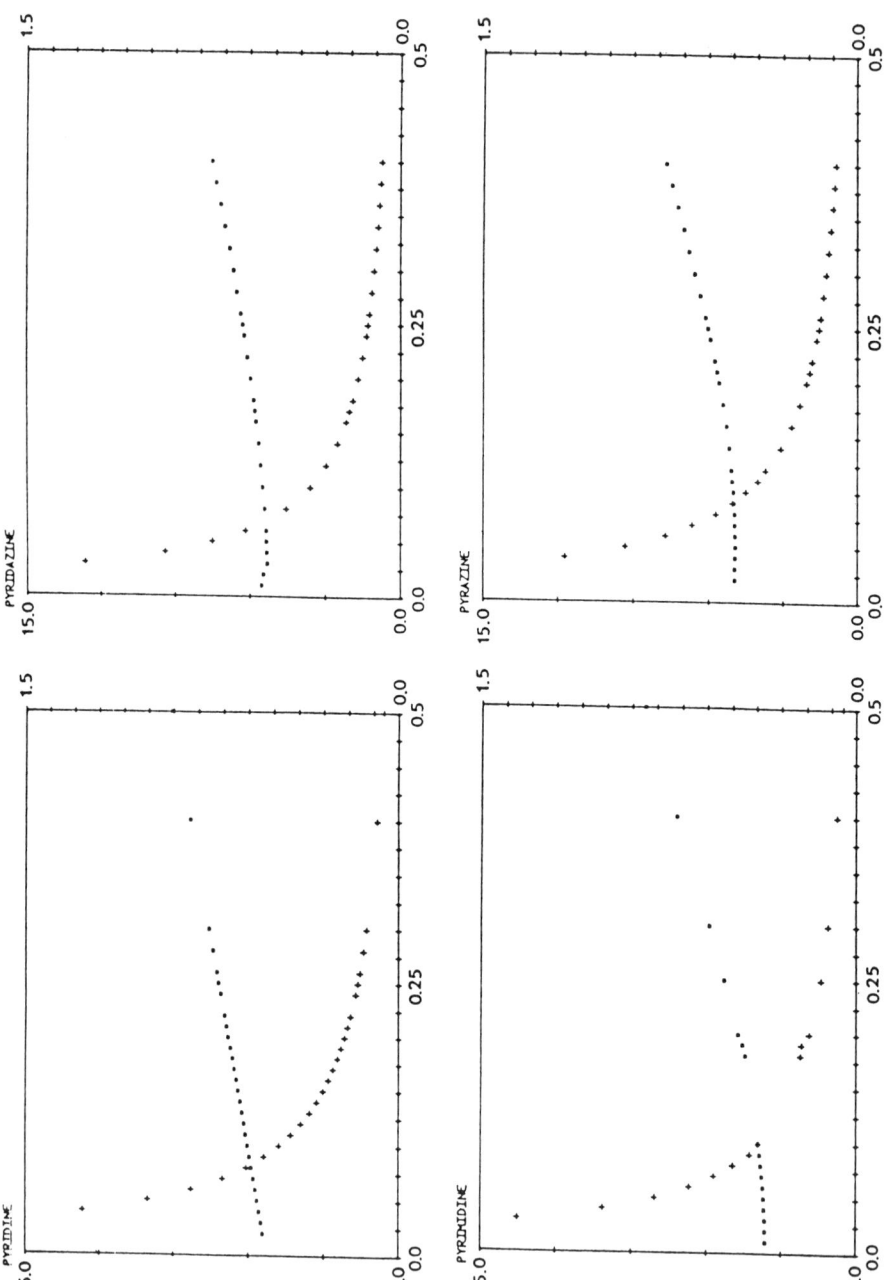

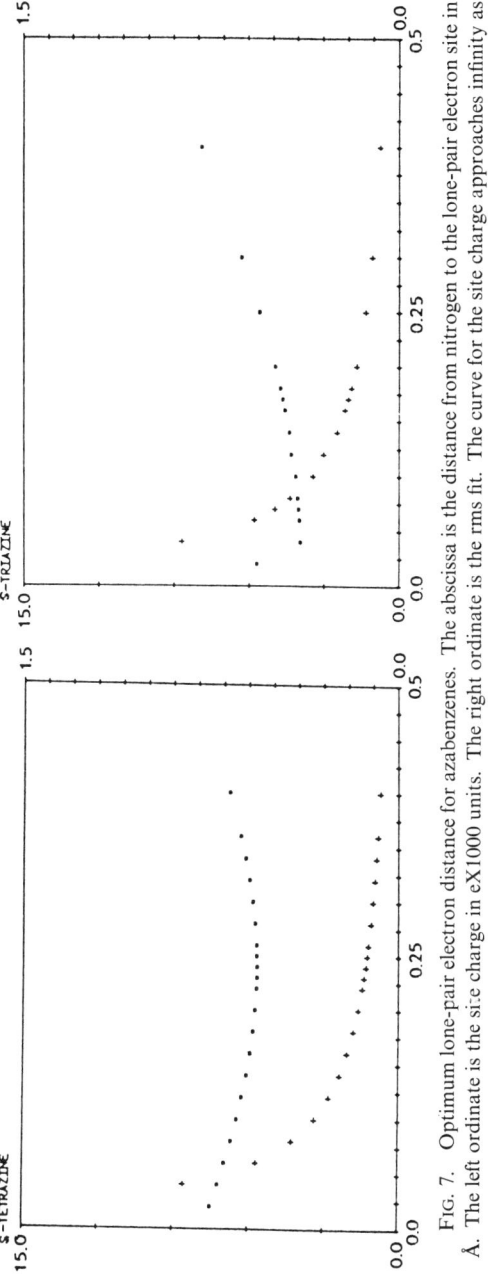

FIG. 7. Optimum lone-pair electron distance for azabenzenes. The abscissa is the distance from nitrogen to the lone-pair electron site in Å. The left ordinate is the site charge in eX1000 units. The right ordinate is the rms fit. The curve for the site charge approaches infinity as the distance approaches zero.

charges. Table III shows PD-LP for pyrimidine and s-tetrazine with the lone-pair site located radially 0.25 Å from nitrogen.

Table III shows the expected behavior of the goodness of fit indices, rms and rrms. The PA charges gave a significantly less accurate description of the electric potential than the PD-AC values. Inclusion of lone-pair sites in the PD-LP further improved the fits to the electric potential. The values of rms and rrms for the PD-LP were generally as good or better than the fits obtained by Cox and Williams (1981) to a variety of smaller nonaromatic molecules.

The molecule which showed the greatest improvement of fit to the calculated electric potential in going from the PD-AC to the PD-LP was s-tetrazine. This was consistent with the above-mentioned early difficulties in predicting the crystal structure of s-tetrazine with PD-AC. Pyridazine also showed a large improvement going from the PD-AC to the PD-LP. Pyridine, pyrimidine, pyrazine and s-triazine were fitted somewhat better by the PD-AC, but nevertheless showed further improved fits with the PD-LP. The fits were uneven by the PA and PD-AC, but it was favorable that all molecules were about equally well fitted by the PD-LP.

TABLE III

PA, PD-AC AND PD-LP FOR PYRIMIDINE AND s-TETRAZINE[a]

Molecule	Quantity	PA	PD-AC	PD-LP
Pyrimidine	q_N	−502	−1046	1301
$u_o = 2.334^b$	q_C	214	1084	−96
	$q_{C'}$	87	894	−31
	$q_{C''}$	−246	−1051	−328
	q_H	227	−41	188
	$q_{H'}$	216	−5	173
	$q_{H''}$	203	322	202
	q_{LP}			−1426
	rms	12.3	1.7	0.5
	rrms	43.9	4.4	1.3
	u_c	3.08	2.94	2.93
s-Tetrazine	q_N	−230	−491	1223
	q_C	181	1167	−175
	q_H	279	−186	237
	q_{LP}			−1254
	rms	19.2	5.9	0.6
	rrms	75.9	16.6	1.6

[a] Site charges are in electronic units $\times 10^3$. Dipole moments are given in Debye units, rms is in kJ/mol, and rrms in percent. The radial distance to the lone-pair site from the nitrogen is 0.25Å.

[b] Blackman, Brown and Burden (1970).

In all of the azabenzene molecules considered except s-tetrazine the electric effect at the nitrogen could be approximated by a point dipole located on the nitrogen, since the optimum lone-pair distance is short and the sum of the nitrogen and lone-pair charges is nearly zero. For s-tetrazine there seemed to be a significant extension of the lone-pair charge sites away from the nitrogens. Bauer and Huiszoon (1982) proposed use of a nitrogen to lone-pair distance of 0.529 Å. In crystal structure energy calculations the distributed dipole model (i.e., the use of a distinct additional charge site) is more convenient, as compared with an atomic point dipole model.

Reynolds (1973) used point dipoles at nitrogen and in the C—H bond. He used a value of 2.2 D obtained from NQR measurement of the electric field gradient at the nitrogen nucleus in pyridine. He obtained a fit for the crystal structure of pyrazine that is comparable to that of the PD–LP of Williams and Weller (1983). Their PD–LP had an effective distributed nitrogen to lone-pair dipole of 1.93 for pyrazine, rather close to the above point dipole value. The distributed dipoles for the other azabenzenes ranged in value from 1.44 to 1.94. Gamba and Bonadeo (1980, 1981) further allowed a variable location for the point dipoles of their model.

The PD–LP have an important advantage of computational simplicity over point multipole models. As an extension of the atom-atom model, the lone-pair site is simply treated as a dummy atomic site having isotropic electrostatic interactions. In contrast, the general formulas for the anisotropic interaction between higher multipoles are very complex (see Burgos and Bonadeo, 1981, for example).

In MPA (Williams, 1972) the intermolecular energy in the crystal is assumed to be a sum of repulsion, dispersion and coulombic components:

$$V(\text{crystal}) = V(\text{repulsion}) + V(\text{dispersion}) + V(\text{coulombic}) \quad (21)$$

Each component is assumed to be a pairwise sum over atoms from different molecules. The corresponding atom-atom nonbonded terms are given by the (exp-6-1) function

$$V_{jk}(\text{exp-6-1}) = B \exp(-C r_{jk}) - A r_{jk}^{-6} + q_j q_k r_{jk}^{-1} \quad (22)$$

where r_{jk} is a nonbonded distance, q_j is a net atomic charge, and A, B, C are adjustable parameters. This function has enjoyed considerable success in modelling crystal structures (Williams, 1981) and has also been used to model molecular clusters (Williams, 1980). The coulombic part of this function requires knowledge of atomic site or nonatomic site charges q_j.

The objective of MPA was to calculate the static equilibrium crystal structure using the observed molecular structure and the observed space group symmetry. The molecule in the crystal was assumed to be rigid, a requirement that is well satisfied by the rigid azabenzene molecules. The success of the MPA model was measured against how well it predicted the observed lattice constants, and the molecular translational and rotational orientation in the cell.

Apparently the only model for any of the azabenzenes which had been tested against the static crystal structure is that of Reynolds (1973) for pyrazine. Rae (1978) performed partial static structure tests for the two crystalline forms of s-triazine: the angle of shear of the unit cell was calculated as well as the molcular rotation angle; the cell edge lengths were held fixed. The less stringent test of predicting only the molecular orientation was used for pyrazine by Sanford and Boyd (1978) and for several azabenzene molecules by Gamba and Bonadeo (1981).

The PD–AC and PD–LP of Williams and Weller (1983) were tested by using them to predict the observed crystal structures of azabenzene molecules. The PD–AC and PD–LP were multiplied by a factor of 0.83 so that they approximately reproduce experimental dipole moments. In the MPA calculations foreshortened (Starr and Williams, 1977) C—H bond lengths were used.

Table IV shows MPA results for the crystal structures of pyrimidine and s-tetrazine. References to the crystal data are given by Williams and Weller (1983). In the MPA calculations all parameters not fixed by the observed space group symmetry were varied simultaneously to find the minimum of the lattice energy. At the final minimum energy of the structures the Hessian (whose elements are the second derivatives of the energy with respect to structural variables) was always positive definite, but not necessarily at the observed structure. To indicate the importance of including the effects of the electric charge distribution in the molecules, no-charge (NC) was subjected to MPA as well as the PA, PD–AC, and PD–LP.

The results for pyrimidine show the poorest fits to the observed crystal structure was given by the NC and PA. The PD–AC gave results of intermediate quality, while the PD–LP gave the best results. Similar patterns of fit were observed with the pyrazine and s-triazine crystal structures.

The crystal structure of s-tetrazine was the most difficult to predict and provided the most stringent test of the model for the molecular charge distribution. The fit of PA and PD–AC were poor; the PD–LP were still not as good as expected. Surprisingly, the NC is about as good as the PD–LP in its prediction of the crystal structure. However, the NC has a negative eigenvalue of the Hessian at the observed structure, and also the calculated energy appears to be too small.

Williams and Cox (1984) further extended the analysis of the crystal structures of azabenzenes and other azahydrocarbons by finding optimum values of nonbonded potential parameters compatible with the PD–LP for azabenzenes. The shifts they obtained in the crystal structure of pyrimidine are indicated in the last column of Table IV. Also, only slight improvements

TABLE IV

Crystal Structures of Pyrimidine and s-Tetrazine[a]

Crystal/obsd. energy	Quantity	NC	PA	PD–AC	PD–LP[b]	PD–LP[c]
Pyrimidine 48.8[d]	Δa	0.95	−1.04	−0.71	−0.01	0.00
	Δb	0.14	0.02	0.40	0.23	0.25
	Δc	−0.27	0.38	0.01	−0.17	−0.05
	θ	19.8	16.9	8.7	2.7	1.7
	t	0.18	0.15	0.18	0.06	0.06
	Energy at obsd. struct.	−35.8	−61.5	−54.7	−55.3	−56.6
	Negative eigenvalues	2	1	1	0	0
	Energy at calc. struct.	−40.5	−64.5	−56.7	−56.4	−57.3
s-Tetrazine	Δa	0.44	−1.31	−1.05	0.39	
	Δb	−0.39	2.49	1.25	0.46	
	Δc	0.36	−0.03	0.44	−0.44	
	$\Delta \beta$	−2.6	5.00	2.9	8.3	
	θ	12.0	39.5	30.9	18.4	
	Energy at obsd. struct.	−28.2	−41.5	−44.7	−58.8	
	Negative eigenvalues	1	1	2	0	
	Energy at calc. struct.	−33.0	−66.0	−61.5	−61.3	

[a] Differences between predicted and observed structures are shown. Δa, Δb, Δc, and $\Delta \beta$ are the predicted shifts in the lattice constants (Å, deg) obtained by minimizing the lattice energy (kJ/mol); θ and t are the predicted molecular rotation (deg) and translation (Å) in the unit cell. The number of negative eigenvalues listed refers to the Hessian matrix of the second derivatives of the lattice energy with respect to the structural variables. The site charges PD/AC and PD/LP were scaled by a factor of 0.83 from the values given in Table III.
[b] Without optimization of nonbonded potential (Williams and Weller, 1983).
[c] With optimization of nonbonded potential (Williams and Cox, 1984).
[d] Nabavian, Sabbah, Chestel and Laffite (1977).

were obtained for the structures of pyrazine and beta-s-triazine. The structure of s-tetrazine was not included in their study.

C. Bond Sites in Fluorocarbons

Electrostatic interactions in fluorocarbon crystals were modelled by Williams and Houpt (1986) by PD-LSF point charges as part of a project to derive MPA values for nonbonded F...F potential parameters. The resulting

F...F nonbonded potential parameterizations were tested by relaxing the lattice constants and molecular positions to find the minimum energy configurations for seven perfluorocarbon crystal structures. The shifted structures were compared with the crystal structures observed by x-ray or neutron diffraction.

Fig. 8 illustrates the molecular structures. Five of the molecules in the data base have only CF groups and the remaining two molecules have only CF_2 groups, besides isolated carbon atoms. The molecular packing in the crystals shows considerable variety. Structures II, V, and VI exhibit the simple case where one molecule is in a general position in the cell and the surrounding molecules are related by the group symmetry operations. In this situation the structural variables are six molecular rotations and translations, in addition to the lattice constants.

In structure III the biphenyl molecules are required to have twofold symmetry in the c lattice direction; the space group also is polar in the c direction. The molecular degrees of freedom are thus reduced to only a rotation about the z axis. In IV the triethanobenzene molecules have unusually high $\bar{6}$ symmetry; only rotation of the molecule about the z axis is symmetry allowed. In VII the substituted cyclooctatetrane molecule has twofold symmetry in the b direction; thus rotation about and translation along the y axis is symmetry allowed.

The most complex molecular packing is shown by the simplest molecule, hexafluorobenzene. This structure was difficult to model because of its increased degrees of freedom. There are six molecules crystallizing per unit cell in a space group of order 4; thus there are one and one-half molecules in the crystallographic symmetric unit. One molecule is placed on an inversion center and another molecule is in a general position. The two crystallographically independent molecules have the same molecular structure within experimental error. In addition to the lattice constants, the structural variables are three rotations of the first molecule, and six rotations and translations of the second molecule. With inclusion of the four lattice constants this structure has thirteen degrees of freedom.

To model the electric potential of these molecules extended or bond site charges were added to atomic sites. The two additional types of sites considered were locations on the extension of the C–F bond axis (PD–EC) or bond charge sites (PD–BC) located at the centers of all C–F bonds. The quadrupole moment of hexafluorobenzene is known (Battaglia, Buckingham and Williams, 1981) so that site charges can be assigned to reproduce this moment.

The split-valence 6-31G basis set was selected to obtain sufficiently accurate electric potentials. However, because of computer limitations the only molecule in the crystal structure data which could be handled with this

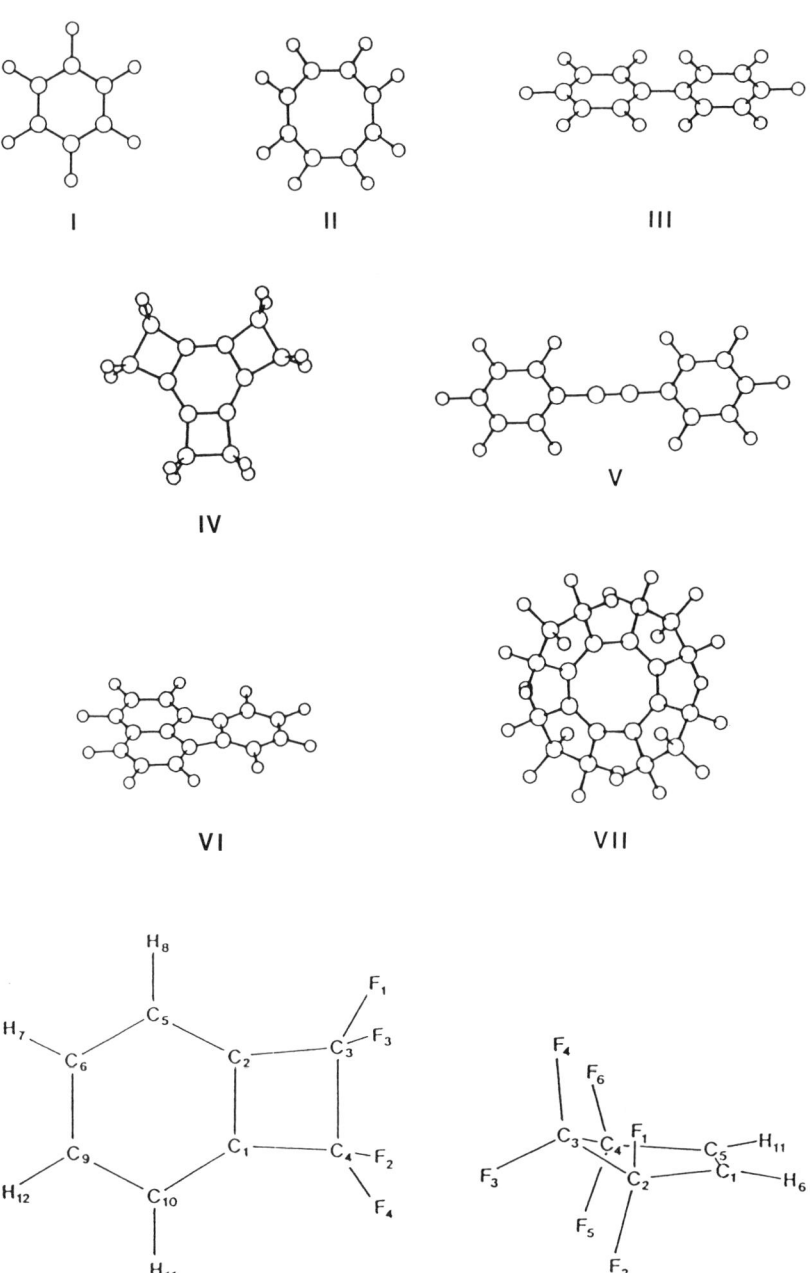

FIG. 8. Molecular structure of perfluorocarbons.

basis set was hexafluorobenzene. The site charge model for hexafluorobenzene was also used to estimate site charges in structures II, III, V, and VI, which contain only CF groups or isolated carbons.

Structures IV and VII contain CF_2 groups, for which site charges also were needed. Molecules VIII and IX (Fig. 8), which looked like fragments of IV and VII, were used as models. The qualitative difference between the two structure types was that IX (and VII) contained extra CF_2 group bonded between adjacent CF_2 groups. Molecules VIII and IX were small enough to permit the calculation of their *ab initio* wavefunctions with the available computational facilities.

The PD-AC, PD-EC and PD-BC were tested with molecules I, VIII and IX. For VIII and IX molecular fragment geometry was obtained from the experimental crystal structures; for I the molecular geometry was taken from an electron diffraction study.

Table V shows fits obtained with the three kinds of models for the three molecules. For I, the PD-AC gave a rrms fit of 11.0%. When an extended site was added at 0.25Å (PD/EC), the rrms fit improved to 4.3%. When a bond center site was added (PD-AC) the rrms fit further improved to 2.1%. Thus, the PD-BC best fitted the surrounding electric potential of I. For VIII and IX, similar comparative results were obtained for the three site charge models. Although PD-BC have the best fit, PD-AC for VIII and IX were relatively better than for I. It was concluded that overall PD-BC were significantly superior to either PD-AC or PD-EC in fitting the electric potential.

Table VI shows values for the site charges. For molecule I the charges of the CF groups always sum to zero by symmetry. In the cases of VIII and IX the sums of the CF_2 group charges, while small, were not exactly zero. Further, in IX there were differences between the charges of the central CF_2 group compared to the adjacent CF_2 groups. In analyzing this situation, it was concluded reasonable to make small adjustments in the charges so that the CF_2 groups had zero charge and so that all CF_2 groups in IX had similar

TABLE V

SITE CHARGE MODELS FOR PERFLUOROCARBONS

	PD-AC		PD-EC		PD-BC	
Structure	rms	rrms	rms	rrms	rms	rrms
I	2.4	11.0	1.0	4.3	0.5	2.1
VIII	1.7	5.3	1.2	3.6	0.9	2.8
IX	1.5	4.5	1.1	3.4	0.9	2.9

TABLE VI

AVERAGED PD-LSF VALUES FOR NEUTRAL GROUP C-F SITE CHARGE

Model/site structure	PD-AC		PD-EC			PD-BC		
	C	F	C	F	E	C	B	F
I	174	−174	95	243	−338	−86	466	−380
VIII	734	−367	308	175	−329	24	405	−417
IX	682	−341	436	3	−221	364	204	−386

charges. All CF_2 groups in both molecules were not required to have similar charges. Table VI shows that somewhat different values were obtained for VIII and IX. The most notable similarity between them is the charge on fluorine in PD-BC. This uniformity of the fluorine charge may be consistent with the superior modelling capability of PD-BC. PD-AC and PD-EC show more charge fluctuations between VIII and IX.

Since dipole moments obtained from PD-LSF charge distributions agree closely with the operator calculated dipole moments, it is reasonable to scale the PD-LSF site charges so that they reproduce observed dipole moments. In a similar way, calculated quadrupole moments may be scaled to observed values. The 6-31G wavefunction for I gave an operator quadrupole moment $Q_{zz} = 48.85 \times 10^{-40}$ Cm^2 vs. an observed $Q_{zz} = 31.7 \times 10^{-40}$ Cm^2 (Battaglia, Buckingham and Williams, 1981). The ratio of Q_{zz} (observed) to Q_{zz} (calculated) is equal to 0.65. Therefore the PD-LSF charges were scaled by this factor in MPA calculations. The practical effect was to essentially make the scaled PD-LSF charges for hexafluorobenzene reproduce the observed quadrupole moment of this molecule. The dipole and/or quadrupole moments of the other molecules were not available for comparison, but the same charge scaling factor was used for them.

Table VII shows the shifts in the crystal structure of hexafluorobenzene obtained by MPA. Since there are thirteen degrees of freedom in this crystal structure, many of the structural shifts were larger than for the other crystal structures. A parallel calculation in which only the lattice constants were varied was carried out. In that case the structural shifts were more comparable to those obtained with the other perfluorocarbon structures.

The improvement in fit obtained with any of the various site charge models was not very great. The conclusion that site charges are unimportant for fluorocarbons in general may be premature, however. Consider a fluorocarbon molecule which has hydrogen atoms. The C-H bonds would have opposite polarity to the C-F bonds. The coulombic interaction between hydrogen and fluorine would be very important in determining the molecular

TABLE VII
Crystal Structure Data of Hexafluorobenzene[a]

Parameter	Site charge model			
	NC	PD–AC	PD–EC	PD–BC
a	−2.1%	−2.6%	−2.8%	−2.6%
$a*$	2.9%	3.4%	3.0%	3.0%
b	10.0%	9.6%	8.4%	7.9%
$b*$	0.3%	0.6%	0.4%	0.4%
c	−3.7%	−2.1%	−1.3%	−0.9%
$c*$	2.4%	2.7%	2.6%	2.5%
β	2.4°	1.7°	2.2°	2.2°
$\beta*$	0.7°	0.4°	0.4°	0.4°
θ_1	12.6°	10.5°	10.3°	9.7°
θ_2	7.7°	8.2°	7.8°	7.4°
t_2	0.35Å	0.35Å	0.39Å	0.37Å

[a] Percentage shifts in a, b, c are shown; other shifts in deg, or Å.
* Molecular rotations and translation held constant.

packing. The coulombic effect would be less important in perfluorocarbons because no such bonds of reverse polarity are present.

The relatively small improvements in fit obtained with the various kinds of site charges confirmed the idea that it was difficult to derive values for site charges from crystal data. It was also found difficult just to derive an overall charge scaling factor from the data, a procedure which was successful in hydrocarbons (Williams and Starr, 1977) and perchlorocarbons (Hsu and Williams, 1980). Fortunately the PD–LSF model allowed an independent estimate of the site charges. Since the perfluorocarbon crystal structures are not extremely sensitive to the choice of the kind of site charge, the improvement in fitting the crystal structures was not as great as the improvement in fitting the electric potential.

D. Water Monomer and Dimer

An oversimplified view of the hydrogen bond is that is can be understood solely on the basis of an electrostatic interaction between atomic site charges. In order to develop more quantitative empirical models for the hydrogen bond, it is desirable to be able to separate out the coulombic intermolecular energy from other components. Water readily forms hydrogen bonded dimer which has been extensively studied both experimentally and theoretically. The empirical form for a hydrogen bond potential energy function V_{HB} for water which is compatible with (exp-6-1) functions for non-hydrogen bonded oxygen and hydrogen atoms is not well established. A clear separation of the coulombic component would assist in defining the empirical form of V_{HB}.

Singh and Kollman (1984) examined PD-AC for water monomer. They obtained a rrms fit of 10.5% to the 6-31G** electric potential. To improve the fit, they added two lone-pair electron sites in the bisector plane at distance d and angle θ from the twofold axis. Instead of finding approximately tetrahedral locations for the lone-pair electron sites, they found an optimum θ of 69.3 deg at distance of 0.465 Å. This places the extra sites on the hydrogen side of the molecule, called "inverted" lone-pair electron sites. The hydrogen charges were 0.489, oxygen -0.544, and the lone-pairs -0.217 $|e|$.

The idea of using a negative charge site on the hydrogen side of the water molecule was earlier used by Popkie, Kistenmacher and Clementi (1973) in connection with an analysis of the calculated Hartree-Fock intermolecular energy of water dimer. This energy was fitted by an empirical model having exponential atom-atom repulsion and site charge coulombic interaction. Charges q on the hydrogen were balanced by charges of $-2q$ on the twofold axes of the molecules. The charges on the twofold axis was moved to an optimum position, not necessarily at the oxygen site. They found $q = 0.670$ $|e|$ and distance of 0.231 Å on the hydrogen side of the oxygen atom. In a later calculation using configuration interaction, Matsuoka, Clementi and Yoshimine (1976) found $q = 0.717$ and $d = 0.268$.

The five charge sites of Singh and Kollman allowed the site on the twofold axis to split into two sites above and below the molecular plane. They also considered four charge sites where the oxygen charge was set to zero as was done in the above intermolecular energy calculation. This simplication of the model led to $q_H = 0.537$ with opposite sign for the "inverted" lone-pair sites. The oxygen to lone-pair distance was 0.279 Å, the angle 65.5 deg, and the rrms fit to the electric potential 7.5%.

Williams and Craycroft (1985) also optimized a similar four charge site for water in connection with a study of water dimer. Using the 6-31G** basis set for the electric potential, they found a rrms fit of 8.3% for the PD-AC; the four charge-site was designated PD-SC rather than PD-LP since it was not clear whether the extra sites could be identified with lone-pair electrons. The O—H distance was shortened by 10% to account for the bonding electron density shift. This electron density center shift for bonded hydrogen atom is well-documented (Starr and Williams, 1977) and leads to a further slight improvement in the PD-SC for water monomer. This gave a rrms fit of 1.0% with $q_H = 0.630$ $|e|$, $d = 0.252$ Å, and $\theta = 48.9$ deg. The rrms fits are not directly comparable to Singh and Kollman (1984) because of a different choice of grid points for the electric potential. But there is no doubt that a significant improvement in rrms occured going from a three charge site PD-AC to a four charge site PD/SC.

These studies agreed that the extra charge sites in the water molecule were on the hydrogen side of the molecule. Since lone-pair electrons are involved in hydrogen bonding, it was of interest to see whether the PD–SC also gave a better description of the electric potential around water dimer, compared to the PD–AC for the dimer. The PD–LSF method may easily be extended to explore site charges in molecular dimers. The dimer wavefunction may be calculated in a straightforward way and the surrounding electric potential found and fitted in a way that is completely analogous to that used for the monomer.

An approximation to the coulombic interaction energy of the dimer can then be obtained by evaluating Coulomb's law directly between the site charges of the different monomers. Further, it was of interest to note whether the charges in the molecular components of the dimer remained essentially unchanged, or if there were significant changes from monomer values as the molecules approached to form the dimer. For this purpose, the change in the site charges going from monomer to dimer were defined as the dimer site charge polarization.

Williams and Craycroft (1985) calculated wavefunctions for 52 configurations of water dimer using the 6-31G** basis set. The geometry of the water molecule was fixed at the experimental monomer values (Benedict, Gailar and Plyler, 1956). Electric potentials were calculated for each dimer configuration.

Fig. 9 shows as examples the linear, bifurcated and cyclic orientations as represented by dimer configurations E39, B13, and C21 of Matsuoka, Clementi and Yoshimine (1976, designated MCY). The figure shows the extra nonatomic charge sites in each molecule.

For the dimer configurations, the PD–AC gave a rms fit to the potential grid points of 1.5 to 2.9 kJ/mole or a rrms fit of 3.2 to 10.4%. Table VIII

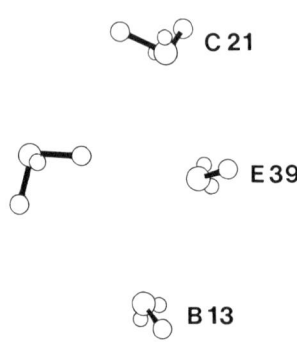

FIG. 9. Charge site placement for water monomer and the configuration of water dimers. The left molecule is fixed and three orientations of the right molecule are shown as specified by MCY: C21 (cyclic), E39 (linear) and B13 (bifurcated).

TABLE VIII
Water Monomer and Dimer[a]

Structure	H1	H2	O1	S1 = S2	H3	H4	O2	S3 = S4	rms	rrms
Monomer(PD–AC)	393	393	−786						2.8	8.3
Monomer(PD–SC)	630	630		−630					0.3	1.0
Linear(PD–AC)	460	409	−868		444	444	−889		1.7	4.1
Bifurcated(PD–AC)	456	456	−883		451	451	−931		1.7	3.9
Cyclic(PD–AC)	369	410	−806		397	399	−769		2.3	9.6
Linear(PD–SC)	681	637		−670	646	646		−635	0.5	1.2
Bifurcated(PD–SC)	654	654		−660	647	647		−641	0.4	1.0
Cylic(PD–SC)	654	642		−654	642	654		−642	0.5	1.9

[a] PD charges in electronic units $\times 10^3$. The O—H distances were foreshortened by 10%.

shows the fits and values for the charges for the near linear, bifurcated and cyclic configurations of the dimer. The rms fits did not vary much between the various configurations, although the bifurcated type B and the linear type E were fitted somewhat better.

In going from the PD–AC to the PD–SC, the rms fits ranged from 0.3 to 0.9 kJ/mol and the rrms fits from 0.9 to 2.3 %. Note that in the PD–SC the out-of-plane sites were required to have equal charge. Test calculations showed that relaxation of this site symmetry restriction gave little further improvement of fit. The uniformly good fit obtained for the various dimer configurations with the PD–SC was quite striking, as compared to the PD–AC.

Kitaura and Morokuma (1976) proposed a method of decomposing the SCF energy into electrostatic, exchange, polarization and charge transfer components. DeWit (1983) recently reported decomposition of water dimer SCF energy into components using the same basis set and the same molecular geometry as MCY. The method of energy decomposition used was described by Smit, Derissen and van Duijneveldt (1979). Table IX shows the coulombic (electrostatic) and polarization (induction) energy components for the three configurations of water dimer as reported by DeWit (designated D).

The intermolecular coulombic energy was estimated for the PD–AC and PD–SC by assiging site charges in the dimer equal to the unpolarized monomer PD–AC or PD–SC values, and summing directly over the charge sites of the different molecules. When the site charges in the dimer were shifted to their polarized PD–AC or PD–SC values, a larger coulombic energy was calculated. The additional energy thus obtained gave a rough estimate of the polarization energy. Note that with this definition of polarization energy no correction was made for any change in the monomer energy in going from monomer-optimized charges to dimer-optimized charges. Such a correction was not possible because the charge sites are not valid for the calculation of intramolecular coulombic energies. Nevertheless, these polarization energies could be compared to those obtained by DeWit since the latter also refer to the unpolarized monomer. Examination of the

TABLE IX

Estimated Coulombic and Polarization Energies (kJ/mol) in Water Dimer

Structure	D		MCY	PD–AC		PD–SC	
	coul.	pol.	coul.	coul.	pol.	coul.	pol.
Linear	−37.7	−3.2	−36.9	−25.1	−8.6	−26.8	−0.4
Bifurcated	−24.8	−1.6	−19.5	−19.3	−11.6	−19.2	0.4
Cyclic	−26.0	−1.6	−31.0	−18.1	1.0	−21.4	−1.3

values showed that parallel trends existed between the components of the D and the corresponding estimates of the MCY, PD–AC, and PD–SC.

Conceptually, the PD–AC and PD–SC were most closely related, since both were obtained directly from the surrounding electric potential and they differ only in the location of charge sites. The PD–SC were the more accurate of the two since they gave better fit to the electric potential. The MCY site charges were obtained from an empirical fit to the intermolecular energy of water dimer, where no explicit provision was made for dispersion energy component. Therefore the MCY coulombic component could have been affected by this neglect of the dispersion component. In the work of DeWit the dispersion energy was indeed evaluated and considered separately.

However, the D values were not obtained from site charges, being based on a classification of the SCF integrals into types. Of the site charge models, MCY gave the best agreement with the D for the coulombic energy components. However, no estimate of the polarization energy was available for the MCY. The PD–AC and PD–SC generally showed smaller magnitudes of estimated coulombic and polarization energy. The SC values were closer to the D values than were the AC values. Note that both D and MCY used different basis set than the calculations for the PD–LSF models.

Fig. 10 shows a plot of the relative change in the charge of the nearest hydrogen atom as the second molecule approaches to form the dimer. The

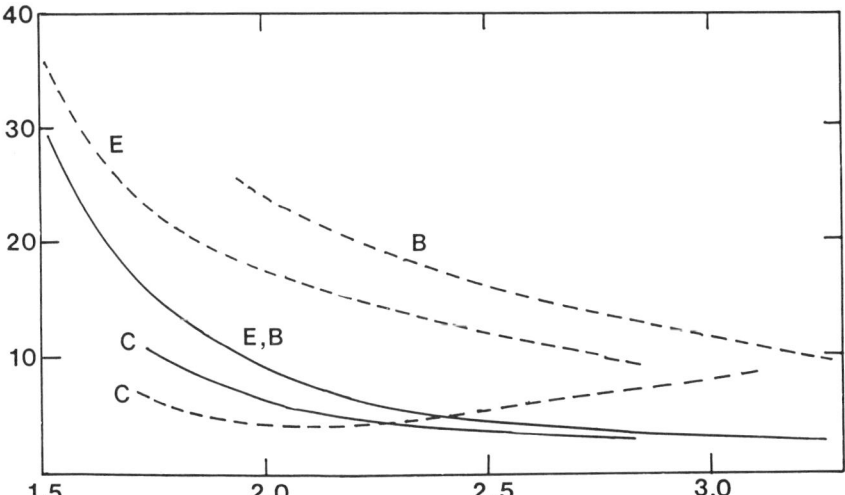

FIG. 10. Relative polarization (in percent) of the charge on the nearest hydrogen by the second molecule, versus the shortest H...O distance (A) in the dimer. The dashed lines show the polarization in the E (linear), B (bifurcated), and C (cyclic) models fitted with PD–AC. The solid lines show similar results for PD–SC.

dashed curves show the PD–AC. The three curves shown are for the B (bifurcated), C (cyclic), and E (linear) cases. For a given H...O distance, the B structure showed the largest site charge polarization, and the C structure the least. The linear E structure was intermediate in its polarization, but since this case was evaluated at the very short H....O distance of 1.51Å, the polarization reached a maximum of 36%.

The solid lines in Fig. 10 show the site charge polarization in the PD–SC. The apparent polarization was smaller, reaching a maximum of 28% at the closest distance for the E structure. The B points no longer showed a larger polarization than the E points; in fact both the B and E points could be plotted on the same curve. In addition, the C points for the PD–SC are closer to the E and B points than they were for the PD–AC. The figure showed that the PD–SC site charge polarization had a more uniform trend vs. the H...O distance than the PD–AC.

V. Conclusion

The PD–LSF point-charge model offers an empirical method for describing the molecular electric potential outside the van der Waals envelope. This model is based on atomic sites, supplemented as necessary by lone-pair sites, bond sites, or other sites placed at optimum locations. The model can be used to quickly estimate the intermolecular energy of electric interaction by using Coulomb's law between sites.

The accuracy of the model depends on how well the electric potential is fitted, and on the accuracy of the electric potential itself. The accuracy of the calculation of the electric potential in turn depends on the quality of the wavefunction and can be quite good for small molecules. The fit of PD–LSF point charges to the electric potential can be as good as desired, depending on the number and location of the charge sites.

The PD–LSF atomic site point-charges may give a sufficiently good fit to the electric potential that is adequate for many applications. For aromatic nitrogen compounds the inclusion of PD–LSF lone-pair electron site point charges is desirable. For perfluorocarbons an improvement of fit occurs when PD–LSF bond site point-charges are included. For water monomer and dimer optimized extra site positions above and below the plane of the molecule, on the hydrogen side, are useful.

The method may also be used to examine charge sites in dimers of small molecules, and the polarization of these charge sites as a function of monomer orientation and intermonomer distance.

ACKNOWLEDGEMENT

This work was supported by research grant GM-37453 from the National Institutes of Health.

REFERENCES

Agresti, A., Bonaccorsi, R., and Tomasi, J. (1979). *Theoret. Chim. Acta* **53**, 215.
Battaglia, M. R., Buckingham, A. D., and Williams, J. H. (1981). *Chem. Phys. Lett.* **78**, 421.
Bauer, G. E. W., and Huiszoon, C. (1982). *Mol. Phys.* **47**, 565.
Benedict, W. S., Gailar, N., and Plyler, E. K. (1956). *J. Chem. Phys.* **24**, 1159.
Binkley, J. S., Frisch, M. J., DeFrees, D. J., Rahgavachari, K., Whiteside, R. A., Schlegel, H. B., Fluder, E. M., and Pople, J. A. (1984). GAUSSIAN 82, available from Department of Chemistry, Carnegie-Mellon University, Pittsburgh, Pennsylvania.
Binkley, J. S., Pople, J. A., and Hehre, W. J. (1980). *J. Am. Chem. Soc.* **102**, 939.
Blackman, G. L., Brown, R. D., and Burden, F. R. (1970). *J. Mol. Spectrosc.* **35**, 444.
Burgos, E., and Bonadeo, H. (1981). *Mol. Phys.* **44**, 1.
Christoffersen, R. E., and Baker, K. A. (1971). *Chem. Phys. Lett.* **8**, 4.
Connolly, M. (1982). Program no. 429, Quantum Chemistry Program Exchange, Indiana University, Bloomington, Indiana.
Cox, S. R. (1980). M. S. Dissertation. University of Louisville, Louisville, Kentucky.
Cox, S. R., Hsu, L. Y., and Williams, D. E. (1981). *Acta. Crystallogr. Sect. A* **37**, 293.
Cox, S. R., and Williams, D. E. (1981). *J. Computational Chem.* **2**, 304.
Crowder, C. D., Alldredge, G. P. and White, H. W. (1985). *Phys. Rev. Sect. B* **31**, 6676.
Davidson, E. R. (1967). *J. Chem. Phys.* **46**, 3320.
DeWit, H. G. M. (1983). Ph.D. Dissertation, University of Utrecht, Utrecht, Netherlands.
Dupuis, M., Rys, J., and King, H. F. (1977). HONDO 76, program no. 336, Quantum Chemistry Program Exchange, Indiana University, Bloomington, Indiana.
Gamba, Z., and Bonadeo, H. (1980). *Chem. Phys. Lett.* **69**, 525.
Gamba, Z., and Bonadeo, H. (1981). *J. Chem. Phys.* **75**, 5059.
Hagler, A. T., Huler, E., and Lifson, S. (1974). *J. Amer. Chem. Soc.* **96**, 5319.
Hall, G. G. (1985). *Adv. Atomic Mol. Phys.* **20**, 41.
Hariharan, P. C., and Pople, J. A. (1972). *Chem. Phys. Lett.* **66**, 217.
Hehre, W. J., Ditchfield, R., and Pople, J. A. (1971). *J. Chem. Phys.* **54**, 724.
Hehre, W. J., Radom, L., Schleyer, P. v. R., and Pople, J. A. (1986). "*Ab Initio* Molecular Orbital Theory." Wiley, New York.
Hehre, W. J., Stewart, R. F., and Pople, J. A. (1969). *J. Chem. Phys.* **51**, 2657.
Hirshfeld, F. L., and Mirsky, K. (1979). *Acta Crystallogr. Sect. A* **35**, 366.
Hsu, L. Y., and Williams, D. E. (1980). *Acta Crystallogr. Sect. A* **36**, 277.
Kitaura, K., and Morokuma, K. (1976). *Int. J. Quantum Chem.* **10**, 325.
Kollman, P. A. (1978). *J. Amer. Chem. Soc.* **100**, 2974.
Krishnan, R., Frisch, M. J., and Pople, J. A. (1980). *J. Chem. Phys.* **72**, 4244.
Lifson, S., Hagler, A. T., and Dauber, P. (1979). *J. Amer. Chem. Soc.* **101**. 5111.
Löwdin, P. O. (1953). *J. Chem. Phys.* **21**, 374.
Lowe, J. P. (1978). "Quantum Chemistry." Academic Press, New York.

Matsuoka, O., Clementi, E., and Yoshimine, M. (1976). *J. Chem. Phys.* **64**, 1351.
Momany, F. A. (1978). *J. Phys. Chem.* **82**, 592.
Mulliken, R. S. (1955). *J. Chem. Phys.* **23**, 1833.
Nabavian, M., Sabbah, R., Chestel, R., and Laffite, M. J. (1977). *J. Chim. Phys.-Chim. Biol.* **74**, 115.
Neumann, D. B., Basch, H., Kornegay, R. L., Snyder, L. C., Moskowitz, J., Hornback, C., and Liebman, P. (1971). POLYATOM, program no. 199, Quantum Chemistry Program Exchange, Indiana University, Bloomington, Indiana.
Pauling, L. (1960). "The Nature of the Chemical Bond." Cornell University Press, Ithaca, New York.
Politzer, P., and Harris, R. R. (1970). *J. Amer. Chem. Soc.* **92**, 6451.
Pollak, M., and Rein, R. (1967). *J. Chem. Phys.* **47**, 2045.
Popkie, H., Kistenmacher, H., and Clementi, E. (1973). *J. Chem. Phys.* **59**, 1325.
Price, S. L. (1985). *Chem. Phys. Lett.* **114**, 359.
Rae, A. I. M. (1978). *J. Phys. Sect. C* **11**, 1779.
Ray, N. K., Shibata, M., Bolis, G., and Rein, R. (1984). *Chem. Phys. Lett.* **109**, 352.
Ray, N. K., Shibata, M., Bolis, G., and Rein, R. (1985). *Int. J. Quantum Chem.* **27**, 427.
Reynolds, P. A. (1973). *J. Chem. Phys.* **59**, 2777.
Sanford, W. E., and Boyd, R. K. (1978). *Mol. Cryst. Liq. Cryst.* **46**, 121.
Schaeffer, H. F. III (1986). *Science* **231**, 1100.
Scrocco, E., and Tomasi, J. (1978). *Adv. Quantum Chem.* **11**, 115.
Singh, U. C., and Kollman, P. A. (1984). *J. Computational Chem.* **5**, 129.
Smit, P. H., Derissen, J. L., and van Duijneveldt, F. B. (1979). *Mol. Phys.* **37**, 521.
Snyder, L. C., and Basch, H. (1972). "Molecular Wave Functions and Properties: Tabulated from SCF Calculations in Gaussian Basis Set." Wiley, New York.
Starr, T. L., and Williams, D. E. (1977). *J. Chem. Phys.* **66**, 2054.
Stone, A. J. (1981). *Chem. Phys. Lett.* **83**, 233.
Tang, A. G., Yang, Z. Z., and Li, Q. S. (1982). "Quantum Chemistry." Academic Publishing House of China, Beijing.
Veillard, A. (1972). IBMOL, available from IBM, San Jose, California.
Wasiutynski, T., van der Avoird, A., and Berns, R. M. (1978). *J. Chem. Phys.* **69**, 5288.
Wasserman, E., and Schaeffer, H. F. III (1986). *Science* **233**, 829.
Williams, D. E. (1972). *Acta Crystallogr. Sect. A* **28**, 629.
Williams, D. E. (1980), *Acta Crystallogr Sect. A* **36**, 715.
Williams, D. E. (1981). *Topics in Applied Physics* **26**, 3.
Williams, D. E., and Cox, S. R. (1984). *Acta Crystallogr. Sect. B* **40**, 404.
Williams, D. E., and Craycroft, D. J. (1985). *J. Phys. Chem.* **89**, 1461.
Williams, D. E., and Houpt, D. J. (1986). *Acta Crystallogr. Sect. B* **42**, 286.
Williams, D. E., and Starr, T. L. (1977). *Computers and Chemistry* **1**, 173.
Williams, D. E., and Weller, R. R. (1983). *J. Amer. Chem. Soc.* **105**, 4143.
Xu, G. X., Li, L. M., and Wang, D. M. (1985). "Quantum Chemistry—Principle and *Ab Initio* Calculation." Academic Publishing House of China, Beijing.
Yáñez, M., Stewart, R. F., and Pople, J. A. (1978). *Acta Crystallogr. Sect. A* **34**, 641.
Yuen, P. S., Lister, M. W., and Nyburg, S. C. (1978). *J. Chem. Phys.* **68**, 1936.

TRANSITION ARRAYS IN THE SPECTRA OF IONIZED ATOMS

J. Bauche, and C. Bauche-Arnoult

Laboratoire Aimé Cotton
Centre National de la Recherche Scientifique
Orsay, France

M. Klapisch

Racah Institute of Physics
The Hebrew University
Jerusalem, Israel

I. Introduction 132
   A. Principles 134
   B. Distribution Moments 135
II. Energy Distribution of Configuration States 137
   A. Elementary Results 137
   B. Use of the Second Quantization Scheme 139
   C. Case of Subconfigurations 141
III. Transition Arrays 142
   A. Average Transition Energy 142
   B. Variance 146
   C. Skewness 148
   D. Spin-Orbit-Split Arrays 150
   E. Absorption Arrays 151
   F. Alternative Types of Arrays 154
IV. Comparisons with Experiment 155
   A. One Gaussian Model 156
   B. Skewed Gaussian Model for $l^{N+1} - l^N l'$ Arrays 158
   C. Spin-Orbit-Split Arrays 161
V. Level Emissivity 164
   A. Line-Strength Sums 164
   B. Emissive Zones of an Atomic Configuration 166
   C. Preferential De-excitation 169
VI. Extension to More Physical Situations 171
   A. Configuration Mixing 171
   B. Breakdown of j-j Coupling 174
   C. Relativistic Corrections 176
   D. Line Intensities 176
VII. Level and Line Statistics 179
   A. Level Statistics 180

    B. Line Statistics 181
    C. Line-Strength Statistics 182
VIII. Conclusion 186
    A. Theory 186
    B. Application of the UTA Model 189
IX. Appendix. 192
    References 192

1. Introduction

In the past decade or so, the community of atomic physics witnessed a deep renewal of the spectroscopy of highly ionized atoms. This renewal was mainly caused by the striving of plasma physicists after ever higher temperatures and longer lifetimes, and by progress in X–UV astrophysics. In the seventies, it was realized that, in magnetically confined plasmas, heavy atomic impurities reach the hot core, where they become very highly ionized, and thus are responsible for dramatic radiative energy losses. Concurrently, powerful lasers were built and spectra of laser-produced plasmas were also studied. This was the beginning of the exploration—both experimental and theoretical—of the spectra of highly ionized atoms, most of which were completely unknown.

One of the most intriguing and frequent findings in these heavy atoms' spectra consisted of intense bands, in some cases regularly spaced, in others, isolated. More often than not, they resisted resolution into lines with the available instruments. It was soon realized that some of these must be ordinary bound-bound transitions between configurations with open d or f shells, which contain a very large number of levels. As a consequence, the lines are so numerous and closely packed that Doppler or other broadening effects suffice to merge them together. However not much more than that could be said, and in many instances these spectra remained unpublished for lack of satisfying interpretation.

We shall adhere to the definition of Condon and Shortley (1935) who, following Harrison and Johnson (1931), have called "transition array" the totality of lines resulting from the transitions between two configurations.

The first published experimental spectra exhibiting patterns of transition arrays can be found in the pioneering work of Edlén (1947), his XUV spectra of transition elements being generated by low inductance vacuum sparks. Since then, spectra with transition arrays were recorded from sparks, exploding wires, tokamaks, laser-produced plasmas, in a spectral range between a few Å and a few hundred Å. A characteristic example is shown in Fig. 1, (Gauthier et al., 1986) reproducing $2p$-$4d$ transitions of highly ionized indium, in a laser-produced plasma.

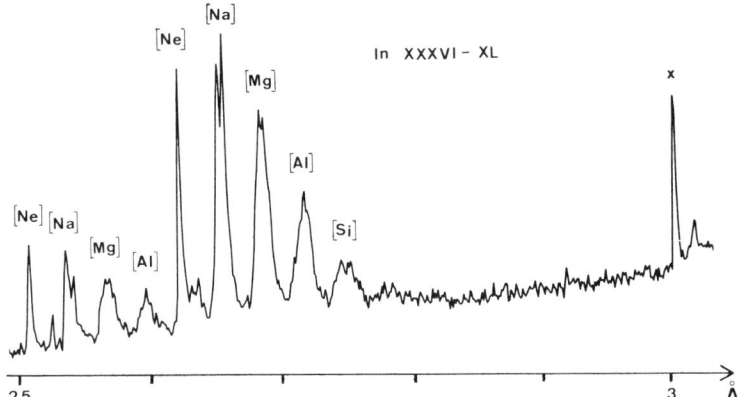

FIG. 1. $2p - 4d$ transitions of highly ionized indium ($Z = 49$). At shorter wavelengths, the Ne-like $2p_{1/2} - 4d_{3/2}$ line and its Na-, Mg-, and Al-like satellites. At longer wavelengths, the Ne-like $2p_{3/2} - 4d_{5/2}$ line, and its Na-, Mg-, Al-, and Si-like satellites. (Gauthier et al., 1986.)

First attempts to interpret these bands as transition arrays were performed by Cowan and coworkers (Burkhalter et al., 1974, Cowan 1977, see also Cowan, 1981, p. 616 & seq.), who computed in detail all the numerous possible lines in these arrays, whose overall width is seldom more than a mere Ångstrom. Indeed, the number of lines, in a transition array of the type $d^{N+1} - d^N f$, or $f^{N+1} - f^N d$, generally amounts to thousands. For the spectral width of a transition array, a simple argument can explain why it appears narrower as the atom gets more ionized. Transition energies E along an isoelectronic sequence behave as Z^2, at least for $\Delta n \neq 0$, whereas configuration spans ΔE behave as Z (neglecting spin-orbit), so that $\Delta E/E$ is proportional to $1/Z$. Moreover, in hot plasmas, as a rule, many ionization stages are present simultaneously, or at least recorded as such in spectra which are not time- nor space-resolved. These simple facts explain qualitatively the characteristic band patterns in these spectra, each band being a transition array of one ionization stage or a superposition of several of these.

Now, it is clear that the detailed computation method requires an enormous amount of computer time, and is thus not easily practicable. Moreover, it is very often unfit and even wasteful, since anyway the experimental data is a nearly structureless band pattern. On the other hand, an interpretation of the latter is unavoidable, since in many instances nothing else is seen on the spectra to support even an elementary identification, let alone diagnostics. In some cases, the bands are so intense that it is of the essence to understand them in order to control the plasma in a useful manner.

A. Principles

In the present paper, after the works of Bauche-Arnoult et al. (1978, 1979, 1982b, 1984, 1985), and of Klapisch et al. (1982, 1983), transition arrays are considered as *spectroscopical objects in their own right*, characterized by their mean energy, spectral width, and other properties. Now, let

$$I_{ab}(E)dE = N_a A_{ab} P(E_{ab} - E)dE \tag{1}$$

be the intensity distribution of an individual transition from level a to level b, expressed as the number of photons of energy E emitted per second in all directions, in the energy interval dE, by 1 cm³ of the source containing N_a atoms on the initial energy level a. A_{ab} is the Einstein transition probability and $P(E_{ab} - E)$ is a profile function depending on the parameters of the plasma, centered at the transition energy E_{ab}, and normalized in the sense that $\int_0^\infty P(E_{ab} - E)dE = 1$. For instance, for non-hydrogenic highly ionized atoms in usual plasmas, a good approximation for $P(E_{ab} - E)$ is a Doppler (Gaussian) profile. Under the assumption that P is the same function for all the transitions of the array, the intensity distribution of the whole array,

$$I(E) = \sum_{ab} I_{ab}(E)$$

is a convolution product

$$I(E) = P(E) * A(E) \tag{2}$$

where

$$A(E) = \sum_{ab} N_a A_{ab} \delta(E_{ab} - E). \tag{3}$$

The moments

$$\mu_n[I] = \frac{\int_{-\infty}^{\infty} I(E) E^n \, dE}{\int_{-\infty}^{\infty} I(E) dE} \tag{4}$$

are the quantities that we shall endeavour to compute. If we assume that $P(E)$ is indeed Gaussian, the following relation between centered moments holds

$$\mu_n^c[I] = \mu_n^c[P] + \mu_n^c[A] \tag{5}$$

Actually, only $\mu_2^c[P]$ is useful. This is just the Doppler width (squared) of an isolated line. It can usually be measured directly on the spectrum.

The remaining problem is then the evaluation of

$$\mu_n[A] = \frac{\int_{-\infty}^{\infty} A(E) E^n \, dE}{\int_{-\infty}^{\infty} A(E) dE} = \frac{\sum_{ab} N_a A_{ab} E_{ab}^n}{\sum_{ab} N_a A_{ab}} \tag{6}$$

It is important to note, at this point, that a clear distinction must be made between what we call the *theory* of transition arrays and the *model* of unresolved transition arrays (UTA).

The former, which addresses the computation of various distribution moments like $\mu_n[A]$, is *exact*. It can be developed in several directions, yielding informations on many properties of transition arrays (Sections II, III, V, VI). Although the concepts, such as distribution and moments, pertain to statistics, the theory does not involve any statistical approximation. Moreover, although the historical developments quoted above involve highly ionized atoms, the theory is valid for any spectrum, whether ionized or not.

The latter (UTA model) is an *approximation* which consists in assuming a specific analytical shape for $I(E)$, such as Gaussian, or skewed Gaussian, in which a few moments $\mu_n[I]$—in this case $n \leq 3$—are taken as equal to $\mu_n[A]$, the others being neglected. This approximation is good and useful if the array is unresolved. Indeed, the loss of resolvability is a result of the smoothing due to the convolution with P. This means that the higher $\mu_n[I]$ are not significant. On the other hand, if the array is resolved, the 2- or 3-moments description may be very crude. A more detailed discussion is left for Section IV.

The problem of the resolvability is outside the scope of this review. It is connected with the statistics of lines and line strengths. These statistics are outlined in Section VII, the only one in this paper where genuine statistical methods are applied.

B. Distribution Moments

In the following sections, many types of moments are computed.

Let us recall that the distribution moment μ_n of order $n = 1$ is the mean value of the distribution, the quantity $\sigma = [\mu_2 - (\mu_1)^2]^{1/2}$ is its mean deviation, the moment μ_3 is used for the description of the asymmetry, and μ_4 for the flattening.

In Section II, the first two moments of the distribution of the quantum state energies in a configuration C are computed in terms of the radial energy parameters. The formula reads

$$\mu_n(C) = \sum_{m \in C} \frac{\langle m|H|m \rangle^n}{g_C} \quad (7)$$

where H is the atomic Hamiltonian, the sum runs over all states m of configuration C, g_C is the total number of states, and $n = 1, 2$. The hamiltonian can be restricted to the sum of the central field hamiltonian, the electrostatic repulsion Coulomb operator

$$G = \sum_{i>j=1}^{N} \frac{e^2}{r_{ij}} \tag{8}$$

and the spin-orbit operator

$$\Lambda = \sum_{i=1}^{N} \xi(r_i) s_i \cdot l_i \tag{9}$$

In Sections III, A and B, the first two moments of the weighted distribution of the energies of the electric dipole (E1) transitions between configurations C and C' are computed through the formula

$$\mu_n(C - C') = \sum_{mm'} \frac{[\langle m'|H|m'\rangle - \langle m|H|m\rangle]^n w_{mm'}}{W} \tag{10}$$

In Eq. (10) the sum runs over all states m and m' of the respective configurations C and C', and the weight $w_{mm'}$ is the E1 strength of the $m - m'$ transition, with

$$w_{mm'} = \langle m|D|m'\rangle^2 \tag{11}$$

$$W = \sum_{mm'} w_{mm'} \tag{12}$$

In Eq. (11), D is the electric dipole operator of the atom, in atomic units:

$$D = -\sum_{i=1}^{N} r_i \tag{13}$$

The third order moment ($n = 3$ in Eq. (10)) has been computed approximately (Section III, C).

In the above equations, the sums over states can be limited to subconfigurations in j-j coupling (Sections II,C and III,D), or to a single Russell–Saunders term (Sections III,E and V,A). They can also be extended to more than two configurations, in case of configuration mixing (Section VI,A). In III,F, the electric dipole operator D is replaced by the general double tensor operator $U^{(\kappa k)K}$. In Section VII,A, other equations are written for the distribution moments of M_J and M_L, the projection quantum numbers. For M_J, the calculation is carried over up to the order $n = 4$.

The form of Eqs. (7) and (10) makes them convenient for calculations with the help of the second quantization formalism (Section II,B). This is also true for the Equations in M_J and M_L, and, in general, for the distribution moments of the eigenvalues of any operator commuting with the hamiltonian.

Now, the question of the adequation of these equations to actual physical problems is essential. First, it is clear that they imply nonrealistic simplifications. This is necessary for ensuring that the mathematical summations yield simple, compact results. These restrictions are removed in Section VI. Second, the calculation of moments opens the way to the modelization of a distribution in terms of the Gram–Charlier series of type A (Kendall and Stuart, 1963, p. 169). If only the first two moments are known, the *best* representation of the distribution is a Gaussian function. In the following, this approximation is generally chosen. Another representation, using Stieltjes moments, has been proposed (Bloom and Goldberg, 1986). But, because it depends on heavy computer work and not on simple compact formulas, it has not yet yielded general results.

In the following sections, we call *transitions* the quantum jumps between atomic states αJM, $\alpha'J'M'$, and *lines* those between levels αJ, $\alpha'J'$. Where two states, or levels, or configurations, appear separated by a hyphen, the first one is the lowest in energy, whether one deals with absorption or emission. Levels or lines are characterized by their energies, not wavenumbers, and, for the variance of a distribution, the notation $\mu_2 - (\mu_1)^2$, or sometimes μ_2, is replaced by v.

II. Energy Distribution of Configuration States

In the nonrelativistic central field scheme of Slater (1929), the basic level set is the electronic configuration. The states belonging to one configuration are degenerate eigenstates of a zeroth-order atomic hamiltonian, in which the physical energy interactions have been replaced by a purely radial one-electron potential. They are the basis states for the matrix M of the perturbing hamiltonian H_p, which is to be diagonalized in the first-order perturbation procedure. Methods for building this matrix have been described long ago (Condon and Shortley, 1935), for the electrostatic repulsion operator G(Eq. (8)) and the spin-orbit operator Λ (Eq. (9)).

The tensor-operator techniques proposed by Racach (1942) are much more efficient, and apply to all types of operators (Judd, 1963) and to all kinds of configurations.

A. ELEMENTARY RESULTS

It is well known that the trace of a hermitian matrix is invariant under any orthogonal transformation of its basic states. Therefore, the sum of the eigenvalues of the M matrix is equal to the sum of the diagonal elements of H_p in any convenient basis scheme. This is why it is such a simple linear combination of radial electronic integrals, whatever the dimension of M and the complexity of its diagonalization. Only the electrostatic operator G contributes. Tables 1^6 and 2^6 in the book by Condon and Shortley (1935) can be used for the explicit computation. The results are conveniently expressed in terms of the average interaction energy $E_{av}(nl, n'l')$ of two electrons, with given nl and $n'l'$ symbols, and averaging over all possible values of the m_s and m_l quantum numbers. Slater (1960) gives numerical tables in his Appendix 21. In the following the corresponding formal expansions,

$$E_{av}(nl, nl) = F^0(nl, nl) - \frac{(2l+1)}{(4l+1)} \sum_{k \neq 0} \begin{pmatrix} l & k & l \\ 0 & 0 & 0 \end{pmatrix}^2 F^k(nl, nl) \qquad (14)$$

and

$$E_{av}(nl, n'l') = F^0(nl, n'l') - \left(\frac{1}{2}\right) \sum_k \begin{pmatrix} l & k & l' \\ 0 & 0 & 0 \end{pmatrix}^2 G^k(nl, n'l'),$$

$$\text{if } n'l' \neq nl, \qquad (15)$$

are of use.

These expansions are, in fact, the first-order moments μ_1 of the energies of the states of the configurations nl^2 and $nln'l'$ respectively (Eq. (7)), with weight 1 for each state. Computing the second order moment μ_2 for the same configurations, with the same weights, is also feasible by elementary means. Indeed, the sum of the squares of *all* the elements of a real hermitian matrix is invariant under any orthogonal transformation of its basis states, because it is identical to the trace of $(M)^2$, the square of the M matrix. The numerical computation being cumbersome, it is more convenient to use the tensor-operator techniques. For instance, the result for the *variance* of the energies of the states of nl^2 is, in usual notations,

$$v(nl^2) = \sum_{k \neq 0} \sum_{k' \neq 0} \left[\frac{2\delta(k, k')}{2k+1} - \frac{1}{(2l+1)(4l+1)} - \begin{Bmatrix} l & l & k \\ l & l & k' \end{Bmatrix} \right]$$

$$\times \frac{(2l+1)^3}{(4l+1)} \begin{pmatrix} l & k & l \\ 0 & 0 & 0 \end{pmatrix}^2 \begin{pmatrix} l & k' & l \\ 0 & 0 & 0 \end{pmatrix}^2 F^k(l, l) F^{k'}(l, l)$$

$$+ \frac{2l^2(l+1)}{(4l+1)} (\zeta_{nl})^2 \qquad (16)$$

It is noteworthy that the total energy variance of a two-electron configuration contains no cross product of electrostatic and spin-orbit radial integrals. This is understood easily in the second quantization scheme.

B. Use of the Second Quantization Scheme

The second quantization method has been introduced in atomic spectroscopy by Judd (1967). It is most convenient in the study of N-particle problems.

The basic concept is that of creation and annihilation operators, denoted $a^\dagger$ and a respectively. These operators are adequate for specializing the one- and two-electron operators

$$F = \sum_{i=1}^{N} f_i = \sum_{\alpha\beta} a^\dagger_\alpha \langle \alpha | f | \beta \rangle a_\beta \tag{17}$$

and

$$G = \sum_{i>j=1}^{N} g_{ij} = \left(\frac{1}{2}\right) \sum_{\alpha\beta\gamma\delta} a^\dagger_\alpha a^\dagger_\beta \langle \alpha_1 \beta_2 | g_{12} | \gamma_1 \delta_2 \rangle a_\delta a_\gamma \tag{18}$$

when they act between states of specific configurations. In these formulas, each Greek letter stands for a set of electronic quantum numbers (n, l, m_l, m_s) (or, sometimes, (n, l, j, m_j)). The corresponding sums run over the infinite orthonormal basis of the one-electron eigenstates of the central field, including the continuum.

The calculation of the electrostatic repulsion contribution to the variance $v(nl^N)$ of the nl^N configuration is a typical example. Eq. (7) is written

$$\mu_2 = \frac{\sum_m \langle m | H | m \rangle \langle m | H | m \rangle}{\binom{4l+2}{N}} \tag{19}$$

In this equation, the sum runs over all states m of the nl^N configuration, whose total number appears in the denominator. It can equivalently be written as the double sum

$$\sum_{mm'} \langle m | H | m' \rangle \langle m' | H | m \rangle, \tag{20}$$

provided the m and m' states are the eigenstates of the H matrix in the space of the nl^N states, i.e., the intermediate coupling states. Then, the electrostatic repulsion part of μ_2 can be written in the following way:

$$\mu_2(G) = \left(\frac{1}{4}\right) \sum_{mm'} \langle m | \sum_{\alpha\beta\gamma\delta} a^\dagger_\alpha a^\dagger_\beta \langle \alpha_1 \beta_2 | g_{12} | \gamma_1 \delta_2 \rangle a_\delta a_\gamma | m' \rangle$$

$$\times \frac{\langle m' | \sum_{\varepsilon\zeta\theta\eta} a^\dagger_\varepsilon a^\dagger_\zeta \langle \varepsilon_1 \zeta_2 | g_{12} | \eta_1 \theta_2 \rangle a_\theta a_\eta | m \rangle}{\binom{4l+2}{N}} \tag{21}$$

The sums over the Greek letters can evidently be restricted to sums over the magnetic quantum numbers m_l and m_s of the (n, l, m_l, m_s) electronic states. Consequently, the sum over m' can be extended to the entire N-electron Hilbert space, without changing the value of $\mu_2(G)$, so that the identity operator

$$\sum_{m'} |m'\rangle\langle m'|$$

can be deleted. If the string of creation and annihilation operators, occurring in the operator

$$O = \sum_{\text{GreekLetters}} a_\alpha^\dagger a_\beta^\dagger a_\delta a_\gamma a_\varepsilon^\dagger a_\zeta^\dagger a_\theta a_\eta \langle \alpha_1\beta_2|g_{12}|\gamma_1\delta_2\rangle \langle \varepsilon_1\zeta_2|g_{12}|\eta_1\theta_2\rangle \quad (22)$$

is transformed in order to bring all the creation operators to the left, it appears to be the sum of a two-, a three- and a four-electron operator.

In conclusion, $\mu_2(G)$ is a multiple of the trace of the operator O in the nl^N subspace. It is a function of n, l, and N only. As it is recalled in the Appendix, the dependence on N of the trace of a k-electron operator in a nl^N subspace is the combinatorial coefficient

$$\binom{4l - k + 2}{N - k}$$

(Uylings, 1984). Then, taking into account the denominator of Eq. (19), it appears that the dependence on N of the μ_2 quantity is a polynomial of degree 4.

In an analogous way, the dependence of the μ_1 quantity is shown to be of degree 2 in N, so that the variance $v = \mu_2 - (\mu_1)^2$ is of degree 4. But this variance is evidently zero for $N = 0$ and 1, because G is a two-electron operator, and also for $N = 4l + 2$ and $4l + 1$, because the *relative* electrostatic energies of the levels are the same in nl^N and nl^{4l-N+2}. In this way, the variance is shown to be a multiple of the product $N(N - 1)(4l - N + 1)(4l - N + 2)$, and, therefore,

$$v_G(nl^N) = \frac{N(N-1)(4l-N+1)(4l-N+2)}{8l(4l-1)} v_G(nl^2) \quad (23)$$

where $v_G(nl^2)$ is the electrostatic part of $v(nl^2)$ in Eq. (16). Equation (23) has first been suggested by Moszkowski (1962) and proven later by Layzer (1963). The proof which is detailed above is the prototype of many others, with operators O different from that in Eq. (22). The complete results for the variances of nl^N and $nl^N n'l'^{N'}$ can be found in Table 1 of Bauche-Arnoult et al. (1979).

The variance $v_G(nl^N)$ is maximum for $N = 2l + 1$, the case of the half-subshell, and this maximum is flat. This is clearly visible on Fig. 2, which is

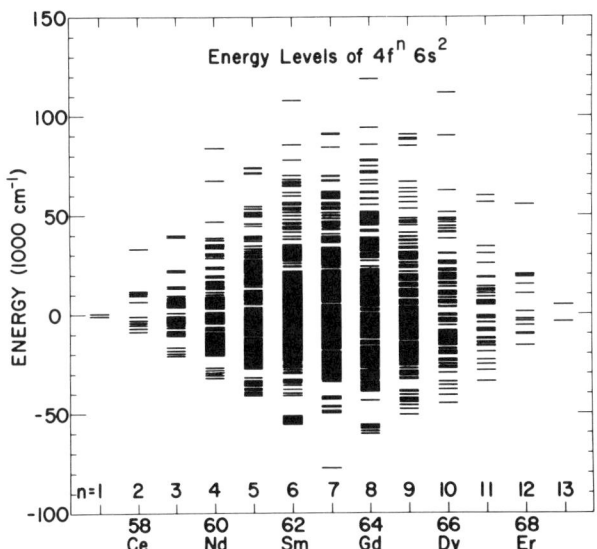

FIG. 2. Energy level structures of the $4f^n6s^2$ configurations of the neutral lanthanides, computed *ab initio* via the HX approximation (Cowan, 1967). The energies are plotted in each case on a scale in which the center-of-gravity energy is zero. (From Cowan, 1973.)

the graph of all the levels of the $4f^N6s^2$ configurations in the neutral lanthanides, computed *ab initio* by Cowan (1973).

The third order moment μ_3, linked with the asymmetry, has a complicated dependence on N. No formula has yet been published. Cowan (1981; p. 623) has given a table of numerical examples. The $4f^N6s^2$ configurations of Fig. 2, all plotted with their center of gravity at zero energy, exhibit large asymmetries.

The second quantization method is also useful for obtaining easily many qualitative results. For example, the absence of cross products of Slater and spin-orbit integrals in $v(nl^2)$(Eq. (16)) comes from the fact that the corresponding scalar operator O, containing only one operator with rank 1 in the spin space, is necessarily a double tensor of the $T^{(11)0}$ type, whose trace, in any configuration, is zero. Thus, in the formalism of Judd *et al.* (1982), G and Λ can be called orthogonal operators.

C. CASE OF SUBCONFIGURATIONS

It is also interesting to compute energy distribution moments for subconfigurations, i.e., when the large predominance of some spin-orbit radial integral(s) leads to the gathering of the levels into well separated ensembles (see Section III,D for the effects on arrays).

The simplest couplings of this type are $nl^N n'l'_{j'}$, and $(nl_j)^N (n'l'_{j'})^{N'}$, where, respectively, one and all open shell electrons have a large spin-orbit interaction. In the operator of the O type (Eq. (22)), some $a^\dagger$ and a operators now create and annihilate monoelectronic (n, l, j, m_j) states. The corresponding variance formulas have been published by Bauche–Arnoult et al. (1985). For the first order moments, the following average interaction energies for j-coupled electrons are available:

$$E_{\mathrm{av}}(nl_j, nl_j) = F^0(nl_j, nl_j) - \sum_{\substack{k=2 \\ \text{even}}}^{2l} \frac{2j+1}{2j} \begin{pmatrix} j & k & j \\ \tfrac{1}{2} & 0 & -\tfrac{1}{2} \end{pmatrix}^2 F^k(nl_j, nl_j) \quad (24)$$

and, if $nlj \neq n'l'j'$,

$$E_{\mathrm{av}}(nl_j, n'l'_{j'}) = F^0(nl_j, n'l'_{j'}) - \sum_k \begin{pmatrix} j & k & j' \\ \tfrac{1}{2} & 0 & -\tfrac{1}{2} \end{pmatrix}^2 G^k(nl_j, n'l'_{j'}) \quad (25)$$

where the sum runs over values of k of the same parity as $l + l'$, and triangular with l and l'. Numerical tables have been published by Larkins (1976).

III. Transition Arrays

The global properties of transition arrays result of the properties of both the configurations and the transitions. Moszkowski (1962) was the first to compute some of them, with simplifying assumptions.

Through systematic use of the second quantization formalism, in the way described in Section II,B, it is possible to evaluate Eq. (10) with the weight being the $E1$ transition strength:

$$w_{mm'} = |\langle m | D | m' \rangle|^2 \quad (26)$$

Because all $w_{mm'}$ quantities are multiples of the same transition radial integral, this integral disappears in the μ_n moment, which is a ratio. In the following we can replace D, for simplicity, by its z component, denoted $-Z$. Therefore, the operator of type O (Eq. (22)) now contains twice the expansion of Z in the form of Eq. (17), together with the expansion of

$$[\langle m' | H | m' \rangle - \langle m | H | m \rangle]^n.$$

A. Average Transition Energy

The most classical transition arrays in atomic spectra are of the types $nl^{N+1} - nl^N n'l'$ and $nl^N n'l' - nl^N n''l''$, denoted $l^{N+1} - l^N l'$ and $l^N l' - l^N l''$, respectively. In the following, their average energies are presented first, and then the general case.

The average energy μ_1 is a linear combination of Slater integrals only, the absence of spin-orbit integrals being easy to explain through an argument like that at the end of Section II,B. It can be computed by means of Eq. (10) with $n = 1$. More simply, some of its properties can readily be found by means of the well-known *J-file sum rule* of Shortley (1935). This rule states that, whatever the coupling of the electronic angular momenta in both configurations, the sum of the strengths $S_{\alpha J \alpha' J'}$ of the lines having a given initial (final) level $\alpha' J'(\alpha J)$ is proportional to $2J' + 1(2J + 1)$, provided that the jumping electron is not equivalent to any other in the final (initial) configuration.

1. The $l^N l' - l^N l''$ Case

In the $l^N l' - l^N l''$ array, the J-file sum rule is valid in both configurations. If the average energy μ_1 is computed through Eq. (10) with $n = 1$, i.e.,

$$\mu_1 = \frac{[\sum_{\alpha J M \alpha' J' M'} \langle \alpha' J' M' | H | \alpha' J' M' \rangle S_{\alpha J M \alpha' J' M'} - \sum_{\alpha J M \alpha' J' M'} \langle \alpha J M | H | \alpha J M \rangle S_{\alpha J M \alpha' J' M'}]}{W} \quad (27)$$

the sum over $\alpha J M M'$ ($\alpha' J' M M'$) in the first (second) part yields a $2J' + 1$ ($2J + 1$) factor. This factor being simply the statistical weight of the relevant level, it is clear that the final result for μ_1 is identical with the difference between the average energies of the configurations themselves, studied in Section II,A. Thus, the average energy T_{av} of the array can be written simply as

$$T_{av}(l^N l' - l^N l'') = E_{av}(l^N l'') - E_{av}(l^N l') \quad (28)$$

2. The $l^{N+1} - l^N l'$ Case

In the $l^{N+1} - l^N l'$ array, the J-file sum rule is only valid in the l^{N+1} configuration. Then, in Eq. (27), only the sum over $\alpha' J' M M'$, in the second part, yields a simple statistical weight. This proves that, when the average energy of the array is written

$$T_{av}(l^{N+1} - l^N l') = E_{av}(l^N l') - E_{av}(l^{N+1}) + \delta E(l^{N+1} - l^N l') \quad (29)$$

the δE quantity comes out from the computation of only the first part of Eq. (27). Two types of Slater integrals are to be considered.

For the (l, l) integrals, the operator of type O (Eq. (22)) is a three-electron operator in the numerator of μ_1, and a one-electron operator in the denominator, both acting in the l^{N+1} configuration. Therefore, the dependence of this part of μ_1 on N is the ratio

$$\frac{\binom{4l-1}{N-2}}{\binom{4l+1}{N}}$$

(see the Appendix), i.e., a multiple of $N(N-1)$. Furthermore, for $N = 4l + 1$, this part is equal to the average energy of the l^{4l+1} subconfiguration, because this subconfiguration only contains one Russell–Saunders term. Therefore, it is identical with the (l, l) part of the average energy of $l^N l'$ for any value of N, and it does not contribute to $\delta E(l^{N+1} - l^N l')$.

For the (l, l') integrals, in an analogous way, the numerator of μ_1 is a two-electron operator, and the μ_1 quantity is the product of N by a linear combination of Slater integrals, which can be computed through tensor-operator techniques for the $l^2 - ll'$ array. The final result is

$$\delta E(l^{N+1} - l^N l') = \frac{N(2l+1)(2l'+1)}{(4l+1)} \left[\sum_{k \neq 0} f_k F^k(l, l') + \sum_k g_k G^k(l, l') \right] \quad (30)$$

TABLE I

Numerical Formulas for the δE Shift in the $l^{N+1} - l^N l'$ Transition Array (Eq. 30)[a]

Transitions	δE
$p \rightleftarrows s$	$\dfrac{N}{4l+1} \left[\dfrac{1}{2} G^1(p, s) \right]$
$d \rightleftarrows p$	$\dfrac{N}{4l+1} \left[-\dfrac{1}{5} F^2(d, p) + \dfrac{19}{15} G^1(d, p) - \dfrac{3}{70} G^3(d, p) \right]$
$f \rightleftarrows d$	$\dfrac{N}{4l+1} \left[-\dfrac{8}{35} F^2(f, d) - \dfrac{2}{21} F^4(f, d) + \dfrac{137}{70} G^1(f, d) - \dfrac{2}{105} G^3(f, d) - \dfrac{5}{231} G^5(f, d) \right]$

[a] Example: the formulas for $f^{N+1} - f^N d$ and $d^{N+1} - d^N f$ only differ by the value of $4l + 1$ (from Bauche et al., 1983).

with

$$f_k = \begin{pmatrix} l & k & l \\ 0 & 0 & 0 \end{pmatrix} \begin{pmatrix} l' & k & l' \\ 0 & 0 & 0 \end{pmatrix} \begin{Bmatrix} l & k & l \\ l' & 1 & l' \end{Bmatrix} \qquad (31)$$

$$g_k = \begin{pmatrix} l & k & l' \\ 0 & 0 & 0 \end{pmatrix}^2 \left[\frac{2\delta(k,1)}{3} - \frac{1}{[2(2l+1)(2l'+1)]} \right] \qquad (32)$$

Numerical applications of Eq. (30) to all cases where $l, l' \le 3$ are given in Table I. Physical applications to the $4d^N - 4d^{N-1}4f$ arrays in a series of praseodymium ions, calculated *ab initio*, are presented in Fig. 3. They show that, for $\delta n = 0$ transitions, the effect of δE can be drastic.

3. General Case

The most general array can be written

$$C - C' = l_1^{N_1+1} l_2^{N_2} l_3^{N_3} l_4^{N_4} \ldots - l_1^{N_1} l_2^{N_2+1} l_3^{N_3} l_4^{N_4} \ldots \qquad (33)$$

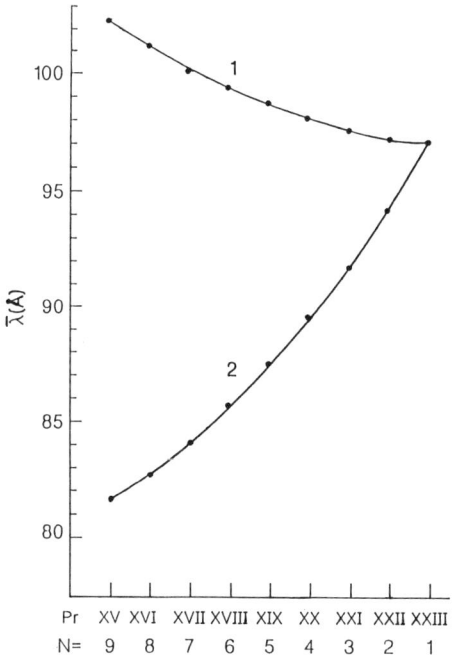

FIG. 3. Mean wavelength $\bar\lambda$ of the $4d^N - 4d^{N-1}4f$ transition arrays, for various ionization stages of praseodymium XV through XXIII, computed *ab initio*. Curve 1: Mean wavelength corresponding to the difference of the average energies of the configurations. Curve 2: mean wavelength including the δE shift of Eq. (30).

with an arbitrary number of "spectator" open subshells $l_3^{N_3}$, $l_4^{N_4}$, etc. Deriving the various dependences in N in the same way as above, it is possible to relate the shift

$$\delta E(C - C') = T_{av}(C - C') - [E_{av}(C') - E_{av}(C)] \quad (34)$$

with simpler shifts derived above:

$$\delta E(C - C') = \frac{(4l_2 - N_2 + 1)\delta E(l_1^{N_1+1} - l_1^{N_1}l_2)}{(4l_2 + 1)}$$

$$\frac{-(4l_1 - N_1 + 1)\delta E(l_2^{N_2+1} - l_2^{N_2}l_1)}{(4l_1 + 1)} \quad (35)$$

The formal relationship $\delta E(C' - C) = -\delta E(C - C')$ is a consequence of Eq. (34). (See the end of Section I,B for the definition.)

Of course, Eqs. (28) and (29) can now be seen as special cases of Eq. (35). The latter shows that, generally, the average energy of a transition array is *not* equal to the difference of the average energies of the two relevant configurations. The only exception is the case where $N_1 = N_2 = 0$, which includes the $l^N l' - l^N l''$ arrays studied in Section III,A,1. The spectator open subshells do not contribute to the $\delta E(C - C')$ quantity.

B. Variance

The variance $v = \mu_2 - (\mu_1)^2$ of the transition energies of an array can be computed by means of Eq. (10) with $n = 2$. The calculation is much more cumbersome than for the average energy. However, the second quantization arguments like that in Section II,B are a great help, in a way which has been described by Bauche-Arnoult *et al.* (1979), for taking into account the intermediate coupling in both configurations. The following is essentially a description of the results.

1. General Case

For the general array $C - C'$ defined in Eq. (33), the variance $v(C - C')$ is simple only in the strict central field assumption, i.e., if any given radial integral has the same numerical value in all the configurations of the spectrum. This assumption is realistic only in the medium or highly ionized atoms. In the neutral or lowly ionized spectra, the spectroscopists obtain much better Slater–Condon interpolations through the (phenomenological) allowance for numerical changes in the Slater integrals values. The relevant variance formulas are described in Section III,B,2.

In the strict central field assumption, $v(C - C')$ is the sum of the electrostatic part

$$v_G(C - C') = \frac{N_1(4l_1 - N_1 + 1)}{(4l_1)} v_G(l_1^2 - l_1 l_2)$$
$$+ \frac{N_2(4l_2 - N_2 + 1)}{(4l_2)} v_G(l_2^2 - l_1 l_2)$$
$$+ \frac{N_3(4l_3 - N_3 + 2)}{(4l_3 + 1)} v_G(l_3 l_1 - l_3 l_2)$$
$$+ \frac{N_4(4l_4 - N_4 + 2)}{(4l_4 + 1)} v_G(l_4 l_1 - l_4 l_2)$$
$$+ \ldots \quad (36)$$

and of the spin-orbit part

$$v_A(C - C') = \frac{(\zeta_1 - \zeta_2)[l_1(l_1 + 1)\zeta_1 - l_2(l_2 + 1)\zeta_2]}{4} + \frac{\zeta_1 \zeta_2}{2} \quad (37)$$

where ζ_i stands for the spin-orbit radial integral of the l_i orbital. The formal expansions of the elementary variances $v_G(l^2 - ll')$ and $v_G(ll' - ll'')$ as linear combinations of squares and cross products of Slater integrals can be found in Tables II and III of Bauche–Arnoult et al. (1979). The numerical values of the expansion coefficients are in Tables IV and V of the same paper. For the spin-orbit part, it is independent of the N_i values.

2. *Special Cases*

For the simple arrays already considered in Section III,A,1 and 2, Eq. (36) reduces to

$$v_G(l^{N+1} - l^N l') = \frac{N(4l - N + 1)v_G(l^2 - ll')}{(4l)} \quad (38)$$

and

$$v_G(l^N l' - l^N l'') = \frac{N(4l - N + 2)v_G(ll' - ll'')}{(4l + 1)} \quad (39)$$

respectively. The former variance is maximum for $N = 2l, 2l + 1$, and the latter for $N = 2l + 1$. The spin-orbit formula (Eq. (37)) remains unchanged, except for the assignment of the l_1 and l_2 quantum numbers.

These variances have been derived first without the assumption that any given radial integral has the same numerical value in all the configurations of the spectrum (Bauche–Arnoult et al., 1979). Then, the parts of the variance which contain Slater integrals over the l orbital are more complicated than in Eqs. (38) and (39). In particular (i) the dependence on N is not the same for all products of Slater integrals (PSI); for instance, in both arrays, it reaches the fourth degree in N for the PSI where the l' and l'' orbitals do not occur; (ii) there appears an N-dependence in the spin-orbit part of both variances. These more general expansions happen to be necessary in the study of emissive zones in configurations (Section V,B).

Because the variance depends on the *relative* energies of the levels, it is easy to prove that the variances of complementary arrays are equal. For instance,

$$v(l^{N+1} - l^N l') = v(l^{4l-N+1} l'^{4l'+2} - l^{4l-N+2} l'^{4l'+1}).$$

3. Numerical Applications

Although the formal Tables (II, III) published by Bauche-Arnoult *et al.* (1979) look very complicated, their numerical applications by use of Tables IV and V of the same paper are easy, especially in the strict central field assumption. The example of the $d^{N+1} - d^N p$ arrays is presented in Table II. In this Table, F^k, F'^k, and G^k are the Slater integrals $F^k(d, d)$, $F^k(d, p)$, and $G^k(d, p)$, respectively. Each integral $F^k(d, d)$ and ζ_d is supposed to have the same numerical value in both configurations.

TABLE II

NUMERICAL EXPANSION OF THE VARIANCE OF THE TRANSITION ENERGIES OF THE $d^{N+1} - d^N p$ ARRAY[a].

$v(d^{N+1} - d^N p) = N(9 - N)[2.385E - 3(F^2)^2 - 8.998E - 4(F^2 F^4)$
$\quad + 1.125E - 3(F^4)^2 - 3.338E - 2(F^2 F'^2) + 6.299E - 4(F^4 F'^2)$
$\quad - 2.440E - 3(F^2 G^1) + 5.240E - 4(F^2 G^3) - 5.879E - 4(F^4 G^1)$
$\quad - 2.538E - 3(F^4 G^3) + 3.272E - 3(F'^2)^2 + 8.820E - 3(G^1)^2$
$\quad + 4.056E - 4(G^1 G^3) + 1.859E - 3(G^3)^2 - 2.551E - 3(F'^2 G^1)$
$\quad - 1.285E - 4(F'^2 G^3)] + 1.5(\zeta_d)^2 - 1.5(\zeta_d \zeta_p) + 0.5(\zeta_p)^2$

---

[a] In the strict central field assumption, with the notations $F^k = F^k(d, d)$, $F'^k = F^k(d, p)$ and $G^k = G^k(d, p)$. $E - 3$ stands for $\times 10^{-3}$. (From Bauche-Arnoult *et al.*, 1979; Table IV).

C. SKEWNESS

The skewness of a distribution is characterized by the asymmetry coefficient

$$\alpha_3 = \frac{\mu_3^c}{(\mu_2^c)^{3/2}} \tag{40}$$

where μ_i^c is the centered moment of order i, with $\mu_2^c = v$, and $\mu_3^c = \mu_3 - 3\mu_2^c\mu_1 - (\mu_1)^3$. The noncentered moment μ_3 can be computed by means of Eq. (10). But, compared to the case of v (Section III,B), the computations turn out to be much more cumbersome.

Actually, the complexity depends upon the type of triple product of radial integrals whose coefficient is desired. Only the case of the $l^{N+1} - l^N l'$ array has been calculated so far (Bauche-Arnoult et al., 1984). It has been found that the triple products of type AAB, where A denotes an $F^k(l, l)$ and B an $F^k(l, l')$ or $G^k(l, l')$ Slater integral, occur in μ_3^c with a coefficient which is a polynomial of the fifth degree in N. But, for some of the other types of triple products, the analogous polynomial does not exceed the third degree in N. Only the coefficients of the latter products have been computed exactly. For the former, an approximate method has been used. If one assumes that the electronic radial functions are all hydrogenic for a common effective nuclear charge Z^*, which is roughly true in highly ionized atoms, one deduces that the part of μ_3^c containing only Slater integrals is the product of $(Z^*)^3$ by a polynomial in N. This polynomial can be determined using explicit diagonalizations and summations for simple values of N.

In Fig. 4, the *ab initio* line-by-line description of the $4d^8 - 4d^7 4f$ array of praseodymium XVI is compared with two global descriptions, with and

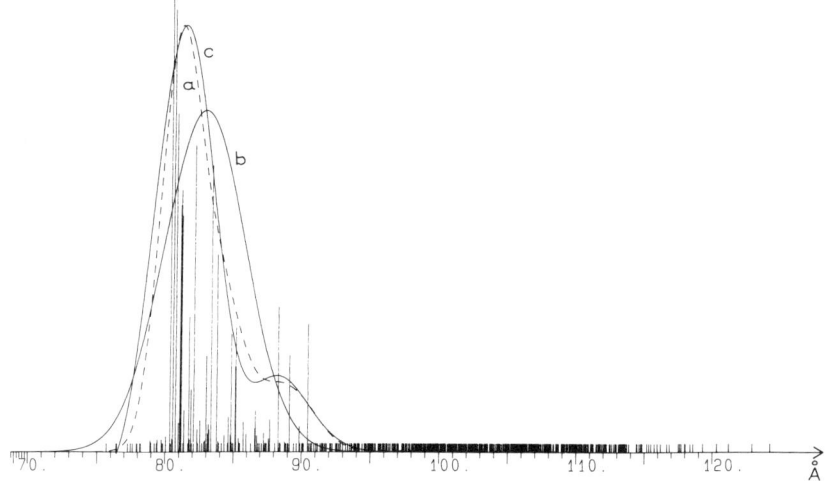

FIG. 4. Calculated $4d^8 - 4d^7 4f$ transition array in praseodymium XVI. Each line is represented with a height proportional to its strength, except those with a strength less than 3% of the highest, which have all, conventionally, been increased to that 3% limit. (a) Envelope calculated by adding the different lines with a given FWHM (0.5 Å). (b) Gaussian curve, using μ_1 and v. (c) Skewed Gaussian curve using μ_1, v, and μ_3. (Bauche-Arnoult et al., 1984.)

without account of the skewness, respectively. The curves are defined in Section IV,A and B.

D. Spin-Orbit-Split Arrays

1. Array Splitting

It is well known (Cowan, 1981, Chap. 19) that, in hydrogenic atoms of nuclear charge Z, the electrostatic repulsion and the spin-orbit integrals are proportional to Z and Z^4, respectively. Consequently, the latter predominate over the former, when the electrons see a large enough effective nuclear charge. The following two cases are typical.

(i) In the X-ray emission spectra of neutral or lowly ionized atoms, the spin-orbit integral $\zeta_{n'l'}$ is often the largest radial integral in the upper configuration $n'l'^{4l'+1}nl\ldots$, if $l' \neq 0$, whatever external subshells are present. Then, the energies of the initial levels depend primarily on the value of the j' angular momentum of the internal hole $n'l'^{4l'+1}$. Their configuration splits into two subconfigurations.

(ii) In highly ionized atoms, the optical transitions occur between external subshells, but these subshells experience large effective nuclear charges. An nl^N configuration splits into two or several subconfigurations, if $l \neq 0$.

Now, the $E1$ selection rule in $j-j$ coupling is

$$|j - j'| \leq 1 \leq j + j' \tag{41}$$

Combined with the rule $|l - l'| = 1$, it leads to the fact that only three lines occur between orbitals nl and $n'l'$, or two if l or $l' = 0$. It is easy to derive that the arrays considered in (i) and (ii) also split into two or three parts.

In general, we call a *subarray* the totality of lines resulting from the transitions between two subconfigurations defined like above (sometimes, between a subconfiguration and an unsplit configuration), with an arbitrary number of spectator orbitals. For example, the $nd^2 - ndn'p$ array splits into two parts, if ζ_{nd} only is very large, but each of the parts is the superposition of two subarrays.

2. Moments of Subarrays

For identification purposes in experimental spectra, it is interesting to compute mean positions and widths of subarrays. Equation (10) can be used, with sums restricted to the relevant subconfigurations. For example, in the case of the $(d_{3/2})^2 - d_{3/2}p_{3/2}$ subarray, the sums run over all the eigenstates of H in the $(d_{3/2})^2$ and $d_{3/2}p_{3/2}$ subspaces, which comprise respectively 6 and 16 states. For the average energy, a δE quantity is defined, much like in Section III,A, by the equation

$$\delta E(X - Y) = T_{av}(X - Y) - [E_{av}(Y) - E_{av}(X)] \qquad (42)$$

where X and Y are subconfigurations. δE vanishes in the first two of the following cases, whose variance formulas have been published (Bauche-Arnoult et al., 1985):

(i) $nl^N n'l'_{j'} - nl^N n''l''_{j''}$, denoted $l^N j' - l^N j''$. The complementary array, denoted $l^{4l-N+2}(j')^{-1} - l^{4l-N+2}(j'')^{-1}$, sharing the same properties, is typical of the atomic X-ray lines.

(ii) $j^N j' - j^N j''$, where all subshells have a large spin-orbit interaction.

(iii) $j^{N+1} - j^N j'$ (same remark as in (iii)). Here, δE is proportional to N.

In all three cases, the occurrence of spectator open subshells has been accounted for. All three are generally useful for describing the subarrays into which an $l^{N+1} - l^N l'$ array splits. The total strength of each subarray can be easily computed (Bauche-Arnoult et al., 1985, Eq. (19); in this paper the numerical applications of Eq. (20) are incorrect by a factor of 2). Formulas are also available for other cases of splitting due to spin-orbit effects, namely $j^{N+1} j^{N'} - j^N j^{'N'+1}$ (the general $j-j$ array), $l^N j' - l^N l''$ (same as in (i) above, except that the spin-orbit interaction for the $n''l''$ orbital is not large), and $l^{N+1} - l^N j'$ (but, in this last case, the formulas are very complicated).

In Fig. 5, the *ab initio* line-by-line description of the $3d^8 4s - 3d^8 4p$ array is compared with its global description in three ionic spectra: krypton X, where the array does not split (Section III,A,2); molybdenum XVI and praseodymium XXXIII, where it splits into two peaks. This Figure shows the changes in the shape of the spectrum when the spin-orbit integrals become predominant.

E. ABSORPTION ARRAYS

The far-UV absorption experiments carried over in many laboratories produce some spectra which contain transition arrays of another type than those studied above: all the lines of an array link one or more low levels of the atomic system with all the levels of an upper electronic configuration, or subconfiguration. The average energy and width of such an array cannot be computed by means of the formal results derived above. Equation (10) remains essentially valid, except that the summation over states m must be restricted, e.g., to the $(2S + 1)(2L + 1)$ states of the Hund's Russell–Saunders (RS) term of an l^N configuration (the (S, L) term with the highest orbital momentum L, among those with highest possible spin S), or to the $(2J + 1)$

states of the lowest J level of this term. The typical cases are those of photoabsorption from the $nd^{10}4f^N$ to the nd^94f^{N+1} configuration ($n = 3, 4$), in triply (or doubly) ionized rare earths in crystals. The complementary array, $4f^{14-N} - 4f^{13-N}nd$, which gives the same results, is easier to compute.

(i) For $n = 3$, the predominance of the spin-orbit ζ_{3d} integral splits the array into two parts (Sugar, 1972a; Bonnelle *et al.*, 1977; Esteva *et al.*, 1983; Sugar *et al.*, 1985).

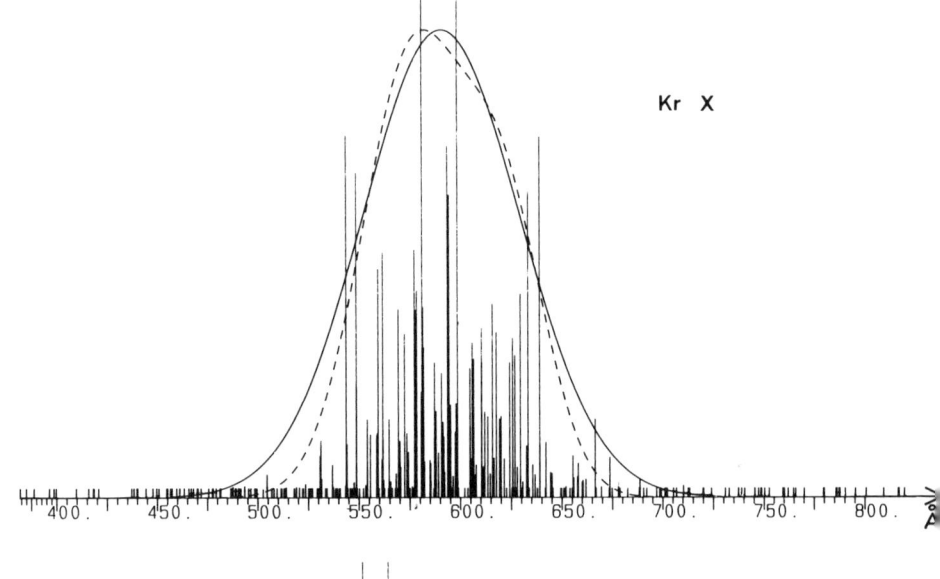

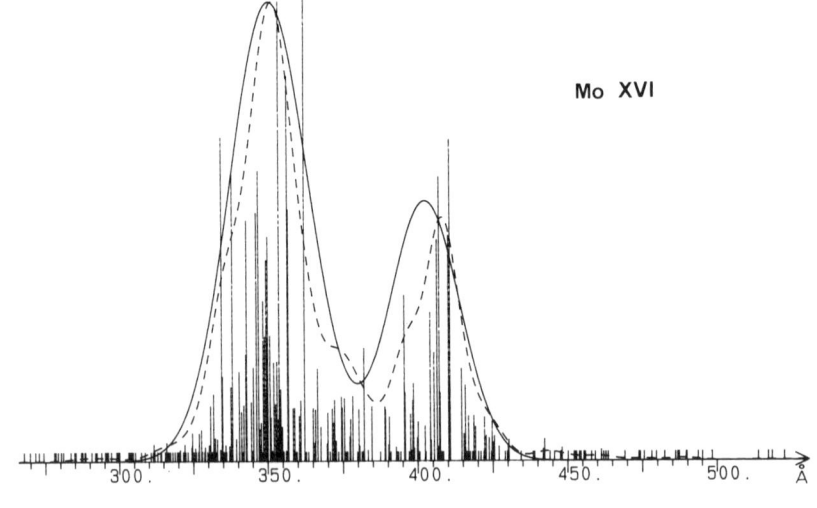

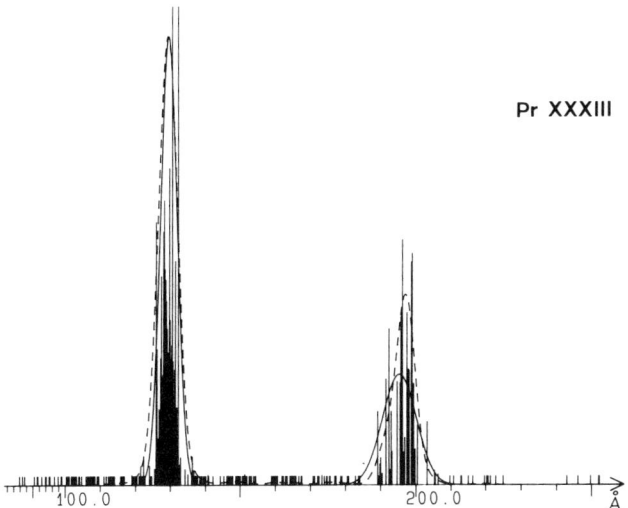

FIG. 5. Examples of calculated spectra in the $3d^84s - 3d^84p$ series. Each line is represented with a height proportional to its strength, except those with a strength less than 3% of the highest, which have all, conventionally, been increased to that 3% limit. The dashed curves are the envelopes of the line spectra for some small linewidth, sufficient for the coalescence of the lines in one or two peaks. The solid curves have been calculated for Kr, in the one-Gaussian model (Section IV,A), and, for Mo and Pr, in the SOSA model (Section IV,C). (From Bauche-Arnoult et al., 1985.)

(ii) For $n = 4$, the ζ_{nd} integral is much smaller, and $G^1(nd, 4f)$ much larger. No subarray appears (Dehmer et al., 1971; Sugar, 1972b).

1. Case of l^{N+1} (Hund) $- l^N l'$

For the (ii) case above, the sum over m in Eq. (10) is restricted to all the states of the Hund's RS term of l^{N+1}, denoted $(S_H L_H)$ below. Only the Slater integrals are of interest.

In the average transition energy, the part with the (l, l) integrals is remarkably simple. As can be derived from Eq. (61), it differs from the (l, l) part of the average energy of the full array by the addition of

$$\Delta\mu_1(l, l) = \frac{-2[E(S_H L_H) - E_{av}(l^{N+1})]}{(N + 1)} \tag{43}$$

Writing the relevant O operator (Eq. (22)), it appears that, in the numerator of μ_1 (Eq. (10)), the part with the (l, l') integrals is the expectation value, in the $(S_H L_H)$ term, of a two-electron operator. First, Eq. (1) of a paper by Racah (1949) can be used for a simple chain calculation when $N \leq 2l + 1$, and, second, the results for $N > 2l + 1$ follow from the properties of

TABLE III

Coefficients of the $F^k(f, d)$ and $G^k(f, d)$ Integrals in the Shift $\Delta\mu_1$ of $f^{N+1} - f^N d$ Absorption Arrays[a].

Hund's term	$F^2(f, d)$	$F^4(f, d)$	$G^1(f, d)$	$G^3(f, d)$	$G^5(f, d)$
$\Delta\mu_1(f^2\ {}^3H)$	−106/1365	−155/9009	+16/455	−34/455	+575/33033
$\Delta\mu_1(f^3\ {}^4I)$	−76/585	−118/3861	−8/1365	−404/4095	+850/99099
$\Delta\mu_1(f^4\ {}^5I)$	−5/39	−46/1287	−6/91	−10/91	−3405/99099
$\Delta\mu_1(f^5\ {}^6H)$	−716/6825	−1696/45045	−304/2275	−2376/20475	−3004/33033
$\Delta\mu_1(f^6\ {}^7F)$	−4/39	−5/117	−8/39	−14/117	−175/1287
$\Delta\mu_1(f^7\ {}^8S)$	−8/65	−2/39	−16/65	−28/195	−70/429

[a] $\Delta\mu_1$ is the shift from the mean energy of the whole $f^{N+1} - f^N d$ array to that of just the transitions from the Hund's RS term $f^{N+1} S_H L_H$. The results are symmetrical with respect to the half-shell. For example, $\Delta\mu_1(f^6\ {}^7F) = \Delta\mu_1(f^8\ {}^7F)$.

complementary subshells l^{N+1} and l^{4l-N+1} (Racah 1942). For example, the coefficients of the (l, l') Slater integrals in the average transition energy of the array $f^{N+1} S_H L_H - f^N d$ are listed in Table III. Of course, these expressions are also valid for the complementary array $d^{10} f^{13-N} S_H L_H - d^9 f^{14-N}$.

2. *Case of* $l^{N+1}(Hund) - l^N j'$

In case (i) above, the j' quantum number of the l' electron marks the subarray of interest, and the $\zeta_{l'}$ parameter is responsible for the main part of the energy difference between subarrays. The sum over m in Eq. (10) is restricted to all the states of the Hund's term of l^{N+1}, and that over m' to all the states of the subconfiguration $l^N j'$. The average energy of the subarray contains the same addition as defined in Eq. (43). For its (l, l') part, the same methods as in Section III,E,1 can be used. But, here, the results depend on j'.

3. *Variances*

In both cases studied above, the expansion of the second-order moment μ_2 involves operators acting on up to five electrons. Its evaluation has not yet been completed.

F. Alternative Types of Arrays

Electric dipole transitions are not the only ones observed in electromagnetic spectra. They are largely predominant in neutral and lowly ionized atoms, but higher-order radiative transitions, e.g., the electric quadrupole E2 transitions, possess large transition probabilities in highly ionized atoms, where the effective nuclear charge seen by the radiating electrons is large (Cowan, 1981, Chap. 15). Therefore, one could consider E2 transition arrays, like $3d^{N+1} - 3d^N 4s$, and call them *forbidden arrays*. However, the importance of such arrays in the spectra is not so large as to motivate by itself the calculation of distribution moments. Another reason stems from the current calculations of the various elementary processes occurring in hot plasmas. Indeed, the results of these calculations may be expanded in terms of tensor operators. The problem of collision strengths is of particular interest (Cowan, 1981, p. 566). Indeed, it can be shown (Klapisch and Bar Shalom, 1987) that in the distorted wave approximation framework, collision strengths can be described by one-electron tensor operators, acting on bound electrons only, the continuum electrons appearing in the related radial integrals. In the $nljm_j$ basis, these operators are simple multipoles, while in the $nlm_l m_s$ basis, they are double tensors. For that reason, the variance has been computed (Bauche, 1986) for the $l^{N+1} - l^N l'$ array, using the following equation for the strength:

$$w_{mm'} = \langle m | U_0^{(\kappa k_1)K}(l, l') | m' \rangle^2 \tag{44}$$

where

$$U_0^{(\kappa k_1)K}(l, l') = \sum_{i=1}^{N+1} [u_0^{(\kappa k_1)K}(l, l')]_i \tag{45}$$

and

$$\langle sl_a \| u^{(\kappa k_1)}(l, l') \| sl_b \rangle = \delta(l, l_a)\delta(l', l_b) \tag{46}$$

The δE quantity, defined like in Eq. (29), is given by Eq. (30), with

$$f_k = (-1)^{k_1+1} \begin{pmatrix} l & k & l \\ 0 & 0 & 0 \end{pmatrix} \begin{pmatrix} l' & k & l' \\ 0 & 0 & 0 \end{pmatrix} \begin{Bmatrix} l & l & k \\ l' & l' & k_1 \end{Bmatrix} \tag{47}$$

$$g_k = \begin{pmatrix} l & k & l' \\ 0 & 0 & 0 \end{pmatrix}^2 \left[\frac{2\delta(\kappa, 0)\delta(k, k_1)}{2k+1} - \frac{1}{2(2l+1)(2l'+1)} \right] \tag{48}$$

IV. Comparison with Experiment

As mentioned in the Introduction, the observable quantity is the array intensity distribution

$$I(E) = P(E) * A(E)$$

In order to compare between theory and experiment, we need theoretical values of the moments of $I(E)$, $\mu_n[I]$. In Sections II and III, the theory of transition arrays is used to obtain analytical expressions for the moments of A, $\mu_n[A]$ for $1 \leq n \leq 3$. We now use these theoretical results in the unresolved transition array (UTA) *model*, assuming that $P(E)$ is very narrow with respect to A, so that, for $n \leq 3$

$$\mu_n[I] \approx \mu_n[A]$$

Indeed, this approximation is justified for highly ionized heavy atoms, in very hot plasmas. The line shape is dominated by Doppler broadening, and the line width is much smaller than that of the array. This can be checked directly on the spectra if there appear resolved individual lines. Actually, even if the individual line shape is not purely Gaussian, as long as it remains approximately symmetrical (i.e., very small $\mu_3[P]$) and nearly the same for all the lines in the array, the above relation is satisfied.

Few papers present recordings of unresolved or partially resolved arrays, probably because of lack of interpretation, although many authors have recorded such arrays in their spectra. The papers quoted below are only meant to be a sample of the material published recently. Earlier publications are quoted by Bauche–Arnoult et al., (1979).

We first discuss the two-moments description of the arrays, i.e., assume that $I(E)$ is a Gaussian, then mention cases where the third moment is important, and describe some spectra with the SOSA model.

A. One-Gaussian Model

1. $l^{N+1} - l^N l'$ Arrays

Formulas obtained in Section III for arrays of this type can be used for the interpretation of many transitions, like $1s - 2p$, e.g., Folkmann et al. (1983), Morita and Fujita (1985); transitions 3-3, e.g., Finkenthal et al. (1985); transitions 3-4, e.g., Kononov (1978), Burkhalter et al. (1980), Klapisch et al. (1982); transitions 4-4, e.g., Sugar and Kaufman (1980), Denne and Poulsen (1981), O'Sullivan and Carroll (1981), Finkenthal et al. (1986), and transitions 5-5, e.g., Carroll and O'Sullivan (1981), Carroll et al. (1984).

Let us describe in more details an example of $3d^{N+1} - 3d^N 4f$ transitions. Mo and Pd spectra (Klapisch et al., 1982) have already been interpreted with the UTA Gaussian model, and the importance of the shift δE has been stressed. Here we show results from heavier atoms. Figure 6 shows a spectrum of La obtained in a laser-produced plasma at Soreq Nuclear Research Center (Zigler et al., 1987). The upper curve is obtained as a superposition of Gaussians defined in the following way, for each $3d^{N+1} - 3d^N 4f$ array from La XXXI to La XXXV.

(i) The mean wavelength corresponds to the first-order moment of the energy distribution (Eq. (29)).

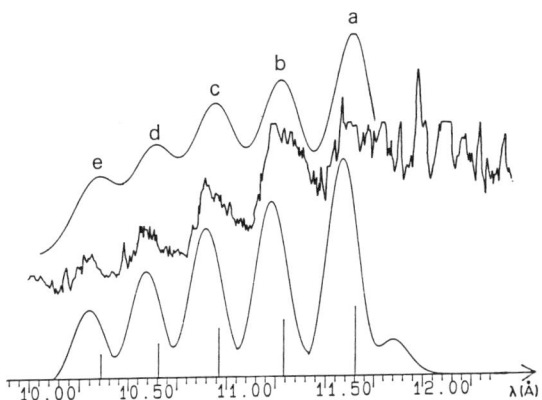

FIG. 6. Spectrum of laser produced lanthanum plasma: (a) La XXXI $3d^9 - 3d^84f$, (b) La XXXII $3d^8 - 3d^74f$, (c) La XXXIII $3d^7 - 3d^64f$, (d) La XXXIV $3d^6 - 3d^54f$, (e) La XXXV $3d^5 - 3d^44f$. Upper curve: theoretical arrays represented as Gaussians. Middle curve: experimental spectrum. Lower curve: theoretical arrays represented as skewed Gaussians. The lines represent the theoretical mean wavelengths of the arrays.

(ii) The full width at half maximum (FWHM) is, in the energy scale, the square root of the variance of the same distribution (Eq. (38)), multiplied by $2\sqrt{2\ln 2} = 2.355$.

The middle curve is the experimental spectrum, in which the ordinate is proportional to the number of photons. The lower curve is explained in paragraph B below.

The array marked a is for Co-like La XXXI, b for Fe-like La XXXII, c for Mn-like La XXXIII, d for Cr-like LaXXXIV, and e for V-like La XXXV. The Slater integrals and configuration energies were calculated *ab initio* by the RELAC program (Koenig, 1972, Klapisch *et al.*, 1977). It is clear that the shapes of the experimental arrays are not Gaussian. Nevertheless, the agreement on the positions of the UTAs is good, at any rate largely sufficient for the identification of the transitions for ionization stages higher than Co-like. Comparison of the FWHMs raises the itchy question of fixing the background in the experimental recording. For the same reason, the intensities of the theoretical arrays were fitted to the peaks and not to the areas. Actually, if the UTA model is adequate, it can yield a way of estimating the area of the experimental array through fitting of the maximum, which is very convenient. This paves the way to quantitative spectroscopy of the UTAs, and to the possibilities of plasma diagnostics.

For the Co-like spectrum, the array is partially resolved. It has been shown recently (Klapisch *et al.*, 1987), in a detailed study of Co-like Xe by means of a 300-level collisional-radiative model, that even at densities typical of laser produced plasmas, $n_e = 10^{20}$ cm^{-3}, some low-lying levels of $3d^84f$ are very much populated, thus explaining the features around 12Å. This means that the approximation involved in the assumption of statistical weight population may become inadequate if there is a systematic—i.e., not random—deviation from it. It is remarkable that for the higher ionization states, where the configurations have more levels, this problem disappears.

2. Transitions with Spectator Electrons

Figure 7 reproduces the above mentioned laser produced Pd spectrum, and Fig. 8 a part of it with the addition to the theoretical curve of satellite transitions of the type $3d^N4s - 3d^{N-1}4s4p$, around 30-34Å, $4s$ being a "spectator" electron. It is noticeable that the agreement on the width of the $3d$-$4f$ arrays in Fig. 7 is better than on the previous case of heavier La. An explanation for that will be proposed in the next paragraph. The $3d$-$4p$ satellites are much wider than the resonant transitions, in accord with the formulas of Section III. It is clear that the satellites contribute to improve considerably the agreement with experiment. The remaining discrepancy is probably connected with the shapes of the arrays not being Gaussian. A comparison with the Mo spectrum (Fig. 1 in the same paper), obtained from a low inductance vacuum spark, shows that in the latter case, satellites are absent. As these doubly excited configurations become more populated at higher electron densities, this demonstrates the possibility of plasma diagnostics by UTA spectroscopy, even with the simple Gaussian model.

B. SKEWED GAUSSIAN MODEL FOR $l^{N+1} - l^N l'$ ARRAYS

Formulas for the third moment were shown in Section III to be cumbersome, and established approximately for transitions of the type $l^{N+1} - l^N l'$ only. When this third moment is taken into account, the array is assumed to have a skewed Gaussian shape

$$I(x) = \left[\frac{1 - \alpha_3(x - x^3/3)}{2}\right] \exp\left(\frac{-x^2}{2}\right) \quad (49)$$

where x is the reduced energy measured in units of $\sigma = \sqrt{v}$ (the r.m.s. deviation), referred to the reduced first moment of the distribution, $x = \mu_1/\sigma$, and α_3 is the coefficient of skewness (Eq. 40). This function, derived from the Gram-Charlier expansion (Kendall and Stuart, 1963, p. 168), has a negative portion that is nonphysical, and which is put to zero here. Figure 4, (Bauche-Arnoult et al., 1984), examplifies the difference between the Gaussian (curve b) and the skewed Gaussian shapes (curve c) for the array

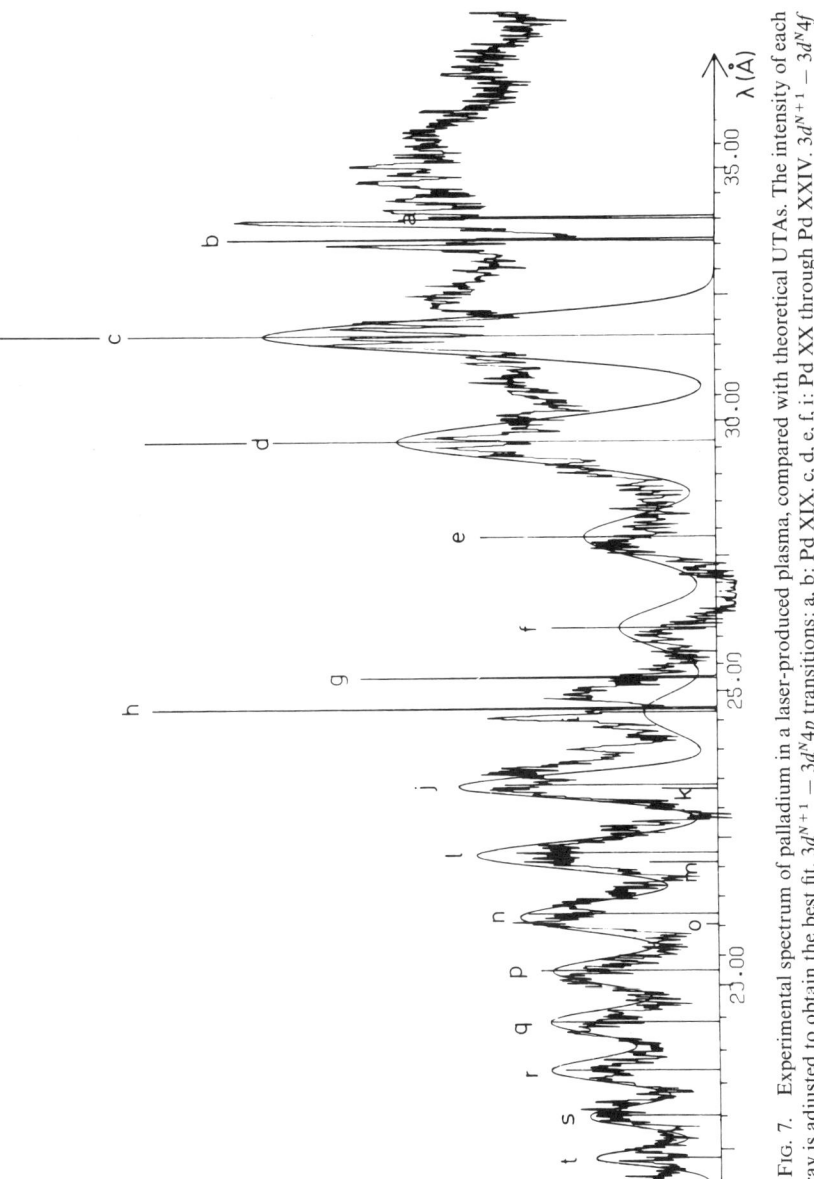

FIG. 7. Experimental spectrum of palladium in a laser-produced plasma, compared with theoretical UTAs. The intensity of each array is adjusted to obtain the best fit. $3d^{N+1} - 3d^N 4p$ transitions: a, b: Pd XIX. c, d, e, f, i: Pd XX through Pd XXIV. $3d^{N+1} - 3d^N 4f$ transitions: g, h: Pd XIX. j, l, n, p, q, r, s, t: Pd XX through XXXVII. (From Klapisch et al., 1982.)

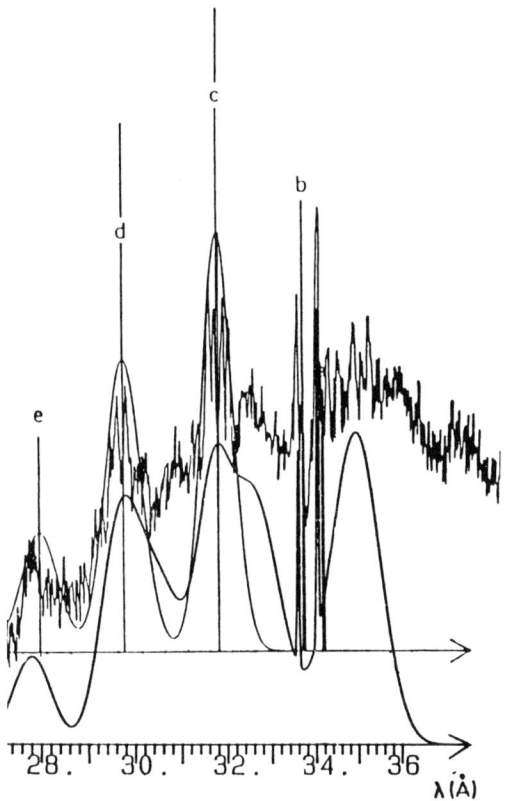

FIG. 8. A portion of the spectrum displayed in Fig. 7, where the satellite arrays of the type $3d^{N+1}4s - 3d^N 4s 4p$ have been included in the lower theoretical curve.

$4d^8 - 4d^7 4f$ in PrXVI, comparing to detailed computations (dashed curve a). The curve a is an envelope obtained by giving to each line a width of 0.5 Å. The conspicuous effects of the introduction of the skewness are a *displacement of the maximum* of the curve with respect to the mean, and *a smaller width at half maximum*, the variance being of course unaffected. The skewed Gaussian agrees better with the detailed computations by diagonalization. The shoulder at right is an artifact due to the functional form.

In order to compare with experiment, we have computed the α_3 parameters for the $3d$-$4f$ transitions of La. The lower curve of Fig. 6 is obtained by plotting the skewed Gaussian profile for each array with the corresponding value of α_3. Although the skewed Gaussian function is unable to reproduce the somehow triangular shape of the experimental arrays, the skewness clearly improves the agreement. The theoretical maxima are closer to the peaks, and because the widths are smaller, the UTAs appear more separated.

On the other hand, the discrepancy for the Co-like ion is larger than with the Gaussian profile. If the explanation in terms of level populations of the previous paragraph is correct, this means that one should be cautious in comparing the theoretical skewness with the observed one when the statistical weight population (SWP) is not ascertained.

An interesting point is that the skewness increases steeply as a function of Z along an isoelectronic sequence. Indeed, Eq. (40) can be written, from the results of Bauche–Arnoult et al., (1984):

$$\alpha_3 = k(N) + \frac{(N\mu^s + \mu^\zeta)}{v^{3/2}}$$

where μ^s includes the parts S_2 to S_5 of Table IV of the quoted paper, with N factored out, and μ^ζ includes S_6 to S_9. Now, the dominant term in μ^s is the product $G^1(3d, 4f)\zeta_{3d}^2$. If we assume hydrogenic behavior for Slater and spin-orbit integrals, this term behaves as Z^9, whereas $v^{3/2}$ is roughly proportional to Z^3, giving a dependence as Z^6. This may explain why the effect of skewness is much smaller in the Mo spectrum.

To conclude this paragraph, let us state that α_3 is worth evaluating for highly ionized heavy atoms, especially if the ratios of the peak intensities (as opposed to integrated areas) are to be used for diagnostic purposes.

C. Spin-Orbit-Split Arrays

In cases where one or several spin-orbit integrals are much larger than the Slater integrals, the coupling becomes nearly j-j, and the arrays appear resolved in subarrays. In that case the formulas for the moments of $l^{N+1} - l^N l'$, while still exact, are no more useful to obtain an adequate description of the observed arrays. For this reason, the spin-orbit-split array (SOSA) formulas have been derived (Section III,D). Only the first and second moments have been obtained, so the intensity distributions $I(E)$ for the subarrays are to be taken as Gaussian.

The experimental spectra corresponding to this situation are mostly transitions 2-3, e.g., Aglitsky et al. (1974, 1979, 1981, 1984), Gordon et al. (1979), and some transitions 3-4 e.g., Zigler et al. (1980), Klapisch et al. (1980), Busquet et al. (1985), Klapisch et al. (1986). Recently, transitions 3-5 have been recorded, e.g., Audebert et al. (1985), Bauche–Arnoult et al. (1986), Zigler et al. (1986), and also transitions 2-4, Gauthier et al. (1986), as shown on Fig. 1.

Figure 9, taken from the work of Audebert et al. (1985), shows the $3d - 5f$ transitions of Ta XLII to Ta XLVI. The Cu-like satellites were computed with all the possible subarrays of the type $3d^{10}nl_j - 3d^9 5f nl_j$, altogether

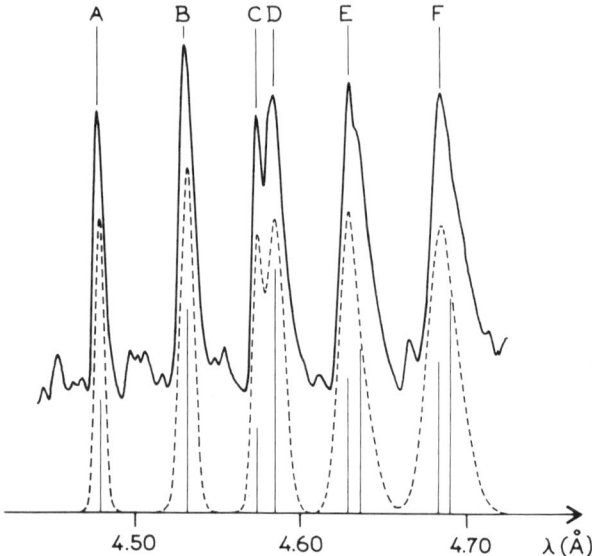

FIG. 9. $3d - 5f$ transitions in the experimental spectrum (solid line) and calculated spectrum (dashed line) of a laser produced plasma of tantalum ($Z = 73$). The mean wavelength and the width of each peak have been computed with allowance for spectator electrons $4s$, $4p$, $4d$, $4f$ only. The intensities are adjusted to obtain the best fit. A: the Ni-like $3d_{3/2} - 5f_{5/2}$ line of Ta XLVI; B, D, and long wavelength parts of E and F: its satellites in Ta XLV through Ta XLII. C: the Ni-like $3d_{5/2} - 5f_{7/2}$ line; short wavelength parts of E and F: its satellites in Ta XLV and XLIV. The calculated spectrum is shifted downwards for clarity. (From Audebert et al., 1985.)

seven subarrays. The same procedure was applied to the Zn-like (28 subarrays). For the Ga-like (82 subarrays) and Ge-like (196 subarrays) satellites, some simplifying assumptions were used.

This is a case where the width of an individual line is not small with respect to the width of the subarrays, and Eq. (5) must be used. Accordingly, the experimental width of 0.0065 Å was quadratically combined to the calculated widths to draw the figure. The agreement obtained is seen to be excellent for the positions as well as the FWHMs of the peaks.

Table IV gives some more quantitative results for mean wavelengths and widths of several subarrays of Ta, W and Re, namely $3d - 5f$, $3d - 6f$ and $3d - 7f$. Here again the agreement between theory and experiment is excellent.

Such results confirm the assignment of the ionic species present in the plasma. In the above examples, they also prove that some autoionizing configurations like $[3d^9 5f$ + spectator electrons$]$ contribute appreciably to the line emission, and thus participate in the population equilibrium of the plasma.

TABLE IV

COMPARISON BETWEEN THEORETICAL AND EXPERIMENTAL MEAN WAVELENGTHS λ AND WIDTHS FWHM FOR 3d – 5f, 6f, 6f, 7f TRANSITIONS IN Ni-, Cu-, AND Zn-LIKE TANTALUM ($Z = 73$), TUNGSTEN ($Z = 74$), AND RHENIUM ($Z = 75$).[a]

Transition		Tantalum				Tungsten				Rhenium			
		Calculated		Experiment		Calculated		Experiment		Calculated		Experiment	
		λ	FWHM	λ	FWHM	λ	FWHM	λ	FWHM	λ	FWHM	λ	FWHM
$3d_{5/2} - 5f_{7/2}$	(Ni)	4.5753		4.573		4.046		4.405		4.2437		4.244	
	(Cu)	4.628	0.0145	4.629	0.014[b]	4.455	0.012	4.456	0.014[b]	4.290	0.011	4.289	0.010
	(Zn)	4.685	0.017	4.684	0.015[b]	4.508	0.015	4.506	0.016	4.342	0.014	4.343	0.013
$3d_{3/2} - 5f_{5/2}$	(Ni)	4.4787		4.478	0.0085	4.3092		4.309	0.006	4.1494		4.149	0.005
	(Cu)	4.533	0.013	4.530	0.012[b]	4.360	0.010	4.357	0.010	4.197	0.010	4.194	0.009
	(Zn)	4.587	0.015	4.584	0.016	4.411	0.013	4.411	Blend	4.245	0.012	4.244	Blend
$3d_{5/2} - 6f_{7/2}$	(Ni)	4.0316		4.032		3.8792		3.878		3.7356		3.737	
	(Cu)	4.091	0.012	4.092	0.012	3.935	0.010	3.932	0.011	3.788	0.009	3.789	0.012[b]
	(Zn)	4.152	0.015	4.149	0.013	3.993	0.013	3.990	0.014[b]	3.843	0.013	3.841	0.013
$3d_{3/2} - 6f_{5/2}$	(Ni)	3.9555		3.956		3.8039		3.804		3.6610		3.661	
	(Cu)	4.013	0.011	4.013	0.011	3.857	0.009	3.856	0.010	3.711	0.008	3.712	0.009
	(Zn)	4.072	0.013	4.068	0.012	3.913	0.011	3.910	0.011	3.764	0.010	3.761	0.009
$3d_{5/2} - 7f_{7/2}$	(Ni)	3.7632		3.763		3.6199		3.620		3.4849		3.485	
	(Cu)	3.824	0.011	3.823	0.010	3.677	0.009	3.676		3.539	0.008	3.540	
	(Zn)	3.887	0.014	{3.885, 3.889}	Blend	3.737	0.012		Blend				

[a] (in Å). The experimental widths (FWHM of isolated Ni-like transitions underlined in the table) are included in the calculations. The estimated experimental accuracies are 0.002 Å for λ, and 15% for FWHM. From Tragin et al., (1988)). [b] Structure perturbed by other lines or UTA's.

V. Level Emissivity

The line strengths are implicitly present in the transition-energy distribution moments, because they are the weights in the summations of Eq. (10). But they can also be studied for themselves, in order to put into evidence some interesting trends in the level emissivity. Can one compute easily the total emissive strength of a level? What is the most emissive part of the upper configuration? Towards which levels do upper levels radiate most? Unfortunately, no exact results can be derived for genuine intensities, which depend on the level populations, and therefore on the overall behavior of the emissive medium. In this Section, like above, only transition *strengths* are considered, denoted $w_{mm'}$ in Eq. (11). All results obtained for emission can also be used in absorption problems.

A. Line-Strength Sums

1. Total Strength of a Transition Array

The first interesting quantity is the sum W of the strengths of all the transitions of an array. For the basic array $l_1^{N_1+1} l_2^{N_2} - l_1^{N_1} l_2^{N_2+1}$, it is easily found to depend on N_1 through the combinatorial coefficient

$$\binom{4l_1 + 1}{N_1}$$

and on N_2 through

$$\binom{4l_2 + 1}{N_2}.$$

If a passive open subshell $l_3^{N_3}$ is added, the W quantity is to be multiplied by

$$\binom{4l_3 + 2}{N_3}.$$

The final formula reads

$$W(l_1^{N_1+1} l_2^{N_2} l_3^{N_3} - l_1^{N_1} l_2^{N_2+1} l_3^{N_3})$$

$$= 2l_> \binom{4l_1+1}{N_1}\binom{4l_2+1}{N_2}\binom{4l_3+2}{N_3}(P_{l_1 l_2})^2 \qquad (50)$$

where $l_>$ is the larger of l_1 and l_2, and

$$P_{l_1 l_2} = \int_0^\infty R_{n_1 l_1}(r) r R_{n_2 l_2}(r) dr \tag{51}$$

is the usual dipolar radial integral.

Equation (50) is, in other notations, Eq. 31.67 of Sobelman (1972) and Eq. 14.92 of Cowan (1981). Actually, it was already known at the time when the book of Condon and Shortley (1935) was published.

2. Generalization of the J-file Sum Rule

Indeed, the quoted book contains the statement of the *J-file sum rule* (Shortley, 1935). A J-file sum $W_{\alpha J}$ is the sum of the strengths of all the transitions connecting a given αJ level with the states of the other configuration. It can be written

$$W_{\alpha J} = \sum_{Mm'} |\langle \alpha JM | D | m' \rangle|^2 \tag{52}$$

Thus, it is the trace, in the αJ subspace, of the operator

$$\sum_{\alpha\beta\gamma\delta} a_\alpha^\dagger b_\beta b_\gamma^\dagger a_\delta \langle \alpha | d | \beta \rangle \langle \gamma | d | \delta \rangle \tag{53}$$

where $a^\dagger$, $a(b^\dagger, b)$ are creation and annihilation operators of an $l_1(l_2)$ electron. Now, the operator string in the sum can be rewritten

$$a_\alpha^\dagger a_\delta \delta(\beta, \gamma) - a_\alpha^\dagger b_\gamma^\dagger b_\beta a_\delta. \tag{54}$$

This is the sum of a one-electron (scalar) operator acting on l_1 electrons in the configuration of the αJ level (say, the $l_1^{N_1+1} l_2^{N_2}$ configuration), and of a two-electron (scalar) operator acting on an (l_1, l_2) pair in the same configuration. Furthermore

(i) In the one-electron operator, $\alpha \equiv \delta$. The sum over α through δ yields the *number operator* (Judd, 1967), whose trace in the αJ subspace is equal to $(N_1 + 1)(2J + 1)$.

(ii) The two-electron operator is an exchange operator of the dipole-dipole type in the orbital space. Its trace is also a multiple of $2J + 1$.

One conclusion is that the $W_{\alpha J}$ sum only depends on αJ by the $(2J + 1)$ factor if this level belongs to a configuration with no l_2 electron ($N_2 = 0$). This is exactly the J-file sum rule, stated in another way. Now, another conclusion stems from the (ii) statement above: if $N_2 \neq 0$, the quantity $W_{\alpha J}$ depends on αJ through the product of $(2J + 1)$ by a linear function of the coefficient $C(G^1; \alpha J)$ of the $G^1(n_1l_1, n_2l_2)$ Slater integral in the energy of the αJ level. Thus, the equation

$$W_{\alpha J} = (2J + 1)\left[\frac{(N_1 + 1)l_>}{(2l_1 + 1)} + C(G^1; \alpha J)\right](P_{l_1l_2})^2 \qquad (55)$$

is the generalization of the J-file sum rule to the case where the αJ level belongs to the $l_1^{N_1+1}l_2^{N_2}$ configuration in the $l_1^{N_1+1}l_2^{N_2} - l_1^{N_1}l_2^{N_2+1}$ array (Bauche et al., 1983). Several consequences have been derived.

(i) Eq. (55) is also valid for the most general array, where an arbitrary number of spectator open subshells are added, like in Eq. (33).

(ii) The sum of the strengths of all the transitions connecting a given RS term with the states of the other configuration reads

$$W_{\alpha SL} = (2S + 1)(2L + 1)\left[\frac{(N_1 + 1)l_>}{(2l_1 + 1)} + C(G^1; \alpha SL)\right](P_{l_1l_2})^2 \qquad (56)$$

in evident notations.

(iii) The sum of the strengths of all the transitions of an array is like in Eq. (50).

(iv) When $W_{\alpha J}(W_{\alpha SL})$ is zero, which may happen only if $N_2 \neq 0$, the relevant level (RS term) is metastable with respect to the $l_2 \to l_1$ transition. This can be used for checking angular calculations for electric dipole strengths. It leads to a drastic reduction of the number of strong lines in arrays where the $G^1(n_1l_1, n_2l_2)$ value is very large (Section VII,C).

(v) If the $G^1(n_1l_1, n_2l_2)$ integral has a relatively large value, the high levels of the $l_1^{N_1}l_2^{N_2+1}$ configuration are the most strongly connected by $E1$ lines to the other configuration.

B. Emissive Zones of an Atomic Configuration

It is stressed in Section III,A,3 that the average energy $T_{av}(C - C')$ of the $C - C'$ transition array is generally not equal to the difference between the average energies of the two relevant configurations.

1. Centers of Gravity

For getting a deeper understanding of this property, it is possible to compute the weighted average q_c of the level energies of configuration C, the weight being for each level the total strength of the lines linked with it. In terms of transitions between atomic states:

$$q_C = \sum_{mm'} \langle m|H|m\rangle \frac{W_{mm'}}{W} \qquad (57)$$

in the notations of Eq. (10). Defining the analogous $q_{C'}$ average, one can write that the average energy T_{av} of the array is simply

$$T_{av}(C - C') = q_{C'} - q_C \qquad (58)$$

It is clear that $T_{av}(C - C')$ is not equal to $E_{av}(C') - E_{av}(C)$ if $q_{C'} \neq E_{av}(C')$ and/or $q_C \neq E_{av}(C)$. Three typical cases can be distinguished, as in Section III,A.

(i) In the $l^N l' - l^N l''$ array, the J-file sum rule applies in both configurations. Then, $q_C = E_{av}(C)$, $q_{C'} = E_{av}(C')$, and $T_{av}(C - C') = E_{av}(C') - E_{av}(C)$.

(ii) In the $l^{N+1} - l^N l'$ array, the J-file sum rule only applies in the l^{N+1} configuration. Then, $q_C = E_{av}(C)$, and $q_{C'} = E_{av}(C') + \delta E(l^{N+1} - l^N l')$, with δE computed by means of Eq. (30).

(iii) In the more general $l_1^{N_1+1} l_2^{N_2} - l_1^{N_1} l_2^{N_2+1}$ array,

$$q_{C'} = E_{av}(C') + \frac{(4l_2 - N_2 + 1)}{(4l_2 + 1)} \delta E(l_1^{N_1+1} - l_1^{N_1} l_2) \qquad (59)$$

with an analogous expression for q_C.

The spectator open subshells do not contribute to the q quantities.

2. Widths

It is also interesting to determine the widths of the active zones of both configurations, i.e., their extensions on both sides of the centers of gravity found just above.

(i) In the $l^N l' - l^N l''$ array, these widths are exactly those of the configurations themselves. This is true even if spectator open subshells are added.

(ii) In the $l^{N+1} - l^N l'$ array, it is intuitive that the variance of the energies of the emitting levels of the upper configuration is that part of the total variance of the array which contains only integrals useful in this upper configuration (Bauche et al., 1983). This is also true if spectator open subshells are added. The relevant variance expansion can be written from the tables published by Bauche-Arnoult et al. (1979, 1982b).

3. Discussion

The main results obtained above are well exemplified in the spectrum of Gd XII. Figure 10 shows series of six low odd configurations, $4d^{10}4f^M 5p^{5-M}$ with $M = 0$ through 5, and of six high even configurations, $4d^9 4f^{M'} 5p^{6-M'}$ with $M' = 1$ through 6. All the latter but one can emit towards two low configurations, resulting in $5p \to 4d$ and $4f \to 4d$ arrays, respectively. In each of these configurations, the two parts which are responsible for the bulk of the

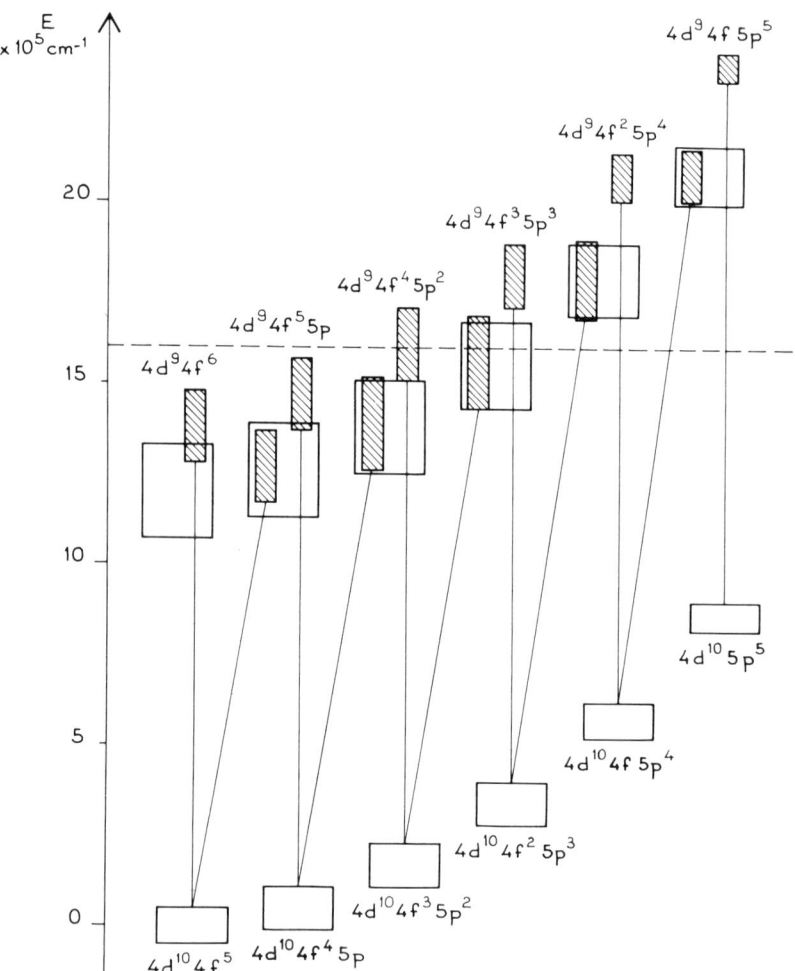

FIG. 10. Low configurations and transition arrays in the spectrum of gadolinium XII. Each rectangle represents a wavenumber distribution. Its mean ordinate is the mean wavenumber; its vertical dimension is equal to 2.355 $(v)^{1/2}$. The broad rectangles correspond to the configurations and the narrow rectangles to the emissive zones relevant to the "vertical" or "oblique" transitions. The dashed horizontal line represents the ionization limit. (From Bauche et al., 1983.)

emission of the two arrays differ in mean energy and in width. They can be called the *emissive zones* corresponding to the two arrays.

The quantities which are displayed in Figure 10 have been evaluated by means of the relativistic parametric potential code RELAC (Koenig, 1972; Klapisch et al., 1977). The spectacular difference between the locations of the two emissive zones comes from the fact that the G^1 Slater integral, whose

TRANSITION ARRAYS IN THE SPECTRA OF IONIZED ATOMS   169

influence predominates in the δE shift (see Table I), is much larger for the (4d, 4f) than for the (4d, 5p) pair. It even appears that the bulk of the $4f \to 4d$ strength may be emitted by levels which lie high above the bulk of the upper-configuration states. Nothing alike happens for the *receptive zones* of the lower configurations, which coincide exactly with these configurations themselves, for all arrays.

C. PREFERENTIAL DE-EXCITATION

Using the concept of emissive (receptive) zones, it is possible to know the origin and the destination of the main part of the intensity radiated in a transition array. But the following information is finer: towards (from) which part of the lower (upper) configuration does a given group of levels of the other configuration mostly emit (receive)?

In most electronic configurations of interest, the open subshell electrons can be classified into *core* and *external* electrons. Then, the largest energy integrals contain core electrons. Two extreme coupling situations can be considered.

1. Russell–Saunders Coupling

In the case of Russell–Saunders (RS) coupling, if only one core subshell, say, nl, is open in both configurations of the array, the $F^k(nl, nl)$ Slater integrals generally predominate over the other radial integrals.

In the $nl^N n'l' - nl^N n''l''$ arrays, the situation is very simple. The S and L angular momenta of the nl^N subconfiguration are good quantum numbers, and the level energies mainly depend on them. Because the radiative jump only concerns the $n'l'$ and $n''l''$ electrons, the core of the atom remains essentially unchanged in the transition. Thus, on one hand, low (high) levels of $nl^N n''l''$ radiate mainly towards low (high) levels of $nl^N n'l'$, and, on the other hand, the array width is small.

In the $nl^{N+1} - nl^N n'l'$ array, the atomic core itself changes in the transition. The simple selection rule applied above does not hold any more, but use can be made of another property: in the transition $nl^{N+1}\alpha SLJ - (nl^N \tilde{\alpha}\tilde{S}\tilde{L}, n'l')\alpha' J'$, the $\tilde{\alpha}\tilde{S}\tilde{L}$ term must be a parent of αSL, in Racah's formalism (Racah 1943). This has quantitative consequences.

The weighted center of gravity of the energies $E_{m'}$ of all the upper states m' linked to a given state m of nl^{N+1} can be written

$$E_{av}(nl^N n'l' \to m) = \sum_{m'} \frac{E_{m'} W_{mm'}}{W} \qquad (60)$$

where the weights $w_{mm'}$ and W are $E1$ strengths (see Eq. (12)), and where the states are

$$|m\rangle = |nl^{N+1}\alpha SLM_S M_L\rangle$$

and

$$|m'\rangle = |(nl^N \tilde{\alpha}\tilde{S}\tilde{L}, n'l')S'L'M'_S M'_L\rangle.$$

As is shown by Bauche–Arnoult et al. (1982a), when only the $F^k(nl, nl)$ integrals are not zero, the sum over M'_S, M'_L, S', L', can first be effected, because $E_{m'}$ does not depend on these quantum numbers. The result is proportional to the square of the coefficient of fractional parentage $(l^N \tilde{\alpha}\tilde{S}\tilde{L}|\}l^{N+1}\alpha SL)$. Then, the final summation on $\tilde{\alpha}\tilde{S}\tilde{L}$ is easily done through the use of Eq. (1) of a paper by Racah (1949), leading to

$$E_{av}(nl^N n'l' \to m) = E_{av}(nl^N n'l') + \frac{(N-1)}{(N+1)}[E_m - E_{av}(nl^{N+1})] \quad (61)$$

In the same way, one obtains

$$E_{av}(m' \to nl^{N+1}) = E_{av}(nl^{N+1}) + \frac{(4l-N)}{(4l-N+2)}[E_{m'} - E_{av}(nl^N n'l')] \quad (62)$$

for the weighted center of gravity of the energies E_m of all the lower states m linked to a given state m' of $nl^N n'l'$. The meaning of each of the Eqs. (61) and (62) is twofold.

(i) If only the core–core interactions are taken into account, there exists a simple linear relationship between the energy of a state and the weighted average energy of the states to which it is linked by transitions.

(ii) Because the coefficient of the above linear relationship is positive, one can say that low (medium, high) levels of the upper configuration de-excite mostly towards low (medium, high) levels of the lower configuration. Through the parentage condition recalled above, this is evidently linked with the fact that, in all l^N configurations, the higher is the spin S, the lower is the energy, as an average.

Eq. (61) has already been tested in a physical case, with allowance for the corrections due to the contributions of the energy integrals neglected in its derivation (Bauche–Arnoult et al., 1982a).

2. $j-j$ Coupling

In the highly-ionized spectra, the core spin-orbit integrals are predominant. The levels tend to gather in $j-j$ terms, whatever the external electrons. It is evident that the low (high) levels of the upper configuration, built with mainly low-j (high-j) core electrons, de-excite mostly towards low (high) levels of the lower configuration.

3. Intermediate Coupling

In addition to the direct derivations given above, there exists an indirect proof of the preferential de-excitation phenomenon: the variance of the transition energies is smaller than those of the level energies in both configurations (Bauche-Arnoult *et al.*, 1978). This statement, and the continuity argument, prove that the preferential de-excitation also exists in intermediate coupling.

Combined with the assumption of a Boltzmann distribution in the upper configuration, the law of preferential de-excitation is the reason why, in any configuration, spontaneous emission generally populates the lower levels more than the higher (Bauche-Arnoult *et al.*, 1982a).

VI. Extension to More Physical Situations

As is stressed in the Introduction, exact formal calculations on global properties of transition arrays can be carried over only if simplifying assumptions make the relevant summations mathematically tractable. Thus, the results obtained in Sections II, III, and V rely on such assumptions. In the present section, we consider the corrections which must be taken into account when dealing with the actual physical situations, where generally some of the above simplifications are not valid.

A. CONFIGURATION MIXING

The concept of transition array (Harrison and Johnson, 1931) relies on that of atomic configuration, which participated in the success of the central field method of Slater (1929). But it has been noticed early that the simple properties of a configuration can be spoiled by its mixing, also called interaction, with another configuration. Chapter XV of Condon and Shortley (1935) is entitled "*Configuration interaction*". This phenomenon can often be neglected in highly ionized atoms. But it cannot, for example, in two typical cases:

(i) At increasing degrees of ionization, the nl orbitals with higher n values tend to become less bound than those with lower n values, whatever the value of l. Thus, in the vicinity of the crossing-over of the energies of such orbitals, there may appear spectra with many low, close configurations (see, for example, Carroll and O'Sullivan, 1982).

(ii) When the ionization degree becomes very high, close configurations, built from quasi-hydrogenic orbitals differing only in their l values, tend to group together to form a complex (Layzer, 1959).

Efficient mathematical formulas for the explicit study of configuration mixing have been available for a long time (Goldschmidt and Starkand, 1970; Goldschmidt, 1971; Sobelman, 1972). A typical phenomenon observed is the cancellation, in the strengths of some lines, of the contributions from different configurations (Cowan, 1970; Aymar, 1973). In the following, we deal with the *global* effects on transition arrays.

1. Example: the $4p^64d^N - (4p^64d^{N-1}4f + 4p^54d^{N+1})$ Mixed Array

Remarkable examples have been studied recently by Mandelbaum et al. (1987) in the spectra of praseodymium XV through XXIII, where the $4l$ orbitals have neighboring energies. Due to the fact that the $4p4f$ and $4d^2$ pairs of orbitals have almost the same energy, the configurations $4p^64d^{N-1}4f$ and $4p^54d^{N+1}$ strongly mix. Both de-excite towards the $4p^64d^N$ ground configuration. For instance, Mandelbaum et al. have computed *ab initio*, by means of the RELAC code, all the line energies and strengths of the *mixed array* $4p^64d^2 - (4p^64d4f + 4p^54d^3)$ in the Sr-like spectrum Pr XXII. The results are presented in Fig. 11, with calculated spectra where the individual lines have been given a small width for ensuring coalescence.

Plot a is a simple (artificial) superposition of the independent arrays $4p^64d^2 - 4p^64d4f$ and $4p^64d^2 - 4p^54d^3$, computed without allowance for the mixing of the upper configurations. The horizontal segments attached to the vertical bars A and B represent the full widths at half maximum (FWHM) of both arrays, respectively. The overlap is strong.

In plot b the lines displayed are those of the mixed array, obtained when the configuration mixing is taken into account in the *ab initio* calculations. This implies that the hamiltonian matrix on the added subspaces of the two upper configurations, including interconfigurational off-diagonal elements, is diagonalized.

The changes from the upper to the lower part of Fig. 11 are spectacular. The array A has been quenched by the array B. What is left of the latter is much narrower than on the upper plot. This example shows that the effects of configuration mixing on transition arrays can be very large. Unfortunately, a direct experimental evidence of the example shown on Fig. 11 is not accessible, because, in a real plasma, all the spectra Pr XV through XXIII have lines in this wavelength range.

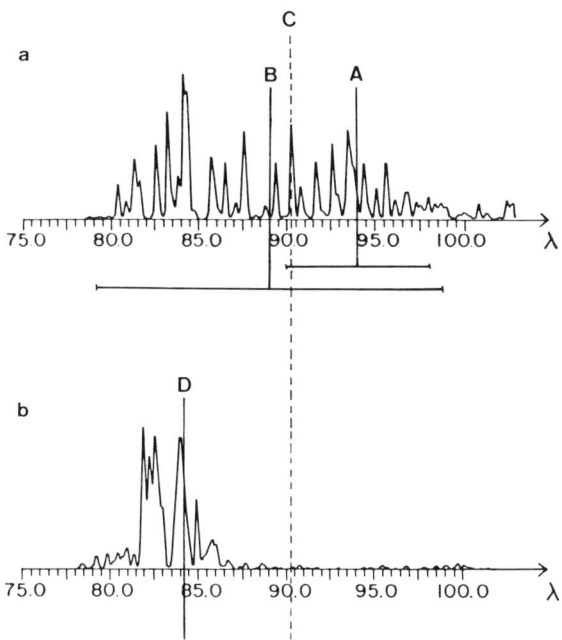

FIG. 11. Calculated spectra of the $4p^64d^2 - 4p^64d4f$ and $4p^64d^2 - 4p^54d^3$ arrays in praseodymium XXII. (a) The pure arrays are superposed. (b) Configuration mixing is accounted for. A, B: centers of gravity of the pure arrays $4p^64d^2 - 4p^64d4f$ and $4p^64d^2 - 4p^54d^3$. C, D: centers of gravity of the superposed and mixed arrays (From Bauche et al., 1987).

2. Interpretation

The results of the explicit computations of Mandelbaum et al. (1987) presented in Fig. 11 correspond to $N = 2$. But they would demand an enormous amount of work, for example, in the case $N = 5$. However, it is possible to obtain the main results more simply (Bauche et al., 1987)

Let the weighted center of gravity μ_1 of the line energies of the mixed array $l'^{4l'+2}l^N - (l'^{4l'+2}l^{N-1}l'' + l'^{4l'+1}l^{N+1})$ be computed by means of Eq. (10), with $n = 1$, and with the sums running over all the states m of the $l'^{4l'+2}l^N$ configuration and over all the states m' of $l'^{4l'+2}l^{N-1}l''$ and $l'^{4l'+1}l^{N+1}$. The denominator is readily derived from Eq. (50) for pure arrays. It reads

$$W = 2 \max(l, l'') \binom{4l+1}{N-1} P_{ll''}^2 + 2 \max(l, l') \binom{4l+1}{N} P_{l'l}^2 \qquad (63)$$

where $\max(l_1, l_2)$ is the larger of l_1 and l_2, and where $P_{l_1 l_2}$ is a transition radial integral (Eq. (51)). For the numerator, it can be split into two parts, the first one only containing energy-matrix elements which involve the integrals $R^k(ll,\ l'l'')$ responsible for the mixing, and the second one containing all the other matrix elements.

(i) The latter part need not be computed, because it cancels out when only the change in μ_1 due to the mixing is expressed.

(ii) For the former part, one can determine, first, the dependence on N through the second quantization method (Section II,B), and, second, the formal expression for $N = 1$ through Racah's methods.

The final result for the change in μ_1 is:

$$\delta\mu_1 = \frac{2N(4l - N + 2)(2l + 1)^2(2l' + 1)(2l'' + 1)\begin{pmatrix} l & 1 & l' \\ 0 & 0 & 0 \end{pmatrix}\begin{pmatrix} l & 1 & l'' \\ 0 & 0 & 0 \end{pmatrix} P_{ll''} P_{l'l}}{3(4l + 1)[N \max(l, l'') P_{ll''}^2 + (4l - N + 2) \max(l, l') P_{l'l}^2]}$$

$$\times \sum_k \left[2\delta(k, 1) - 3 \begin{Bmatrix} l' & l & 1 \\ l'' & l & k \end{Bmatrix} \right] \begin{pmatrix} l & k & l' \\ 0 & 0 & 0 \end{pmatrix} \begin{pmatrix} l & k & l'' \\ 0 & 0 & 0 \end{pmatrix} R^k(ll, l'l'') \quad (64)$$

It is noteworthy that this equation is the only one, in the present paper, where some radial integrals appear in the denominator.

Numerical applications are shown in Fig. 11. In plot a, the vertical bars A and B represent, respectively, the centers of gravity of the arrays $4p^6 4d^2 - 4p^6 4d4f$ and $4p^6 4d^2 - 4p^5 4d^3$, calculated like in Section III,A for pure arrays. The dashed line C represents the center of gravity of the pure arrays A and B, namely, the part of μ_1 considered in (i) just above. In plot b, the bar D represents the center of gravity of the mixed array: it is deduced from C through the shift $\delta\mu_1$ on the energy scale. Without knowing the results of the explicit computation, it is clear from the conspicuous position of the bar D, namely, far to the left of B, that the array A is almost totally quenched. Work on the narrowing effect is in progress.

In cases where the configuration mixing is weaker, its major effect is the partial quenching of one of the arrays. The spectra of ionized copper (Sugar and Kaufman, 1986) are typical examples (Bauche et al., 1987).

B. Breakdown of $j - j$ Coupling

The assumption of $j - j$ coupling is crucial to the application of the results of Section III,D. Mathematically, it means that, in the calculation of moments, the off-diagonal matrix elements of the Coulomb operator G between states belonging to different subconfigurations are negligible. Of course, this can never be perfectly valid in nature. The consequent deviations are of two kinds: the line energies are changed, and the subarray strengths are changed. In the occurrence of configuration mixing, the changes of the strengths are the most important (Bauche et al., 1987). A similar behavior is predicted in the present problem, which can be called subconfiguration mixing, or relativistic-configuration mixing. It has already been observed by Klapisch et al. (1986).

The corresponding calculations are in progress (Oreg, 1986). The deviations in the subarray strengths are clearly visible in many experimental spectra. For example, the spectrum of highly ionized tantalum (Audebert et al., 1984) presented in Fig. 12 shows, among others, the $3d^{10} - (3d^9)_j nf_{j'}$ lines of Ni-like Ta XLVI, for $n = 4$, 5, 6, and $(j, j') = (\frac{3}{2}, \frac{5}{2})$, $(\frac{5}{2}, \frac{7}{2})$. In the nonrelativistic approximation, the strength of the $(\frac{3}{2}, \frac{5}{2})$ line is 70% of that of $(\frac{5}{2}, \frac{7}{2})$. The relativistic corrections, approximated through the use of relativistic P_{3dnf} radial integrals (Section VI,C), decrease this ratio by a factor 0.97 for $n = 4$, and closer to 1 for $n = 5, 6$. If no other phenomenon occurred, the $(\frac{3}{2}, \frac{5}{2})$ line, at shorter wavelength, should be the weaker for any value of n. This is not observed on Fig. 12, where the tilted segments on top of the relevant lines make it clear that the $(\frac{3}{2}, \frac{5}{2})$ peak is by far stronger than the $(\frac{5}{2}, \frac{7}{2})$ peak for $n = 4$, and that both are roughly equal for $n = 5$. These observations are in good conjunction with the fact that the values of the $(3d, nf)$ Slater integrals, which are responsible for the breakdown of the $j - j$ coupling, rapidly decrease for increasing n. It is also noteworthy that, at the same time, the interpretation of the positions and widths of the $(3d, 5f)$ subarrays (see Fig. 9) has been a success, even without allowance for breakdown of $j - j$ coupling. This stresses the fact that the intensities are much more sensitive to a slight change in the wavefunctions than the energies.

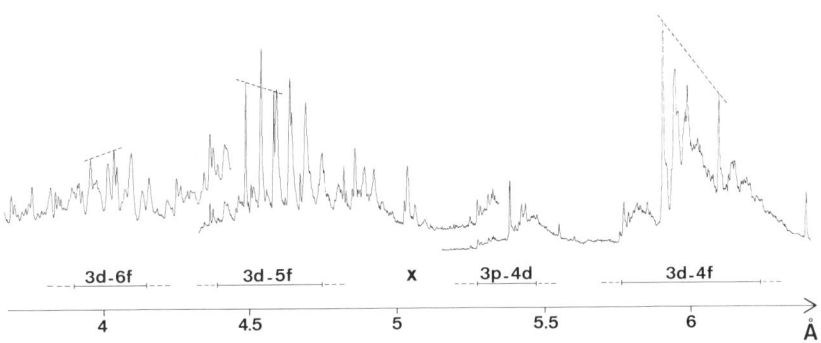

FIG. 12. Experimental spectrum of a laser produced tantalum plasma (the three parts of the spectrum correspond to different recordings) (Audebert et al., 1984). Tilted dashed lines are drawn just on top of the pairs of Ni-like lines $3d_{3/2} - nf_{5/2}$ and $3d_{5/2} - nf_{7/2}$, for $n = 4, 5, 6$. The lines denoted X are sulfur lines for calibration. The $3d - 5f$ part is analyzed in Fig. 9, and the numerical results for $3d - 5f$ and $3d - 6f$ are presented in Table IV.

C. Relativistic Corrections

All the equations and results reviewed in the above sections are written in the nonrelativistic scheme. Relativity is not completely absent, though, in the sense that the spin-orbit operator Λ (Eq. (9)) is a nonrelativistic effective operator of relativistic origin. But it should be taken into account in several other ways.

(i) The orbital radial functions delivered by an *ab initio* relativistic code depend on the j quantum number, as well as on n and l.

(ii) The Coulomb operator G(Eq. (8)) should be corrected for relativity (Armstrong, 1966, 1968), and the Breit operator added (Bethe and Salpeter, 1957).

(iii) The relativistic expansion of the $E1$ transition operator (Feneuille, 1971; Armstrong and Feneuille, 1974) should be used.

The j-dependence of the radial functions (i) has implications in the transition matrix elements (iii), and in the level energies. In the latter, it leads to the diversification of the Slater integrals in the expectation value of the electrostatic repulsion operator. This diversification can readily be taken into account in the formulas for the $j^n j' - j^n j''$ and $j^{n+1} - j^n j'$ subarrays (Section III,D,2). But, for $l^n j - l^n j'$, and for all *arrays*, an important problem is: from a given set of relativistic Slater integrals, how can one derive, first, the *best* values of the Slater integrals to be used in the nonrelativistic energy expansions, and, second, the values of the ζ_{nl} spin-orbit integrals? Solutions have been published by Larkins (1976), and Bauche *et al.* (1982). The corrections to G and the Breit operator (ii) have also been accounted for by the latter authors.

The relativistic corrections to the $E1$ transition operator (iii) have not yet been introduced in the formalism of the transition arrays. Only, the j-dependence of the radial functions can easily be accounted for in the transition integral (Eq. (51)), if needed for the evaluation of relative total strengths (Section V,A). An example of application is given in Section VI,B.

D. Line Intensities

In the above sections, the line strengths are used in place of the line intensities, wherever an exact mathematical treatment is attempted. But, in the experimental context, only the line intensity is relevant. Neglecting geometrical factors specific to the experiment, it is the product of the strength by the population density of the initial level, and, in the case of spontaneous emission, by the well-known v^3 factor.

Although the population problem deserves most of the attention, it can be avoided easily, in a first approximation. In the assumption of statistical weight population (SWP), the populations of the J levels are proportional to $2J + 1$, so that the populations of all αJM states are equal. Therefore, the latter can be omitted in the calculation of the moments. In this approximation, all the results described in Sections III, V, VI,A–C are valid. Other situations must now be considered.

1. Case of Boltzmann Equilibrium

Local thermodynamic equilibrium (LTE) is often chosen as being close to the actual physical conditions in plasmas, e.g., in arcs and sparks. If it is assumed in one of the relevant configurations, say, C', for temperature T, the ratio of the population densities of emitting *states* m and m' reads

$$\frac{N_{m'}}{N_m} = \exp\left[\frac{-(E_{m'} - E_m)}{kT}\right] \tag{65}$$

which tends to 1 for $T \to \infty$. It is clear that introducing the exponential population densities N_m multiplying the strengths in summations such as Eq. (10) prevents any exact calculation to be carried over. However, this equation reads, for the case of emission,

$$\mu_n(C - C') = \frac{\sum_{mm'}[\langle m'|H|m'\rangle - \langle m|H|m\rangle]^n N_{m'} w_{mm'}}{\sum_{mm'} N_{m'} w_{mm'}} \tag{66}$$

and one can write

$$N_{m'} = N_0 \exp\left[\frac{-(E_{m'} - E_0)}{kT}\right] \approx N_0 \left[1 - \frac{(E_{m'} - E_0)}{kT}\right] \tag{67}$$

to the first order in $(E_{m'} - E_0)/kT$, where N_0 is the population density of some reference level of energy E_0. Equation (67) is sufficiently accurate if the energy range of the configuration C' is appreciably smaller than kT. This is often true in the observed ionic spectra, whose ionization potentials are only a few times kT (Cowan, 1981). The denominator of Eq. (66) now reads

$$N_0 W + \frac{N_0}{kT} \sum_{mm'} (E_0 - E_{m'}) w_{mm'}$$

The sum on the right vanishes if the reference level is the center of the emissive zone, with energy $E_0 = q_{C'}$ as given by Eq. (59). Then, in the assumption of Boltzmann equilibrium, μ_n is changed by the increment

$$\Delta\mu_n(C - C') = \frac{1}{kT} \sum_{mm'} [\langle m'|H_p|m'\rangle - \langle m|H_p|m\rangle]^n$$
$$\times [E_0 - \langle m'|H|m'\rangle] \frac{w_{mm'}}{W} \quad (68)$$

This equation is easily evaluated if the formal expansions of μ_n and $\mu_{n+1}(C - C')$ are known in full detail. At the present time, this is the case only for $n < 2$. Thus, one obtains for $n = 1$

$$\Delta\mu_1(C - C') = -\left(\frac{v_{\text{upper}} + \frac{v_{\text{cross}}}{2}}{kT}\right) \quad (69)$$

where v_{upper} is the part of the variance $v(C - C')$ which only contains integrals of the upper configuration, and where v_{cross} is the part with cross-products of integrals of both configurations. These parts can be found in the tables published by Bauche-Arnoult et al. (1979, 1982).

2. Other Physical Situations

In general, the populations of the atomic levels in a plasma can be described in a collisional-radiative model. The SWP and LTE cases are the limits of this model at very high temperatures and very high densities, respectively. But some others are also interesting.

In the assumption of extreme complexity, it can be supposed that the populations of the *states* of the upper configuration are stochastic numbers, obeying, for instance, a Gaussian distribution. Through the explicit computation of some transition-energy variances it has been observed that this stochasticity distorts very little the SWP results. But this is not true when the line strengths themselves are supposed to be stochastic, even in the SWP assumption: the specific correlation between the quantum strengths and energies is the master characteristics of any transition array.

In absorption spectra, only one level, or RS term, is initially populated. The mean energy and variance of the corresponding array are studied in Section III,E.

In the coronal equilibrium (CE) (see Cowan, 1981, p. 545), the populations of the levels of the upper configuration are proportional to the cross section for collisional excitation from the ground level. The consequences for the distribution moments can be studied only through the detailed collisional-radiative model.

3. The v^3 Factor

The v^3 factor relevant to spontaneous emission can be expanded to the first order in the vicinity of the average transition array

$$T_{av}(C - C') = E_{av}(C') - E_{av}(C) + \delta E \tag{70}$$

in the notations of Section III. Then, the v^3 factor can be accounted for approximately, for the emission from state m' to state m, by the factor

$$1 + \frac{3[\langle m'|H|m'\rangle - \langle m|H|m\rangle - T_{av}(C - C')]}{T_{av}(C - C')}$$

It is straightforward to deduce the corresponding increment to the moment μ_n:

$$\Delta\mu_n = \frac{3\mu_{n+1}}{T_{av}(C - C')} - 3\mu_n \tag{71}$$

where the μ_n moments are computed as in Section III.

VII. Level and Line Statistics

All the above sections are devoted to the study of the weighted distributions of level or line energies, with the weights being the line strengths of some radiative type, mostly the electric dipole type $E1$. But the strengths themselves obey a specific distribution function, which is a property of the array. The distribution of $E1$ strengths is an important piece of data, e.g., in opacity problems. In a given array, denoted $C - C'$, its moments can be written

$$\mu_n = \frac{\sum_{\alpha J \alpha' J'} (S_{\alpha J \alpha' J'})^n}{L(C - C')} \tag{72}$$

where $S_{\alpha J \alpha' J'}$ is the $E1$ strength of the $\alpha J - \alpha' J'$ line, and $L(C - C')$ the total number of lines in the array. There appears in this equation an important difference with all those written above: the sum runs over the relevant *levels*, not over the *states*. The $L(C - C')$ quantity in the denominator is also not a sum over states. Its evaluation can be made through application of the $E1$ selection rule

$$|J - J'| \leq 1 \leq J + J' \tag{73}$$

Krasnitz (1984) has established an algorithm for obtaining directly an *exact* value of $L(C - C')$ when, in both configurations, all the electrons are nonequivalent. In the present *statistical* approach, with no condition on C

and C', the distributions of the levels of both configurations according to their J or J' value must first be determined.

A. Level Statistics

Although the quantum number J has a major physical importance in atomic spectroscopy, its distribution in a configuration is not simple to study. This is linked with the fact that J is the eigenvalue of no simple operator. As it is stressed in Section II,B, this prevents the use of the second quantization formalism for an easy evaluation of the moments of the J distribution.

Consequently, an approximate method has been proposed, which goes through some genuine statistical steps (Bauche and Bauche–Arnoult, 1987).

(i) Because the projection quantum number M_J is the eigenvalue of J_z, an operator commuting with the hamiltonian H, its moments μ_n in any configuration C can easily be computed, up to the order $n = 4$.

(ii) The discrete M_J distribution can be approximated, in the limit of very complex configurations, by a Gram–Charlier series of the fourth order (Kendall and Stuart, 1963, p. 168).

(iii) Because the number of levels with angular momentum J is equal to the number of states with $M_J = J$ minus the number of states with $M_J = J + 1$, the J distribution is essentially the derivative of the function introduced in (ii). More precisely, the approximate formula for the number of the levels of configuration C whose total angular momentum has value J reads

$$N(C; J) = \frac{g_c(2J+1)}{v_c\sqrt{8\pi v_c}} \left\{ 1 + \frac{\alpha_4(C) - 3}{24} \left[15 - \frac{5(2J+1)^2}{2v_c} + \frac{(2J+1)^4}{16v_c^2} \right] \right\}$$
$$\times \exp\left[-\frac{(2J+1)^2}{8v_c} \right] \quad (74)$$

where g_c and v_c are respectively the total number of states, and the M_J variance of configuration C, and where $\alpha_4(C) = \mu_4(C)/v_c^2$ is the coefficient of kurtosis (Kendall and Stuart, 1963).

Another approximate formula is available for the total number of levels in configuration C:

$$N(C) = \frac{g_c}{\sqrt{2\pi v_c}} \left[1 - \frac{1}{24v_c} + \frac{\alpha_4(C) - 3}{8} \right] \quad (75)$$

if C contains an even number of electrons, and

$$N(C) = \frac{g_c}{\sqrt{2\pi v_c}} \left[1 - \frac{1}{6v_c} + \frac{\alpha_4(C) - 3}{8} \right] \qquad (76)$$

for an odd number of electrons.

The numerical applications of these equations compare nicely with the exact values (Bauche and Bauche-Arnoult, 1987). It can be proven that $\alpha_4(C)$ is almost equal to 3, the limit for Gaussian distributions. In the above equations, the corresponding terms are small corrections.

Other formulas are available for the numbers of Russell–Saunders terms with specified S and L values, and for the numbers of levels in a given $j - j$ term.

B. Line Statistics

Equations (73) and (74) can readily be combined for the approximate determination of the total number $L(C - C')$ of the $E1$ lines between configurations C and C'. The final formula reads:

$$L(C - C') = \frac{3 g_c g_{c'}}{8v \sqrt{\pi v}} \left[1 - \frac{1}{2v} + (\alpha_4 - 3) \left(\frac{5}{16} - \frac{35}{96v} \right) \right] \qquad (77)$$

with $v = (v_c + v_{c'})/2$ and $\alpha_4 = [\alpha_4(C) + \alpha_4(C')]/2$.

In the assumption $\alpha_4 = 3$, the value of $L(C - C')$ is invariant in the following transformations:

(i) The C and C' configurations can be exchanged.

(ii) In C and C', one can replace simultaneously *all* the subshells by their complementary subshells, e.g., nl^N by nl^{4l-N+2}.

(iii) The $C - C'$ array, denoted in general $l_1^{N_1+1} l_2^{N_2} \ldots - l_1^{N_1} l_2^{N_2+1} \ldots$, possesses exactly as many lines as its *semicomplementary* array, $l_1^{4l_1-N_1+2} l_2^{N_2} \ldots - l_1^{4l_1-N_1+1} l_2^{N_2+1} \ldots$. In this notation, the points represent a string of spectator open subshells, the same one in all four places. It is noteworthy that this correspondence occurs between an array with integer J values and an array with noninteger J values. It is a property of the transition arrays, independently of the statistical approach.

Numerical applications of Eq. (77) to transition arrays of various complexities are compared in Table V with the exact numbers of lines. The relative discrepancy is smallest, as an average, for the most complex arrays. Other examples have been published by Bauche and Bauche-Arnoult (1987).

TABLE V

Numbers $L(C - C')$ of Electric Dipole Lines
in Some $C - C'$ Transition Arrays.[a]

Array	Number of lines		Relative error
	Stat.[a]	Exact	
$p^2 - pd$	36	34	$+5.9\%$
$d^{10}f - d^9fp$	126	130	-3.1%
$d^9s - d^8sf$	310	308	$+0.6\%$
$d^8 - d^7p$	468	466	$+0.4\%$
$d^7 - d^6f$	2 859	2 825	$+1.2\%$
$d^7s - d^6sf$	11 016	10 870	$+1.3\%$
$d^{10}f^3 - d^9f^3p$	33 198	33 150	$+0.1\%$
$d^8f - d^7f^2$	35 539	35 433	$+0.3\%$
$d^7f - d^6f^2$	147 642	146 856	$+0.5\%$

[a] Equation (77), with rounding off.

Another formula is available for evaluating approximately the number of $E1$ lines for vanishing (but nonzero) spin-orbit interactions. This coupling situation is the meaning of *RS coupling* in the following.

C. Line-Strength Statistics

More statistical assumptions are needed for outlining the distribution of line strengths in an array, because the numerator of its moment formula in Eq. (72) cannot be evaluated exactly when $n > 1$. But the way has been opened years ago by nuclear physicists, who have worked in the field of radiation widths. Many of the corresponding publications have been gathered in a book by Porter (1965).

The crucial assumption was made by Porter and Thomas (1956). In view of the complexity of its calculation, they found it reasonable to assume that the *line amplitude* $\langle \alpha J \| D^{(1)} \| \alpha' J' \rangle$ has a Gaussian distribution, with a null average value. The validity of this assumption in atomic transition arrays can be checked through explicit calculations. On this basis, the atomic physicist can now take advantage of simple properties specific to the transition arrays.

(i) In the frequent case where the J-file sum rule (Section V,A) is valid in, at least, one of the configurations, the sum of the strengths of the lines connected to a given J level of this configuration is proportional to $2J + 1$. Thus, it is useful to treat the sets of lines linked to all such levels with this J

value as separate statistical ensembles. These sets are called J sets in the following.

(ii) In extreme coupling cases, many line strengths vanish. The most common example is that of RS coupling, which is often close to the physical coupling observed. In addition to the selection rule on J (Eq. (73)), the strengths obey others on S and L:

$$S = S' \quad \text{and} \quad |L - L'| \leq 1 \leq L + L' \tag{78}$$

The reduction in the number of lines is spectacular. For example, in the $d^4 - d^3 p$ array, 1718 lines reduce to 637, and 1081 lines vanish.

1. Case of Russell–Saunders Coupling

Then, if properties (i) and (ii) just above are valid, it is reasonable to assume that the distribution of line amplitudes in any given J set (defined in (i)) is Gaussian, with a null average value, provided that the lines vanishing in RS coupling are not included in it. For the strengths, this results in a χ-squared distribution (Kendall and Stuart, 1963), which is written:

$$P_J^{RS}(y) = (2\pi y)^{1/2} \exp\left(\frac{-y}{2}\right) \tag{79}$$

where $y = S_{JJ'}/S_{av}(J)$ is the reduced strength, with $S_{av}(J)$ being the average of the nonzero strengths of the considered J set.

The example of the $3d^4 - 3d^3 4p$ array of the FeV spectrum is presented in Fig. 13 and in the first two columns of Table VI.

(i) Figure 13 is a histogram of the line strengths computed explicitly in the intermediate coupling determined by Ekberg (1975). This coupling is close to RS coupling. The ordinates $P^{IC}(S)$ of the points, at increasing abscissas, are the numbers of lines with a strength S smaller than 3, between 3 and 6, 6 and 9, etc., in arbitrary units. The solid curve smoothes off the staggering. It shows in a striking way the existence of a huge number of very weak lines.

(ii) Each entry of Table VI is the number of the strongest lines whose total strength amounts to a given percentage of the array strength. The first column gives the results for the explicit computation (see (i)). The second column gives those of the statistical evaluation, using Eq. (79) for each relevant J set. The agreement between the values in the two columns is a good test of the statistical assumptions made above.

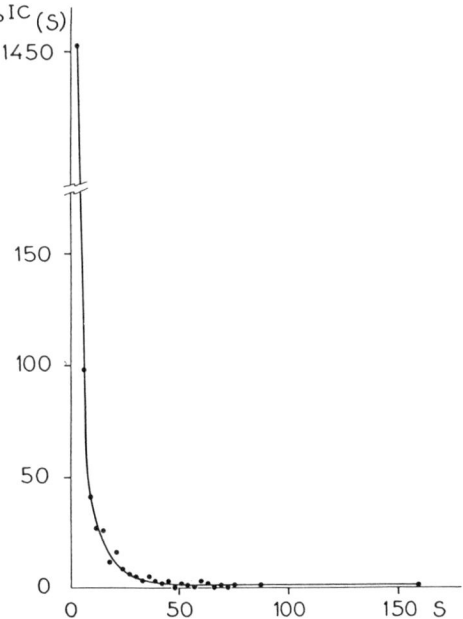

FIG. 13. Distribution of the strengths S of the $3d^4 - 3d^34p$ transition array in the intermediate coupling of iron V (Ekberg, 1975). The point ordinates are the numbers of lines whose strengths are in the successive ranges 0–3, 3–6, 6–9, etc. (From Bauche and Bauche–Arnoult, 1987).

2. Intermediate Coupling

In the most general coupling case, termed *intermediate*, the levels of different RS terms mix. Then, no line has a null intensity. However, in no known case does it lead to the levelling of most lines to an average value. The following examples are typical.

(i) In the $d^4 - d^3p$ array, the 90 strongest lines (i.e., about 5% of the total number of lines) add up, *in any coupling*, to more than one half of the total array strength (for example, 62% in Fe V).

(ii) In the $3d^84s - 3d^84p$ array of Kr X, Mo XVI, and Pr XXXIII (see Fig. 5), the number of the lines which are stronger than 3% of the most intense one is found to be 93, 98, and 89 respectively, (i.e., about 25% of the total number of lines), although the coupling evolves drastically from Kr X to Pr XXXIII. The case of Pr XXXIII is presented in the right part of Table VI, in the same way as FeV (see above, Section VII,C,1). The fair agreement between the explicit and statistical evaluations, in a case so far from RS coupling, shows that the strength distribution is nearly independent of the coupling.

TABLE VI

COMPARISON BETWEEN EXPLICIT AND STATISTICAL EVALUATIONS OF THE NUMBERS OF STRONGEST LINES WHOSE TOTAL STRENGTH AMOUNTS TO SPECIFIC PERCENTAGES OF THE TOTAL ARRAY STRENGTH.[a]

	$d^4 - d^3p$		$d^8s - d^8p$	
	Fe V[b]	stat[c]	Pr XXXIII[b]	stat[c]
10%	5	3	3	1
20%	13	9	6	3
30%	23	20	10	7
40%	38	35	15	11
50%	58	55	21	17
60%	83	82	29	24
70%	118	118	38	34
80%	174	171	48	48
90%	276	256	65	70

[a] The total number of lines is equal to 1718 in Fe V $3d^4 - 3d^34p$ and 401 in Pr XXXIII $3d^84s - 3d^84p$.
[b] Result of the explicit calculation of the strengths through diagonalization.
[c] Computed by using Eq. (79) for each J set.

3. Sundry Extreme Couplings

There exists extreme coupling other than RS, which produce the vanishing of a large fraction of the line strengths (Bauche and Bauche–Arnoult, 1988).

(i) The vanishing of the Slater integrals leads to $j-j$ coupling, with selection rules on the j quantum numbers.

(ii) In the $l^N l' - l^N l''$ array, the vanishing of the spin-orbit and (l, l') and (l, l'') interactions would result in transition selection rules on the l^N core.

(iii) In the $l^{N+1} - l^N l'$ array, the vanishing of all radial integrals except $G^1(l, l')$ generally leads to a larger fraction of vanishing lines than the standard RS coupling (Roth, 1972).

There exist experimental situations akin to the cases (i) and (iii), where the spin-orbit and G^1 integrals largely predominate, respectively. In such situations, the approximate distribution described in Section VII,C,1 is still valid, with only an adequate change of the number of vanishing lines.

VIII. Conclusion

This review paper is devoted to the systematic exposition of the transition array theory, and to its use in the unresolved transition array model. The underlying idea is that in many instances of the physics of highly ionized heavy atoms, when the configurations possess numerous levels, it is useful to depart from the usual level-by-level description of the atom. The latter would lead to a waste of efforts involved in computing very many features that are not observable because of the very conditions in which these atoms exist and are observed. In the theory of transition arrays, the discrete distributions of level energies in the configurations, and of transition energies in the arrays, are considered. Various moments of these distributions (up to order 3, for some cases) are obtained, by using second quantization, as closed formulas involving exact angular coefficients and Slater integrals for all types of configurations and transitions. These moments can be used in two ways: first, in the UTA model, for reproducing the intensity distribution of the actual transition array by a Gaussian or skewed Gaussian function; second, for finding some finer properties of radiative emission and absorption between configurations.

We shall now point out some remarkable results of the theory and discuss possible improvements and extension to different arrays, before describing the applications of the UTA model to plasma spectroscopy.

A. THEORY

1. Conspicuous Results

Although the formulation of the theory was initially motivated by the necessity of modelling the spectra of highly ionized atoms, it should be stressed that it is valid for any atom, whether ionized or neutral, and for any array, resolved or unresolved. The development of this theory leads to some results which are of interest by themselves, shedding some light on facts that were heretofore noticed only empirically.

The radial parts involved in the energy moments are products of as many integrals as the order of the moment, i.e., the integrals themselves for the mean transition energy, products of two integrals for the variance, products of three integrals for the third order moment, and so on. As each such product has a different type of angular coefficient, the complexity of the resulting formulas increases rapidly with the order. This is one of the reasons why the skewness formula has not been completed for all transition types. On the other hand, the dipole radial integrals do not appear in the results, owing to the cancellation with the denominator, except in the cases when configuration interaction is considered.

The first moment (mean transition energy) was shown *not* to be equal, in the case where the jumping electron belongs to a group of equivalent electrons, to the difference of the average energies of the two configurations. The shift, δE, may be very significant for $\Delta n = 0$ transitions. Cowan (1968) described first this phenomenon, in $p^{N+1} - p^N d$ arrays: he observed that *practically all of the strong lines arise from the high energy levels of the excited configuration*. Thus, this δE leads to the concept of emissive zone, i.e., a part of the configuration from which most of the emission takes place. This concept was shown (Mandelbaum et al., 1987) to be useful to understand the perturbation of the emitting configuration by another one which crosses it in an isoelectronic sequence. It is not the crossing of the averages which produces the strongest perturbations on the observed spectrum, but the crossing of the emissive zones.

Most of the arrays considered here are dipolar electric. Because of this, a part of the energy arising from tensors of rank 1, namely the exchange integral G^1 (and R^1) involving the jumping electrons, plays a special role. It gives a prominent contribution to the shift δE, to the exchange-exchange part of the variance, to the skewness, to the configuration-mixing effects and, occasionally, to the statistics of line strengths. It is also remarkable that the coefficient of G^1 in the energy expression is of the essence in the generalized J-file sum rule.

The statistical distribution of the J values in a configuration can be described fairly, except for the very simple configurations, without taking into account the Pauli principle. The total number of $E1$ lines in an array is deduced within a 1% accuracy.

2. Possible Improvements

Like any theory which investigates new concepts, the theory of transition arrays has gathered a momentum of its own, and some questions may be asked, independently of the comparison with experiment.

The developments of Section III have implicitly postulated the dipole-length formula for the oscillator strength. What would change if one used the dipole-velocity formulation? The question is probably not relevant to highly ionized atoms, where the central field approximation is excellent, and in which case, as is well known, the two formulations give the same results. However it is surely worth investigating for lowly ionized atoms, although the results will clearly be more complicated, due to energy terms in the denominator.

Besides some coefficients of the empirical α parameter of Trees (Bauche-Arnoult *et al.*, 1979), and the explicit study of nearby perturbing configurations in the quenching phenomenon (Section VI,A) the theory is based on first-order perturbation theory. On the other hand, it is well known that correlation effects can manifest themselves by additional effective operators and by large deviations between computed and observed Slater integrals. Do they have specific effects on the moments of the arrays, in addition to the effective modification of the Slater integrals? In the recent work of Mandelbaum *et al.* (1987), it is suggested that a significant screening factor applied to the G^1 integral would greatly improve the position of the $4d - 4f$ transition in Pr XVI. This deserves further interest.

Likewise, the transition probability operator considered is not relativistic. For short wavelength transitions, retardation and other terms which do not have a rank one behavior will contribute, and give some additional terms. This has not yet been investigated.

3. Possible Extension of the Theory

Possible extension of the theory to other types of radiative transitions does not present any special difficulty. Formulas have been obtained for arrays of *forbidden*, non electric dipole, transitions. There may not be many applications for these arrays. However, their interest stems from the fact that collisional excitation strengths can be shown to behave like one-electron tensor operators acting only on bound electrons, the effect of the continuum electron being absorbed in a radial integral containing the sum over partial waves (Klapisch and Bar Shalom, 1987). The concept of emissive zones can thus be immediately transposed to "most probably populated zones" by electron impact.

Another extension of the theory could be in the direction of Zeeman and hyperfine patterns. It suffices in these cases to replace the G and Λ operators by the corresponding Hamiltonians. Since the latter are one-body operators with respect to the atomic electrons, the formulas should be much simpler.

At the opposite, one can consider arrays of arrays, or "super arrays". These could occur in LTE plasmas involving highly ionized heavy atoms, so that each resonant transition is accompanied by satellites corresponding to different repartitions of holes between the possible subshells. Some work has already been done for the case where the width of each array is negligible with respect to the overall width, the latter originating from the energy shift between the arrays (Bar Shalom *et al.*, 1987). The formulas come out to be very simple and yield a model which is more flexible and more accurate than the average atom model.

On the other hand, it would be very interesting to consider Auger transitions, arrays of which can be studied experimentally in the so-called electron-energy spectra. Here the transition operator is the non-diagonal Coulomb interaction connecting a doubly excited configuration with one possessing an electron in the continuum. The formulas are bound to be quite complicated.

B. Application of the UTA Model

1. Identification of Spectra

a. Comparison with experimental profile. Comparison with experimental data has shown that agreement is very good. Although the shapes of the experimental arrays are usually not Gaussian, the Gaussian model is sufficient in many cases to identify the spectra in such a way as to point out when a satellite transition or an ionization stage has been forgotten. The fact that even some quantitative information can be obtained from the intensities of the arrays in many cases with such a simple model is particularly gratifying and establishes definitely the physical existence of UTAs as spectroscopical objects of their own. It must be emphasized that although the method for obtaining the moments pertains to abstract theory, the final results can be presented as tables of universal coefficients, in such a way that the mean energy and the width of the arrays can be obtained *without the help of a computer*, once values are obtained for Slater and spin-orbit integrals (see Tables IV and V in the paper of Bauche-Arnoult *et al.*, (1979)).

b. Use of higher moments. It has been shown in Section IV that in some cases, specially $d - f$ transitions, the skewness is important. Its introduction in the model through the skewed Gaussian profile does not give a more realistic shape to the arrays, though. However, it does yield better values for the energy of the transition peak, at the condition, of course, that one has at hand a reliable way of estimating the energies of the configurations. Also, an important reduction of the FWHM occurs, which influences the exploitation of the intensities, through the normalization of the profiles. An example is shown in Fig. 6. This also leaves less doubts on the presence or absence of satellites.

However, it is not advisable to try and obtain higher moments. First, as mentioned above, the formulas for these become increasingly cumbersome.

Second, it is clear that at some stage, the fact that the populations of the levels are not proportional to the statistical weights must influence the shape of the arrays. However, computing these in complex spectra is not an easy task. Recently, a level-by-level model has been made and applied to the Co-like Xe $3d^9 - 3d^84f$ (Klapisch et al., 1987). It shows clearly that, even at laser-produced plasma densities $n_e = 10^{20}$ cm^{-3}, there are some overpopulated quasi-metastable levels, causing the array to appear "double-peaked." Thus, attempting to obtain shapes, without knowing how to account for the population repartition, is an illusion.

c. Resolvability. The knowledge of the fact that the arrays are resolved or not is interesting in the case where emission spectra contain several superimposed arrays or ionization stages (Mandelbaum et al., 1987). Moreover it is of vital importance when considering absorption spectra and the question of opacity of plasmas, because of the nonlinear character of absorption. As mentioned in the introduction, the resolved–unresolved character is dependent on the interplay of $P(E)$, the individual line profile, and of $A(E)$, the discrete distribution studied in the theory. This implies that nothing can be said about the resolvability without considering explicitly $P(E)$. This has not been done yet. Actually, even a precise, operational, definition of resolvability is missing. This is certainly a direction of research that should be pursued in the future.

2. Use of UTA Radiation

Now that we have at hand these UTAs appearing in the spectra of highly ionized heavy atoms, and that they are well understood, the question arises: what are they good for?

a. Intense narrow band emission source. One possible answer is the one of Carroll et al. (1984), and Finkenthal et al. (1986), especially relevant to $4d - 4f$ transitions in ionized rare earths, namely to provide an intense souce of soft X-rays that is easily controllable. Indeed it has been shown (Mandelbaum et al., 1987) that, owing to array quenching, many ionization stages radiate their UTA's at about the same wavelength, varying from 60 Å to 100 Å for different atoms, with a width not exceeding 5 Å in most cases. Although this wavelength is somehow too long for lithography, this inconvenient is compensated by the fact that they can be obtained in a very cheap source, a low inductance vacuum spark, or a small laser. Moreover, the width is large enough to provide for easy coincidence with many atomic transitions, thus enabling a convenient optical pumping device for many experiments. This scheme has been indeed proposed by Mathews et al. (1983) for $3d - 4f$ transitions in Au.

b. *Plasma diagnostics.* It has recently been proposed (Zigler, et al. 1987) to use the $3d - 4f$ UTA's in heavy atoms for plasmas diagnostics, instead of classical line-intensity ratio measurements. Indeed, the fact that the UTAs corresponding to different ionization stages are well separated in this case enables the design of very compact, low spectral resolution spectrometers. The latter can thus benefit from a high acceptance, leading in turn to the possibility of simultaneous space and time resolution. This is very useful in laser-produced plasmas that have a short lifetime, and in tokamaks, which have a low emissivity by unit volume. In the latter case, the injection of chosen impurities for that purpose has been considered. Density measurements can be performed by monitoring the ratio of resonant to satellite arrays, like in the Pd $3d - 4p$ transition in Fig. 8, whereas temperature sensitive information can be obtained by following the ratio of homologous transitions, e.g. $3d - 4f$, for different ionization states. Evidently, the interpretation of the results relies on a proper model for configuration populations, but this is not different than for the line-ratio method. For a first try, one can assume LTE. This has been done on the spectrum of La displayed in Fig. 6, and has yielded temperatures somewhat too low compared with results of more sophisticated methods and measurements. Alternatively, coronal equilibrium can be assumed. Curves for relative abundances of various ion states in coronal equilibrium were published (De Michelis and Mattioli, 1984). However, coronal equilibrium is not quite a good approximation for laser-produced plasmas.

Now, a complete collisional-radiative model, taking into account all the possible ionization stages, which is at present out of reach for a detailed level-by-level description of heavy atoms, should be feasible for UTAs since all the transition probabilities, radiative and collisional, are then averaged over configurations, thus reducing drastically the number of "levels" to be included in the model. A step in this direction was recently proposed, for excitation-autoionization, by Pindzola et al. (1986). However, a step further can now be envisioned, by using for collisional transitions the transition-array theory described at length in this review for radiative transitions. One should then obtain a much more realistic description of the spectra than the configuration average approximation, without the prohibitively long computer times of the full level-by-level description.

Thus, it has been made clear in this review, that a new field of research in the physics of highly ionized atoms is open, a field vast enough for experimental and theoretical research to thrive for a deeper understanding of atomic processes in plasmas.

IX. Appendix

Dependence on N of the trace of a k-electron operator in nl^N

Let a k-electron operator be written as a sum over $2k$ Greek letters of a string of k creation and k annihilation operators, multiplied by a k-electron matrix element K, depending on the Greek letters (see the case of G in Eq. (21)). The trace to be computed reads

$$T = \sum_{ab\ldots n} \langle \{ab\ldots n\}| \sum_{\alpha\beta\ldots\mu} a_\alpha^\dagger a_\beta^\dagger \ldots a_\lambda a_\mu \, K(\alpha, \beta, \ldots, \lambda, \mu)|\{ab\ldots n\}\rangle$$

where $\{ab\ldots n\}$ is an N-electron Slater determinant, with $a, b, \ldots, n$, and the Greek letters, being sets of one-electron quantum numbers. In each nonzero matrix element contained in T, all the Greek letters indexing the annihilation operators also index the creation operators. Let a packet of k such letters, denoted $(\alpha, \beta, \ldots, \kappa)$, be selected. The relevant part in T is equal to

$$\binom{4l - k + 2}{N - k}(k!)^2 K(\alpha, \beta, \ldots, \beta, \alpha)$$

The first combinatorial factor is the number of Slater determinants of nl^N containing the sets $\alpha, \beta, \ldots, \kappa$, and $k!$ is the number of ways in which these sets can be ordered for indexing, for instance, the creation operators. The N-dependence only appears in the first factor.

REFERENCES

Aglitsky, E. V., Boiko, V. A., Krokhin, O., N., Pikuz, S. A., and Faenov A. Ya. (1974). *Sov. Quant. Electron.* **1**, 1067.
Aglitsky, E. V., Golts, E. Ya., Levykin, Yu. A., Livshits, A. M., Mandelstam S. L., and Safronova A. S. (1979). *Opt. Spektr.* **46**, 1043.
Aglitsky, E. V., Golts, E. Ya., Levykin, Yu. A., and Livshits, A. M. (1981). *J. Phys. B: At. Mol. Phys.* **14**, 1549.
Aglitsky, E. V., Antsiferov, P. S., Mandelstam, S. L., and Panin, A. M. (1984). *Can. J. Phys.* **63**, 1924.
Armstrong, L. Jr. (1966). *J. Math. Phys.* **7**, 1891.
Armstrong, L. Jr. (1968). *J. Math. Phys.* **9**, 1083.
Armstrong, L. Jr., and Feneuille, S. (1974). *Adv. At. Mol. Phys.* **10**, 1.
Audebert, P., Gauthier, J.-C., Geindre, J.-P., Monier, P., Bauche-Arnoult, C., Bauche, J., Luc-Koenig, E., Pain, D., Wyart, J.-F., Busquet, M., and Chenais-Popovics, C. (1984). Rapport GRECO-ILM.
Audebert, P., Gauthier, J.-C., Geindre, J.-P., Chenais-Popovics, C., Bauche-Arnoult, C., Bauche, J., Klapisch, M., Luc-Koenig, E., and Wyart, J.-F. (1985). *Phys. Rev. A* **32**, 409.
Aymar, M. (1973). *Nuclear Inst. and Meth.* **110**, 211.
Bar Shalom, A., Klapisch, M., and Oreg, J. (1987). Unpublished.

Bauche, J. (1986). Unpublished.
Bauche, J., and Bauche-Arnoult, C. (1987). *J. Phys. B: At. Mol. Phys.* **20**, 1659.
Bauche, J., and Bauche-Arnoult, C. (1988). To be published.
Bauche, J., Bauche-Arnoult, C., Luc-Koenig, E., and Klapisch M. (1982). *J. Phys. B: At. Mol. Phys.* **15**, 2325.
Bauche, J., Bauche-Arnoult, C., Luc-Koenig, E., Wyart, J.-F., and Klapisch, M. (1983). *Phys. Rev. A* **28**, 829.
Bauche, J., Bauche-Arnoult, C., Klapisch, M., Mandelbaum, P., and Schwob, J.-L. (1987). *J. Phys. B: At. Mol. Phys.* **20**, 1443.
Bauche-Arnoult, C., Bauche, J., and Klapisch, M. (1978). *J. Opt. Soc. Am.* **68**, 1136.
Bauche-Arnoult, C., Bauche, J., and Klapisch, M. (1979). *Phys. Rev. A* **20**, 2424.
Bauche-Arnoult, C., Bauche, J., and Ekberg, J., O. (1982a). *J. Phys. B: At. Mol. Phys.* **15**, 701.
Bauche-Arnoult, C., Bauche, J., and Klapisch, M. (1982b). *Phys. Rev. A.* **25**, 2641.
Bauche-Arnoult, C., Bauche, J., and Klapisch, M. (1984). Phys. Rev. A **30**, 3026.
Bauche-Arnoult, C., Bauche, J., and Klapisch, M. (1985). *Phys. Rev. A* **31**, 2248.
Bauche-Arnoult, C., Luc-Koenig, E., Wyart, J.-F., Geindre, J.-P., Audebert, P., Monier, P., Gauthier, J.-C., and Chenais-Popovics, C. (1986). *Phys. Rev. A* **33**, 791.
Bethe, H. A., and Salpeter, E. E. (1957). "Quantum Mechanics of One- and Two-Electron Atoms." Springer, Berlin.
Bloom, S. D., and Goldberg, A. (1986). *Phys. Rev. A* **34**, 2865.
Bonnelle, C., Karnatak, R. C., and Spector, N. (1977). *J. Phys. B: At. Mol. Phys.* **10**, 795.
Burkhalter, P. G., Feldman, U., and Cowan, R. D. (1974). *J. Opt. Soc. Am.* **64**, 1058.
Burkhalter, P. G., Reader J., and Cowan, R. D. (1980). *J. Opt. Soc. Am.* **70**, 912.
Busquet, M., Pain, D., Bauche, J., and Luc-Koenig, E. (1985). *Phys. Scripta* **31**, 137.
Carroll, P. K., and O'Sullivan, G. (1981). *Phys. Lett. A* **84**, 59.
Carroll, P. K., and O'Sullivan, G. (1982). *Phys. Rev. A* **20**, 275.
Carroll, P. K., Costello, J., Kennedy, E. T., and O'Sullivan, G. (1984). *J. Phys. B: At. Mol. Phys.* **17**, 2169.
Condon, E., U., and Shortley, G., H. (1935). "The Theory of Atomic Spectra." Cambridge Univ. Press, London and New York.
Cowan, R. D. (1967). *Phys. Rev.* **163**, 54.
Cowan, R. D. (1968). *J. Opt. Soc. Am.* **58**, 924.
Cowan, R. D. (1970). *J. Phys. Colloq.* **34**, C4, 191.
Cowan, R. D. (1973). *Nucl. Instrum. Methods* **110**, 173.
Cowan, R. D. (1977). Los Alamos Report LA 6679-MS.
Cowan, R. D. (1981). "The Theory of Atomic Structure and Spectra." University of California, Berkeley.
Dehmer, J. L., Starace, A. F., Fano, U. Sugar, J., and Cooper, J. W. (1971). *Phys. Rev. Letters* **26**, 1521.
De Michelis, C., and Mattioli, M. (1984). *Rep. Prog. Phys.* **47**, 1233.
Denne, B., and Poulsen, O. (1981). *Phys. Rev. A* **23**, 1229.
Edlen, B. (1947). *Physica* (Utrecht) **13**, 545.
Ekberg, J. O. (1975). *Physica Scripta* **12**, 42.
Esteva, J.-M., Karnatak, R. C., Fuggle, J. C., and Sawatzky, G. A. (1983). *Phys. Rev. Letters* **50**, 910.
Feneuille, S. (1971). *Physica* (Utrecht) **53**, 143.
Finkenthal, M., Stratton, B. C., Moose, H. W., Hodge, W. L., Suckewer, S., Cohen, S., Mandelbaum, P., and Klapisch, M. (1985). *J. Phys. B: At Mol. Phys.* **18**, 4393.
Finkenthal, M., Lippmann, A. S., Huang, L. K., Yu, T. L., Stratton, B. C., Moos, H. W., Klapisch, M., Mandelbaum, P., Bar Shalom, A., Hodge, W. L., Phillips, P. E., Price, T. R., Porter, J. C., Richards, B., and Rowan, W. L. (1986). *J. Appl. Phys.* **59**, 3644.

Folkmann, F., Mann, R., and Beyer, H. F. (1983). *Phys. Scripta.* **T3**, 88.
Gauthier, J.-C., Geindre, J.-P., Monier, P., and Tragin, N. (1986). Private communication.
Goldschmidt, Z. B. (1971). *Phys. Rev. A* **3**, 1872.
Goldschmidt, Z. B., and Starkand, J. (1970). *J. Phys. B: At. Mol. Phys.* **3**, L141.
Gordon, H., Hobby, M. G., Peacock, N. J., and Cowan, R. D. (1979). *J. Phys. B: At. Mol. Phys.* **12**, 881.
Harrison, G. R., and Johnson, M. H. Jr. (1931). *Phys. Rev.* **38**, 757.
Judd, B. R. (1963). "Operator Techniques in Atomic Spectroscopy." McGraw-Hill, New York.
Judd, B. R. (1967). "Second Quantization and Atomic Spectroscopy." Johns Hopkins Press, Baltimore, Maryland.
Judd, B. R., Hansen, J. E., and Raassen, A. J. J. (1982). *J. Phys. B: At. Mol. Phys.* **15**, 1457.
Kendall, M. G., and Stuart, A. (1963). "The Advanced Theory of Statistics." Charles Griffin, London.
Klapisch, M., and Bar Shalom, A. (1987). Unpublished.
Klapisch, M., Schwob, J.-L., Fraenkel, B. S., and Oreg J. (1977). *J. Opt. Soc. Am.* **67**, 148.
Klapisch, M., Bar Shalom, A., Mandelbaum, P., Schwob, J.-L., Zigler, A., Zmora, H., and Jackel, S. (1980). *Phys. Lett. A* **79**, 67.
Klapisch, M., Meroz, E., Mandelbaum, P., Zigler, A., Bauche-Arnoult, C., and Bauche, J. (1982). *Phys. Rev. A* **25**, 2391.
Klapisch, M., Bauche, J., and Bauche-Arnoult, C. (1983). *Physica Scripta* **T3**, 222.
Klapisch, M., Mandelbaum, P., Zigler, A., Bauche-Arnoult, C., and Bauche, J. (1986). *Physica Scripta* **34**, 51.
Klapisch, M., Bar Shalom, A., and Cohn, A. (1987). Unpublished.
Koenig, E. (1972). *Physica* (Utrecht) **62**, 393.
Kononov, E. Ya. (1978). *Phys. Scripta* **17**, 425.
Krasnitz, A. (1984). M. Sc. Thesis (Hebrew Univ. Jerusalem). Unpublished.
Larkins, F. P. (1976). *J. Phys. B: At. Mol. Phys.* **9**, 37.
Layzer, D. (1959). *Ann. Phys.* (N.Y.) **8**, 271.
Layzer, D. (1963). *Phys. Rev.* **132**, 2125.
Mandelbaum, P., Finkenthal, M., Schwob, J.-L., and Klapisch, M. (1987). *Phys. Rev. A* **35**, 5051.
Matthews, D. L., Campbell, E. M., Ceglio, N. M., Hermes, G., Kauffman, R., Koppel, L., Lee, R. L., Maus, K., Rupert V., Slivinsky, V. W., Turner, R., and Ze, F. (1983). *J. Appl. Phys.* **54**, 4260.
Morita, S., and Fujita, J. (1985). *Nucl. Instr. Meth. B* **9**, 713.
Moszkowski, A. (1962). *Prog. Theor. Phys.* **28**, 1.
Oreg, J. (1986). Private communication.
O'Sullivan, G., Carroll, P. K. (1981). *J. Opt. Soc. Am.* **71**, 227.
Pindzola, M. S., Griffin, D. C., and Bottcher, C. (1986). *Phys. Rev. A* **33**, 3787.
Porter, C. E. (1965). "Statistical Theories of Spectra: Fluctuations." Academic Press, New York.
Porter, C. E., and Thomas, R. G. (1956). *Phys. Rev.* **104**, 483.
Racah, G. (1942). *Phys. Rev.* **62**, 438.
Racah, G. (1943). *Phys. Rev.* **63**, 367.
Racah, G. (1949). *Phys. Rev.* **76**, 1352.
Roth, C. (1972). *J. Res. Nat. Bur. Stand.* **76B**, 61.
Shortley, G. H. (1935). *Phys. Rev.* **47**, 295.
Slater, J. C. (1929). *Phys. Rev.* **34**, 1293.
Slater, J. C. (1960). "Quantum Theory of Atomic Structure," Vol. 2. McGraw-Hill, New York.
Sobelman, I. I. (1972). "Introduction to the Theory of Atomic Spectra." Pergamon, New York.
Sugar, J. (1972a). *Phys. Rev. A* **6**, 1764.
Sugar, J. (1972b). *Phys. Rev. B* **5**, 1785.
Sugar, J., and Kaufman, V. (1980). *Phys. Rev. A* **21**, 2096.

Sugar, J., and Kaufman, V. (1986). *J. Opt. Soc. Am. B* **3**, 704.
Sugar, J., Brewer, W. D., Kalkowski, G., and Paparazzo, E. (1985). *Phys. Rev. A* **32**, 2242.
Tragin, N., Geindre, J.-P., Monier, P., Gauthier, J.-C., Chenais-Popovics, C., Wyart, J.-F., and Bauche-Arnoult, C. (1988). *Physica Scripta* **37**, 72.
Uylings, P. H. M. (1984). *J. Phys. B: At. Mol. Phys.* **17**, 2375.
Zigler, A., Zmora, H., Spector, N., Klapisch, M., Schwob, J.-L., and Bar Shalom, A., (1980). *J. Opt. Soc. Am.* **70**, 129.
Zigler, A., Klapisch, M., and Mandelbaum, P. (1986). *Phys. Lett. A* **117**, 31.
Zigler, A., Givon, M., Yarkoni, E., Kishinevsky, M., Goldberg, E., Arad, B., and Klapisch, M. (1987). *Phys. Rev. A* **35**, 280.

PHOTOIONIZATION AND COLLISIONAL IONIZATION OF EXCITED ATOMS USING SYNCHROTRON AND LASER RADIATIONS

F. J. WUILLEUMIER

Laboratoire de Spectroscopie Atomique et Ionique
and
Laboratoire pour l'Utilisation du Rayonnement
Electromagnique (LURE), Université Paris Sud, B350
Orsay Cedex, France 91405

D. L. EDERER

Radiation Physics Division
National Bureau of Standards
Gaithersburg, MD 20899

J. L. PICQUÉ

Laboratoire Aimé Cotton, CNRS
Université Paris Sud, B505
Orsay Cedex, France 91405

I. Introduction	198
II. Experimental Techniques	201
A. Recombination Radiation	201
B. Photoabsorption by the Excited Atom	202
III. Theoretical Background	209
IV. Photoionization of an Outer Electron in Excited Atoms	210
A. The Rare Gases	210
B. Photoionization of Excited Alkali Atoms	217
C. Photoionization of Excited Alkaline Earths	226
V. Results from Synchrotron Radiation Ionization of Laser-Excited Atoms . .	228
A. General Background	228
B. Results from Continuum Photoabsorption Experiments in Laser-Excited Atoms	229
C. Laser-Synchrotron Combination: the Feasibility Experiment	231
D. Resonant $3p$ Cross Section in Laser-Excited Sodium Atoms: Determination of Oscillar Strengths for Inner-Shell Excitations . . .	238
E. Inner Shell and Outer Shell Photoionization in Laser-Excited Atomic Barium	251

VI. Collisional Ionization of Laser-Excited Atoms 261
   A. General Background. 261
   B. Experimental Conditions 263
   C. The Show Case of Sodium. 268
VII. Conclusion 278
   Acknowledgement. 279
   References. 279

I. Introduction

This article is devoted to a description of various ionization processes which may be studied in excited atoms. Some of these processes are: photoionization, autoionization, and collisional ionization. The development of this field has been accelerated by combining tunable lasers, which excite the atomic species, with synchrotron radiation, which probes the excited atoms. We will concentrate on the two step photoionization in free atoms where one electron is promoted to an outer orbital and another is promoted either to a highly excited state which subsequently autoionizes, or to a continuum state. The goal of the present article is to cover works achieved by the end of 1985. For other multiphoton processes the reader is urged to consult recent reviews and bibliographies on the subject (Eberly and Karczewski, 1970; Eberly and Gallagher, 1979 and 1981; Eberly et al., 1982; Kimura, 1985).

The photoionization process is an important aid to study the geometrical and dynamical properties of electron motions (Kelley and Simons, 1973; Chang and Poe, 1975 and Johnson et al., 1980). The Coulomb potential is known and the charge of the nucleus dominates the force on the electrons, whose correlated motion provides a many body problem that is tractable through perturbation theory (Starace, 1982). The many body effects studied this way through atomic (and molecular) photoionization provides an important insight towards the solution of other problems such as those encountered in nuclear matter (Wuilleumier, 1976). The photon energy range from threshold to several hundred electron volts above threshold provides the most fruitful region to study these many body interactions. In this spectral region, synchrotron radiation is the most useful source with which to probe the dynamical properties of these systems. The continuing progress in the development of theoretical methods, (Seaton, 1966; Amusia et al., 1971; Kelley and Simons, 1973; Burke and Taylor, 1975; Johnson et al., 1980), has gone hand in hand with development of experimental techniques (Codling, 1979; Koch and Sonntag, 1979; Krause, 1980; Wuilleumier, 1980, 1981a, 1981b; Sonntag and Wuilleumier, 1983), to map out photoionization cross sections and other parameters that are determined by the correlated motion of electrons over a broad energy range. These other parameters such as angular distribution parameters, partial cross sections of the ejected photoelectrons, (Siegbahn and Karlsson, 1982), and the spin polarization, (Heinzmann, 1980) of the photoelectron have provided additional information needed to describe the characteristics of the wavefunctions that are a measure of the potential that governs the photoionization process.

The information obtained from these measurements, while having great value, is somewhat restrictive because of dipole selection rules. Only a limited class of states can be probed and the initial state is often an ensemble of nearly degenerate levels. Photoexcitation from excited states provides access (Eshevick et al., 1976; McIlrath and Lucatorto, 1977) to states of parity opposite to that of the ground state. The excited state can be prepared with a well defined set of quantum numbers (Arnous, et al., 1973; Leuchs et al., 1979; Smith et al., 1980), and the photoionization dynamics of these precisely defined states can be studied as a function of photon energy. The angular distribution of the electrons photoionized from a specifically prepared excited state will reflect the geometry and dynamical properties of that state and yield phase shift information (Lampropoulos, 1976; Kollath, 1980; Kaminski et al., 1980) about the continuum wavefunctions which cannot be obtained from total cross section measurements. The energy dependence of the angular distribution parameter for the different continuum channels is predicted by Lahiri and Manson (1982), to have multiple minima. The measurement of the spin polarization parameters is another means of providing the missing link (Heinzmann, 1980) for a determination of the continuum wave function phase shifts.

The total potential of an atom whose atomic number Z is near that corresponding to the binding of a shell is very sensitive to changes in the electron orbits. The promotion (or ionization) of a valence electron to an excited state modifies dramatically the effective potential of the core electrons. The change in correlation can shift high orbital momentum states closer to the nucleus (Connerade, 1978). This effect produces a radical redistribution of the oscillator strength density as has been observed recently, (Lucatorto et al., 1981), in the $4d$ inner shell ionization of barium neutral, singly and doubly ionized atoms. In the case of barium, the $4d$ cross section is dominated by $4d$ to continuum transitions that have f-wave symmetry. The discrete autoionizing $4d$ to $4f$ transition is weak (Wendin and Starace, 1978) in neutral barium but as the effective nuclear charge increases in Ba^{++}, the $4d$ to $4f$ transition appears as a very strong resonance, with a diminution in the continuum oscillator strength density. A dramatic effect has been observed (Hill et al., 1982) in the xenon isoelectronic sequence where photoionization cross section behavior of the autoionizing resonances converging to $5p^5\ ^2P_{1/2}$ state of the ion is strongly dependent on the nuclear charge. This change in the spectral cross section behavior demonstrates the dramatic effect a rather small change in the core potential can have on the electron-electron correlation.

There are many applications that require information about oscillator strengths and cross sections from these excited states. Some of these applications are: resonance ionization spectroscopy (RIS), (Hurst et al., 1977), which has the sensitivity of single atom detection and has been suggested (Hurst et al., 1977, 1979; Nayfeh, 1980) as a tool for quantitative analysis; XUV lasers (Willison et al., 1980; Caro et al., 1984) where highly energetic metastable excited atoms and ions form the basis for obtaining a population inversion; radiative transfer in laboratory plasmas, and stellar atmospheres (Spitzer, 1968; Aller, 1984). Minimizing VUV radiation transfer in laboratory plasmas is a major factor in producing more efficient plasma heating.

We are witnessing a revolution in the development of techniques to study photoionization from excited states. Multicolor powerful pulsed lasers are available to selectively excite states in atoms (Kaminski et al., 1980) and molecules (White et al., 1982; Pratt et al., 1983) and to study the photoionization products by photoelectron spectroscopy. Often startling discoveries have been made because the production of an ensemble of excited atoms by laser irradiation for a photoionization experiment has led to new insights in the understanding of collisional interactions between excited atoms (Lucatorto and McIlrath, 1976; Bizau et al., 1982). The intimate relationship between these two areas of research is due in part to similar equipment requirements. For the past five years, cw dye lasers and synchrotron radiation have been used over a broad energy range to explore the photoionization and collisional interaction of atoms in excited states. Some of the recent developments in this work (Bizau et al., 1982) will be discussed in this article.

We first recall the experimental techniques in Sect. II and the theoretical background in Sect. III. Next we describe experiments in which lasers and other sources were used to photoionize the excited atoms. The results provided deal exclusively with photoionization of the excited outer electron, and cover a very narrow photon energy range within a few eV above the binding energy of the excited electron. In Sect. V we discuss the new experiments that have been used to explore the photoionization dynamics of excited atoms using synchrotron radiation, giving examples of measurements of oscillator strengths and photoionization cross sections for outer-shell and inner-shell electrons over an extended energy range (15–150 eV). In section VI we will present results that show how electron spectroscopy and synchrotron radiation can be utilized to study collision processes of laser-excited atoms.

II. Experimental Techniques

Experimental studies of the photoionization of excited states require the combination of the usual techniques developed for the photoionization of ground state atoms plus special methods to prepare the atoms in a specific excited state. In this section these techniques will be described.

A. RECOMBINATION RADIATION

The study of the photoionization of excited state atoms began indirectly over fifty years ago through observations (Mohler and Boeckner, 1929 and Mohler, 1929) of the continuum radiated by recombining cesium ions. The process is shown schematically in Fig. 1 for the case of sodium. An electron with velocity v is captured by an ion. The initial state is noted by the symbol ●, and the final state by the symbol ○. The radiated photon traverses the

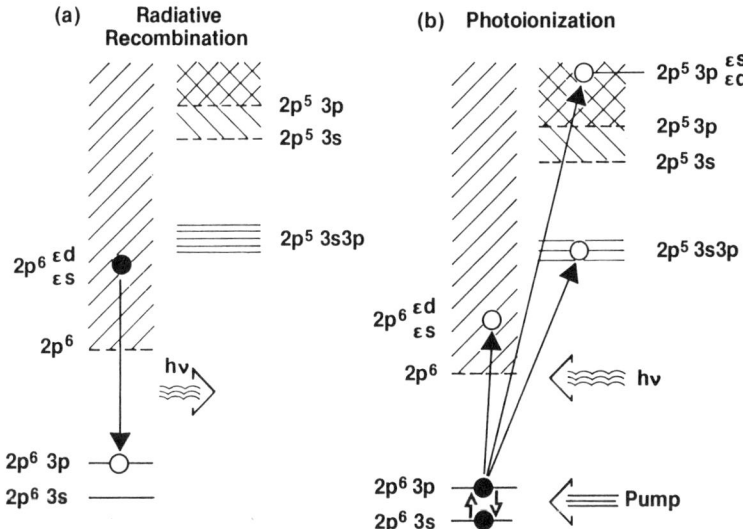

FIG. 1. Schematic representation of excited state processes. Sodium is used to illustrate these processes: a) Radiative recombination: a continuum electron with a kinetic energy ε in an εs or εd continuum channel is captured by ionized sodium and radiates a photon of energy $hv = I(\varepsilon) - E(3p)$, where $E(3p)$ is measured from the ground state of neutral sodium. The final state of excited sodium is $1s^2 2s^2 2p^6 3p$. b) Absorption by the excited atom: A pump mechanism (a laser or an electron beam for example) is used to excite neutral sodium to the $1s^2 2s^2 2p^6 3p$ state. Subsequently a UV photon of energy hv promotes the excited electron to a continuum εs or εd channel or a core electron to an autoionizing level, $2p^5 3s3p$ or a continuum state $2p^5 3p\varepsilon s$ or $2p^5 3p\varepsilon d$.

optically thin plasma which contains electrons that are assumed to be in thermal equilibrium. Measurements like these were carried out for the alkali atoms of lithium and sodium (Roth, 1969 and 1971). A carrier gas was seeded with the alkali and then the plasma was created by shock heating. The intensity $I(v)$ of the recombination radiation is related to the cross section $\sigma(v)$ for recombination by an electron of velocity v as follows;

$$I(v)\frac{dv}{hv} = n_e \cdot n_+ \cdot v \cdot f(v) \cdot \sigma(v) \cdot dv, \tag{1}$$

where n_e and n_+ are the electron and ion densities respectively. The quantity $f(v)$ is the Boltzman distribution for thermal electrons of rms velocity v. The recombination cross section is related to the excited state photoionization cross section $\sigma_p(\lambda)$ through detailed balancing as follows:

$$g_+ \cdot \sigma(v) \cdot (mv)^2 = g_n \cdot \sigma_p(\lambda) \cdot \left(\frac{h}{\lambda}\right)^2, \tag{2}$$

where g_+ and g_n are the statical weights of the states in the ion and the neutral respectively. These measurements are difficult because an absolute flux determination must be made and the ion and electron densities must also be measured. These measurements are made on a transient medium whose parameters change from shot to shot. It was not long before high power pulsed lasers provided a more direct approach.

B. PHOTOABSORPTION BY THE EXCITED ATOM

A more straightforward way to study excited state photoexcitation is to produce a population of atoms in the excited state and then irradiate them with photons before they decay. This process is shown schematically in Fig. 1. Atomic sodium is used again as an example. In this case the ground state is pumped by some means, an electron or laser beam for example. Equilibrium is reached in the excited state, and then a photon beam induces transitions in the excited atom to states that have the same parity as the ground state.

Pulsed lasers with an energy of a few mJ in a band width of about 0.1 cm^{-1} can easily saturate a transition with an oscillator strength of 10^{-4}. This capability led naturally to the technique of obtaining the cross section by ionization saturation.

1. Ionization Saturation

This technique was proposed (Ambartzumian et al., 1976) to eliminate the difficult experimental problems of measuring the absolute value of the density of excited atoms. This measurement is replaced by one more tractable, an absolute measurement of the ionization flux.

The idea evolves from the concept that the probability, $P(v)$ an atom will absorb a photon of frequency v is given by:

$$P(\lambda) = \sigma(v) \cdot I(v) \cdot \tau,$$

where $\sigma(v)$ is the cross section for absorption of a photon of frequency v, in a pulse of intensity $I(v)$ photons/(cm^2-s) and of duration τ sec. Furthermore, if the population of the excited state is not changing rapidly with time, then the excition rate from the bound excited state of number density $N(v, t)$ is given approximately by:

$$\frac{d}{dt} N(\lambda t) = \frac{-NP(\lambda)}{\tau} = -N \cdot \sigma(v) \cdot I(v) \tag{3}$$

Integrating both sides of Eq. (3) we obtain

$$N(v, \tau) = N(v, o)\exp[-\sigma(v)I(v)] \tag{4}$$

and the charge, $J(v)$ collected per pulse for photoionization from the excited state is just

$$J(v) = N(v, o) \cdot \{1 - \exp[-\sigma(v)I(v)]\} \tag{5}$$

The cross section can be obtained from Eq. (5) without a knowledge of $N(o)$; just $J(v)$ and $I(v)$ must be measured.

This technique was first used by Ambartzumian et al. (1976) to obtain the photoionization cross section for the 6 $^2P_{1/2}$ and 6 $^2P_{3/2}$ states in rubidium at two photon energies (3.57 eV and 1.78 eV). Heinzmann et al. (1977) used this technique to obtain the photoionization cross section for the 7 $^2P_{3/2, 1/2}$ states in cesium. The method was further exploited by Smith et al. (1980), to obtain the absolute photoionization cross section of the 4d ^2D and 5s ^2S states of sodium. The authors of this paper recognize that the simplified idea reflected in Eq. (5) is somewhat misleading, because the excited state is assumed to be unaligned. For an isotropic system the total cross section reduces to a rather simple form, being a sum over initial states and an average over final states. The general form of the cross section is well documented (Fano, 1953 and 1957; Duncanson et al., 1976; Hanson et al., 1980; Hellemuth et al., 1981; Klar and Kleinpoppen, 1982). In general it yields cross section values that are different for different magnetic sublevels. The calculations show that to obtain correct cross section values both the spatial alignment of the intermediate state and the temporal inhomogeneity of the ionizing radiation must be taken into account. In the experiment described by Smith et al. (1980), the intermediate state was optically pumped into a well defined level using two stabilized, circularly polarized dye lasers. By choosing the appropriate polarization for the pumping and ionizing lasers, $\sigma(^2D \to \varepsilon f)$ can be obtained for an initial excited state with known quantum numbers. Uncertainties of 10–15% are claimed to be achievable by the saturation method.

2. Pump and Ionization

This method is illustrated in Fig. 1b. A schematic view of a generalized experiment is shown in Fig. 2. The pumping excitation can be provided by resonance lamp (Klucharev and Ryazanov, 1972a, Klucharev and Sepman, 1975), by laser, (McIlrath, 1969; Bradley, 1970), or by an electron beam, as shown Fig. 2. If electrons are used to pump, electrostatic plates Q are provided to sweep the ions from the atomic beam produced by an oven or gas jet. A cell may-be used too (Klucharev and Ryazanov, 1972a; Klucharev and Sepman, 1975; and Ambartzumian et al., 1976). The ionizing radiation, hv, laser, (Smith et al. 1980, Heinzmann et al., 1980), monochromatized lamps, (Nygaard et al., 1975), pulsed continuum, (McIlrath and Lucatorto, 1977), or synchrotron radiation, (Bizau et al., 1982), photoionizes the excited species (and the ground state as well in some cases). Ions or electrons are recorded by detector, D ion chamber (Nygaard et al., 1975; Ambartzumian et al., 1976), photoelectron spectrometer, (Hanson et al., 1980; Bizau et al., 1982; and Siegel et al., 1983). If a cell is used, absorption of the photon beam hv by the excited atoms may be detected using a photon spectrometer (McIlrath and Lucatorto, 1977).

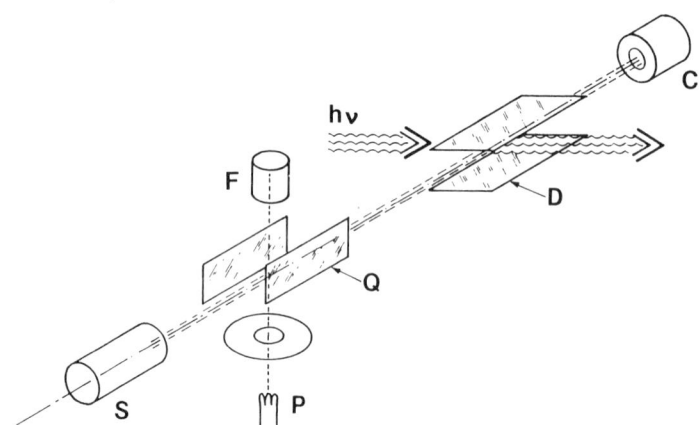

FIG. 2. Schematic representation of an apparatus to observe photoions from an excited beam. The symbols denote the essential components of the apparatus: S, atomic beam source; P, pump (electron beam in this case); F, detector to measure the pump beam intensity (a Faraday cup in this case); Q, electrodes to sweep ions from the atomic beam; hv, ionizing photon beam; D, detector for ions produced by the ionizing radiation hv (ion plates in this case); and C, detector to measure the atomic beam intensity (hot wire for example).

The spectral range available with tunable lasers restricts most of the work reported to alkalies or alkaline earths, because the excited states of these elements can be produced by lasers tuned to the resonance transitions located in the accessible spectral range of the laser. Electron bombardment (Stebbings et al., 1973; Rundel et al., 1975) has been used to produce the highest state of excitation (10–20 eV) at this time. Up to now most of the experiments have probed the photoionization continuum a few eV above the ionization threshold because tunable dye lasers and lamps are generally available only for the visible and the near UV. When an absorption cell is used, it is possible to use a pulsed flash lamp-pumped dye laser, (McIlrath and Lucatorto, 1977), ($1 MW/cm^2$ in $1 \mu s$) to provide a large transient population of excited states or of ions. This excited vapor column can then be exposed to the continuum radiated by a pulsed discharge source, (Balloffet et al., 1961), or a laser produced plasma, (Carroll et al., 1978). The absorption spectrum can be studied in the VUV using spectrometers with photographic registration or recently developed multichannel detectors (Cromer et al., 1985). A typical apparatus for the experiments is shown in Fig. 3.

The inner shell (VUV) photoabsorption spectra of excited state and singly ionized lithium and sodium have been obtained (Lucatorto and McIlrath, 1976; McIlrath and Lucatorto, 1977; and Sugar et al., 1979), as well as spectra in multiply ionized barium (Lucatorto et al., 1981 and Hill et al., 1982). Spectra for ionized magnesium, (Esteva and Mehlman, 1974), lithium, (Carroll and Kennedy, 1977), and beryllium, (Mehlman and Esteva, 1974) have been obtained using pulsed discharges or laser produced plasmas. These pulsed continuum sources of photons have enough intensity to permit the observation of the photoionization continuum and enough stability to obtain in some cases a value of the cross section.

These pulsed sources provide the means to probe excited state atoms in the VUV spectral range. Synchrotron radiation is a source of radiation that extends over a spectral range extending from x-rays to microwaves. Because of its intensity, synchrotron radiation is ideally suited for the observation of photoionization in gaseous samples. Several reviews of this work have been mentioned (Codling, 1979; Koch and Sonntag, 1979; Krause, 1980; Wuilleumier, 1980, 1981a, 1981b). The output intensity of existing monochromators is high enough and the present photoelectron spectrometers are sensitive enough so that angle resolved photoelectron spectra in gaseous samples can be obtained in the ground state (Parr et al., 1984). Instrumentation is also adequate to obtain measurements of the emission of spin polarized photoelectron by the absorption of the polarized synchrotron radiation (Heckenkamp et al., 1984). An excited state population in excess of 10^{11} atoms/cm^3 can be achieved using conventional cw dye lasers (Bizau et al., 1982).

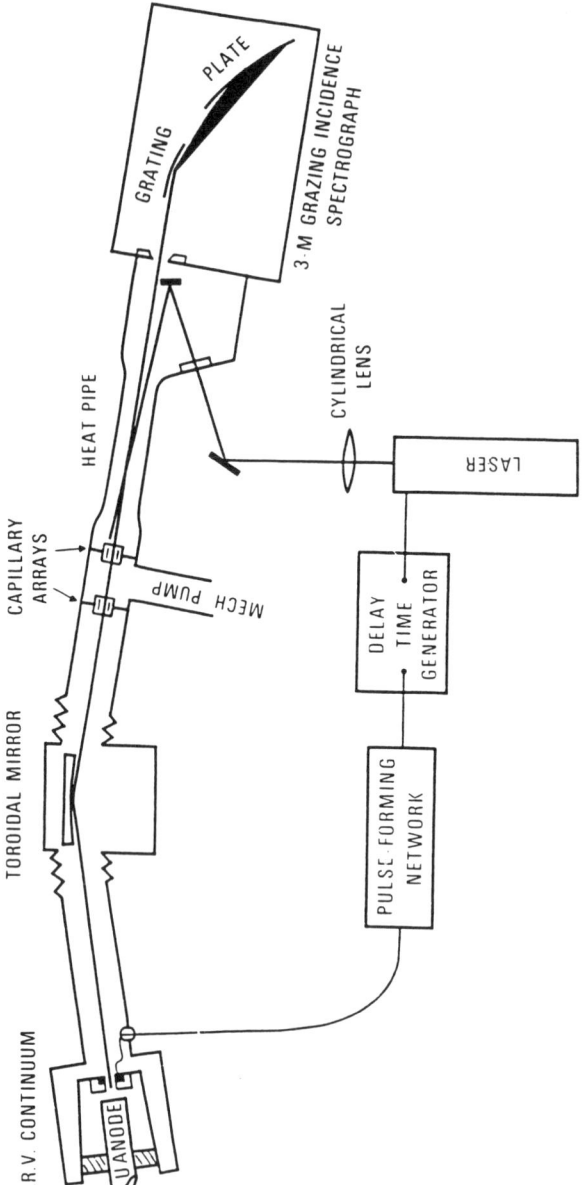

FIG. 3. Schematic of an apparatus to measure the absorption spectrum of excited atoms. A laser tuned to an atomic transition pumps the atomic vapor contained in a heat pipe oven (atomic density $\sim 10^{15}/cm^3$). Pulsed VUV continuum radiation is focused by a toroidal mirror on the entrance slit of a grazing incidence spectrograph. The spectrograph is used to record the VUV absorption spectrum of the excited atomic column (Figure taken from Lucatorto and McIlrath, 1980).

Synchrotron radiation and photoelectron spectroscopy have been combined to study core excitation and outer shell processes in excited atoms over an extremely broad energy range (Bizau et al., 1982). Experiments to study collisional processes have been developed simultaneously with the experiments to study photoionization processes. Photoelectron spectroscopy is the tool of choice when cross sections of subshell electrons are to be measured over a wide energy range, because the technique permits one to select electron associated with ionization from each subshell of interest. Figure 4 is a

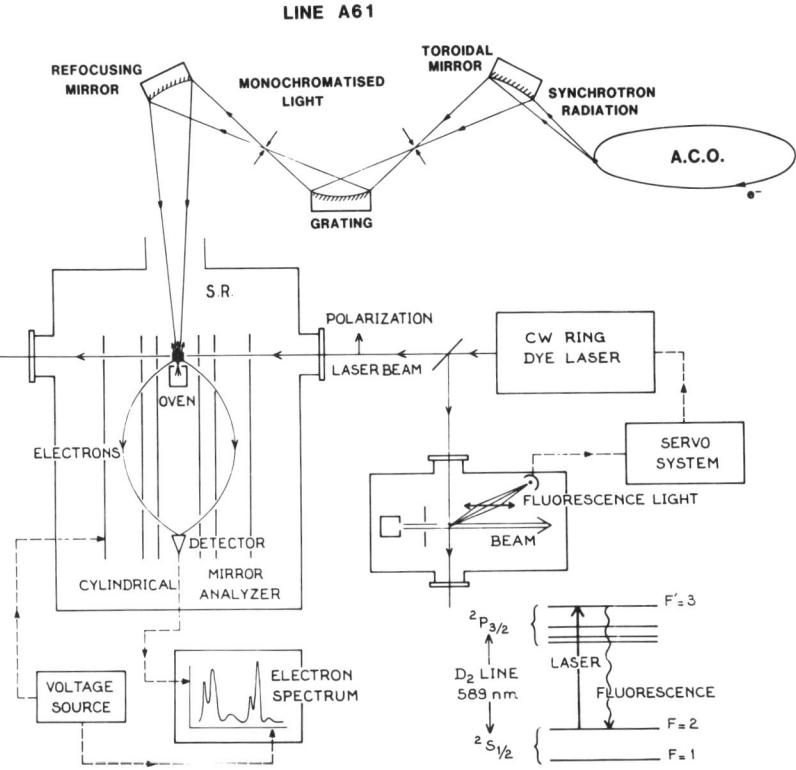

FIG. 4. Schematic diagram of apparatus to study photoionization of excited atoms by synchronotron radiation. Synchrotron radiation from the storage ring Anneau de Collision d'Orsay (ACO) at the University of Paris Sud at Orsay is monochromatized by a toroidal grating monochromator (10^{12} photons/s in a 1% band pass). A toroidal mirror focuses the monochromator output onto the interaction volume of a cylindrical mirror electron energy analyzer (CMA) with a kinetic energy resolution $\Delta E/E = 0.9\%$. A cw dye laser (power of about 10W/cm^2) is locked to the Na $3s\ ^2S_{1/2}(F=2) \rightarrow$ Na $3p\ ^2P_{3/2}(F'=3)$ transition in sodium and pumps the sodium atoms in the interaction volume of the CMA. Electrons of the correct kinetic energy traverse the CMA and are collected by the detector. The kinetic energy spectrum of the electrons is obtained by scanning the voltage between the inner and outer cylinders.

schematic diagram of the experimental apparatus used in the first experiments combining cw dye lasers with synchrotron radiation to probe the photoionization cross section and collisional cross sections of excited atoms. These were performed in Orsay using the synchrotron radiation emitted by the ACO storage ring. A number of pieces of experimental hardware must work simultaneously, the laser, the storage ring, the monochromator, the high temperature furnace, and the electron spectrometer. A toroidal grating monochromator (Larssen et al., 1982) delivers a flux of 10^{12} photons/s in a 1% bandpass for photon energies between 15 and 150 eV. This flux is focused by a toroidal mirror on the active volume of the cylindrical mirror analyzer

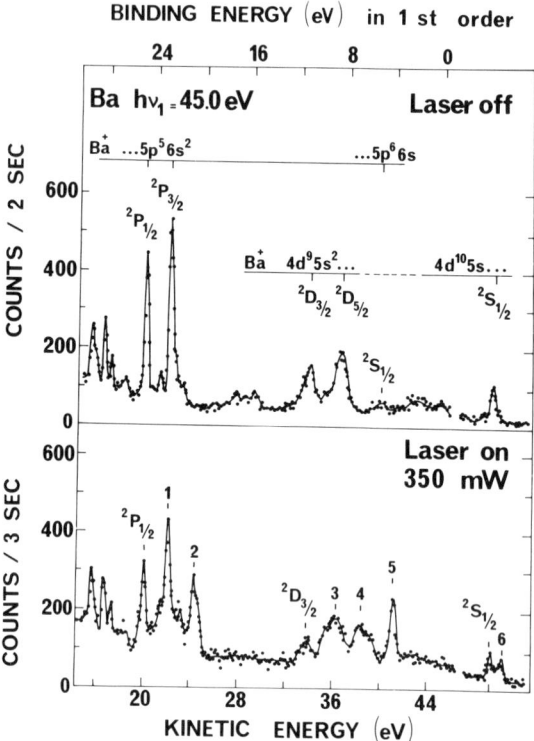

FIG. 5. Top: a photoelectron spectrum of Ba taken at $hv_1 = 45.0$ eV with the laser turned off. From left to right, the various electron peaks are due to: ionization of the 5p-shell electrons (final state $5p^56s^2$ Ba$^+$) by photons of energy hv_1; ionization of the 4d-shell electrons ($4d^95s^2$) by photons of energy equal to three times hv_1 or 135.0 eV; ionization of the 6s-electrons ($^2S_{1/2}$) by photons of energy hv_1; ionization of the 5s-shell electrons ($4d^{10}5s$) by photons of energy equal to twice hv_1 or 90.0 eV. Bottom: a spectrum taken at $hv_1 = 45.0$ eV with the laser turned on: peak 5 is due to photoionization of the 5d excited electron by photons of energy hv_1, while peaks 1–2, 3–4, and 6 are due to photoionization of the 5p, 4d, and 5s inner-shell electrons, respectively, in the laser-excited barium atoms. (Figure taken from Bizau et. al., 1986.)

(CMA) electron spectrometer (Wuilleumier et al., 1977). An oven mounted on the axis of the CMA and of the VUV beam, emits a weakly collimated beam of metal atoms (ground state density $\lesssim 10^{13}$ atoms/cm^3). The laser beam irradiates the atomic beam in a direction that is at right angles to the CMA axis. The beam waist of the laser radiation is adjusted to fit the size (about 4 mm) of the source volume of the spectrometer which constitutes the interaction region of the sample. The laser beam is produced by an argon-ion pumped ring dye laser which has an intensity of up to about ten watts/cm^2, and is linearly polarized in the horizontal plane. It is locked and stabilized to the transition between the ground state and the first excited state of the atoms under investigation by observing the fluorescence emitted from the excited atoms in an auxilliary atomic beam. Details of the use of the laser with an electron spectrometer with an atomic beam will be given in Sect. VI.B.2. The VUV beam is also partially polarized in the horizontal plane at right angles to the laser beam polarization, which is along the CMA axis.

The power of electron spectroscopy to aid the experimenter to unravel complex spectra is illustrated by Fig. 5. This figure is an example of a complex spectrum of electrons ejected from Ba atoms in the ground state (upper frame) and in an excited state (lower frame). Without going into detail, it is evident that photoelectrons ejected from the $4d$, $5s$, $5p$, $6s$, and $5d$ subshells can be differentiated only by measuring their different kinetic energies. Ion detection would provide only the total ionization yield, information similar to that obtained in a photoabsorption experiment. We now turn our attention to the theoretical background which provides the basis for understanding the experiments which have yielded cross sections for the photoionization from atoms in an excited state.

III. Theoretical Background

For an axially symmetric system (i.e. one in which the linear polarization vectors of the photon beams lie in the same plane) with randomly oriented atoms or molecules in the initial state, the angular dependence of the two photon photoionization cross section is given by Arnous et al. (1973), Duncanson et al. (1976), Hanson et al. (1980), Hellemuth et al. (1981), and Klar and Kleinpoppen (1982). This cross section has the form:

$$\frac{d\sigma}{d\Omega} = A_0(t) + A_2(t)P_2[\cos(\theta)] + A_4(t)P_4[\cos(\theta)]. \tag{5}$$

The quantity, θ, is the angle between the polarization direction of the photon beams and the direction of the ejected electron. The functions $P_n[\cos(\theta)]$ are the Legendre polynomials of order n and the quantities $A_n(t)$ are coefficients which are products of appropriate photoexcitation cross sections, Clebsh-Gordon coefficients and a time dependent factor that accounts for the mixing and beating of the intermediate and/or final states. Legendre polynomials of fourth order appear because anisotropy is transferred by the radiation to the atom or molecule in two steps. In general, if the intermediate state is a superposition of hyperfine levels which occurs if the bandwidth of the light producing the intermediate state is broader than the hyperfine levels, the coefficients $A_n(t)$ are time dependent. For cw laser excitation, generally only one hyperfine level is populated and the quantum beats disappear. The resulting alignment in the intermediate state is only modified by collisions. For one photon ionization from a nonaligned intermediate state or an optically pumped intermediate state, the result can be expressed in terms of the usual dipole allowed photoionization cross section (Yang, 1948; Cooper and Zare, 1969):

$$\frac{d\sigma}{d\Omega} = \frac{\sigma}{4\pi}(1 + \tfrac{1}{2}\beta\{P_2[\cos(\theta)]\}). \qquad (7)$$

For most of the cases described in the following sections, the total ion current or the absorption of the photon beam was measured. There are only a few cases in two-step photoionization where the angular distribution of the ejected photoelectrons has been measured (Hanson et al., 1980; Hellemuth et al., 1981; Siegel et al., 1983).

The general techniques for photoionization calculations have been mentioned in Sect. I (Seaton, 1966; Amusia et al., 1971; Johnson et al., 1980). Specific examples of calculations for excited state photoionization are dispersed throughout the text. We mention several in this section for completeness. Salzmann and Pratt (1984) have studied the sensitivity to variations in screening in the photoionization of inner shell electrons from excited atoms. Chang and Kim (1982a) made such a study for sodium, and Wendin (1985) studied barium. These two calculations will be discussed in more detail in Sect. V.C. and in Sect. V.E.1.

IV. Photoionization of an Outer Electron in Excited Atoms

A. THE RARE GASES

1. Helium

As a first example of photoionization from an excited state we wish to describe one of the first cross section measurements of photoionization from an excited state by Stebbings et al., (1973) and Rundel et al., (1975). An atomic beam of helium $2\,^1S$ and $2\,^3S$ was produced by electron impact. The $2\,^1S$ state could be quenched by radiation from a helium discharge lamp and the resulting beam of $2\,^1S$ or $2\,^1S$ plus $2\,^3S$ atoms was photoionized by a frequency doubled pulsed dye laser.

The photoionization cross section for the $2\,^1S$ and the $2\,^3S$ states of helium is shown in Fig. 6. The closed circles are the experimental results for the $2\,^1S$ state; the open circles are the results for the $2\,^3S$ state. The solid line, the

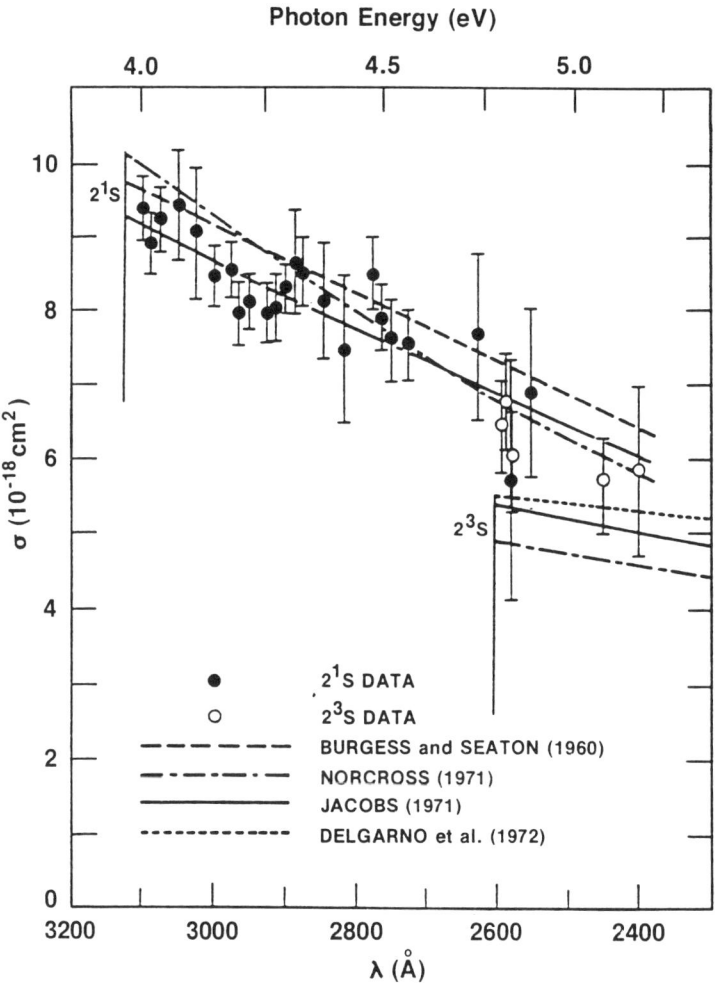

FIG. 6. The photoionization cross section of He $1s2s\,^1S$ and He $1s2s\,^3S$ excited states as a function of wavelength. The data points are the measurements from Stebbings et al. (1973). Several calculations are identified in the figure. (Figure taken from Hurst et al., 1979.)

dash-dot line, and the dashed line are calculations by Jacobs (1971), Norcross (1971), and Burgess and Seaton (1960) respectively. The calculation of Dalgarno et al. (1972), for the $2\,^3S$ cross section is represented by the short dashed line. A more recent calculation by Aymar and Crance (1980), is not shown but is in good agreement with the experiments and the other calculations. Each calculation was different: Jacobs (1971) used a close-coupling formalism, Norcross (1971) used a frozen core Hartree–Fock approximation, Burgess and Seaton (1960) used a quantum defect approach, and Aymar and Crance (1980) used a central field model potential. The different calculations yield essentially the same cross sections for $2\,^3S$, $2\,^1S$, and the $2\,^3P$ and $2\,^1P$ partial cross sections.

Dunning and Stebbings (1974b) continued this work and measured the photoionization of helium atoms in the $n\,^{1,\,3}P$ states at one wavelength. These results are in good agreement with calculated cross sections using the quantum defect method (Burgess and Seaton, 1960) and the close coupling approach (Jacobs, 1971).

One can conclude that there is nothing mysterious in these two electron systems and that the experiment substantiates the calculations for this system.

2. Neon

There have been two recent experiments in neon by Ganz et al., (1982, 1983) and by Siegel et al. (1983), that involve excitation of an excited electron either to autoionizing states between the $2p^5\,^2P_{3/2}$ and $2p^5\,^2P_{1/2}$ ion states or to the continuum near the ionization thresholds. In the first experiment (Ganz et al., 1982, 1983) the autoionizing resonances could not be reached by single photon transitions from the ground state but metastable neon atoms in the $2p^5\,3p\,^2P_2$ state were laser-excited to the $2p^5\,3p\,^1D_3$ state and another tunable cw laser was used to excite electrons to autoionizing levels classified as $2p^5\,^2P_{1/2}\,14s'$, $12d'$. In earlier VUV ionization measurements (Radler and Berkowitz, 1979), these resonances were observed directly but the resonance widths were overestimated because the bandwidth of the spectrometer was 100 times broader than the resonance widths, predicted (Johnson and Le Dourneuf, 1980) to be 3.8 GHz for the $2p^5(^2P_{1/2})\,12d'$ state and 10.5 GHz for the $2p^5(^2P_{1/2})\,14s'$ state. Using a tunable dye laser of 4 GHz bandwidth to study the profile of these autoionizing resonances, Ganz et al. (1983), deconvoluted their measurements and obtained a resonance width of 6.0(5) GHz for the $2p^5\,^2P_{1/2}\,14s'$ resonance and 2.0(5) GHz for the $2p^5\,^2P_{1/2}\,12d'$ resonance in good agreement with the widths predicted by Johnson and Le Dourneuf (1980). The high resolution profiles of these

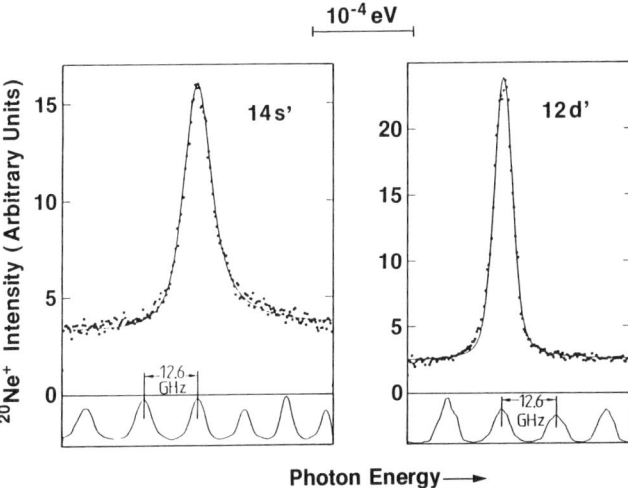

FIG. 7. High resolution measurements of the two lowest autoionizing resonances $Ne2p^5 14s'$, $J = 1$ and $Ne2p^5 12d'$, $J = 3$ in the photoionization of laser-excited $Ne(2p^5 3p\ ^1D_2)$ atoms with a narrow-band (FWHM about 4GHz) cw dye laser. The $12d'$ resonance contains small contributions from $J = 2$ levels. The smooth lines through the data points are fitted Beutler-Fano profiles (Fano (1961)) with widths of $r(14s') = 6$ GHz and $r(12d') = 2$ GHz and shape parameters $q(14s') = 15$ (uncertainty ± 5) and $q(12d') = 100$ (uncertainty ± 30). The calculated profiles have been convoluted with Gaussians chosen to approximately represent the effective laser bandwidth. The lower parts of the figures show frequency markers from a solid etalon (free spectral range 12.6 GHz, finese about 15) recorded simultaneously with the spectra. The laser intensity was essentially constant over the scan range. (Figure taken from Ganz et al., 1983.)

resonances are shown in Fig. 7. In neon the widths of the d' resonance are smaller than the s' resonances, which is just the opposite of that observed (Beutler, 1935; Fano, 1935), for similar resonances in the other rare gases.

Neon is a small atom; its core is about half the radius of argon, krypton, or xenon. Most of the electrostatic interaction occurs within the core and the d wavefunction amplitude increases slowly with increasing radial distance. The overlap of the wavefunction of the excited d electron with the core is small in neon, but much larger for these other rare gases. Therefore, the interaction of the d electron will be much stronger for the heavier rare gases. On the other hand the s electron penetrates the ion core and has an approximately constant quantuum defect as Z increases. One would expect a dramatic change in the width of the d' resonance as the size of the atom increases which changes the effective potential and interaction experienced by the d electron. This phenomenon was also observed in the isoelectronic sequence Xe, Cs^+, and Ba^{++}, where the s' autoionizing resonance changed little in width but the d' autoionizing resonances decreased dramatically as the nuclear charge increased from Xe to Ba^{++}. This increase caused the radius of the atomic core to decrease and thus the overlap of the nonpenetrating d' wavefunction with it.

The measurements of the width of the autoionizing resonances were further refined (Ganz et al., 1984), by using a dye laser stabilized to 50 MHz. The autoionizing resonance $2p^5(^2P_{1/2})14s'(\frac{1}{2}) J = 1$ was populated from several intermediate states: $2p^5(^2P_{3/2})3p\ J = 1, 2$. The profile index, Fano (1961), was found to depend on the intermediate state changing from $q = +29$ for the $2p_5\ ^1P_1$ (Paschen notation) intermediate state to $q = -13$ for the $2p_8\ ^3D_2$ intermediate state. The value obtained for the autoionizing resonance width, Γ, was 4.5 GHz in good agreement with the earlier measurement of Ganz et al. (1983). A MQDT calculation is in semiquantitative agreement with the observations. The autoionizing level is expected to have a width independent of the intermediate state because Γ connects only the final resonance level with the continuum levels through the Coulomb operator.

In the second experiment (Siegel et al., 1983) metastable Ne $2p^5(^2P_{3/2})3s\ ^3P_2$ neon atoms were laser-excited to the Ne $2p^5(^2P_{3/2})3p\ ^3D_3$ level. This excited state was ionized by the uv beam of an argon-ion laser to continuum states between the Ne $2p^5\ ^2P_{3/2}, P_{1/2}$ ionization threshold and 0.7 eV above it. Photoelectron spectroscopy was used to separate the $^2P_{3/2}$ and $^2P_{1/2}$ ionization channels. Ganz et al. (1983) found an upper limit of 10^{-3}, for the cross section ratio $\sigma(2p^5\ ^2P_{1/2})/\sigma(2p^5\ ^2P_{3/2})$. This low value of the branching ratio suggests that core switching transitions are negligible for transitions from the Ne $2p^5(^2P_{3/2})3p\ ^3D_3$ excited state. In the analysis presented by Siegel et al. (1983), the photoionization of the excited state was carried out in jj coupling, a fixed core angular momentum, and no spin-orbit interaction in the continuum. Because the pumping laser induces an alignment of the intermediate state, the correct form of the angular distribution of the photoelectron is given by Eq. 6, where the coefficients A_n do not have a time dependence since equilibrium was established by the cw laser that was used to pump the intermediate state and photoionize it. The partial cross sections $\sigma(3p \to \varepsilon s)$, $\sigma(3p \to \varepsilon d)$ and the phase difference between the d and s continuum state $\delta = \delta d - \delta s$ could be derived from the coefficients $A_n(n = 0, 2, 4)$ and the geometrical factors. These coefficients were determined by measuring the angular distributions for judiciously chosen combinations of the polarization of the pump laser relative to the ionization laser.

The measured phase difference between the d and s continuum wave functions is shown in the upper frame of Fig. 8 as ○. These can be compared with the calculations by Chang (1982), (shown as ∆) and the phase difference calculated from the quantum defects, Siegel et al. (1983), (shown as —). Chang's many body formalism demonstrated that correlations play a minor role and modify the cross section by less than 15%. Siegel et al. (1983) suggest

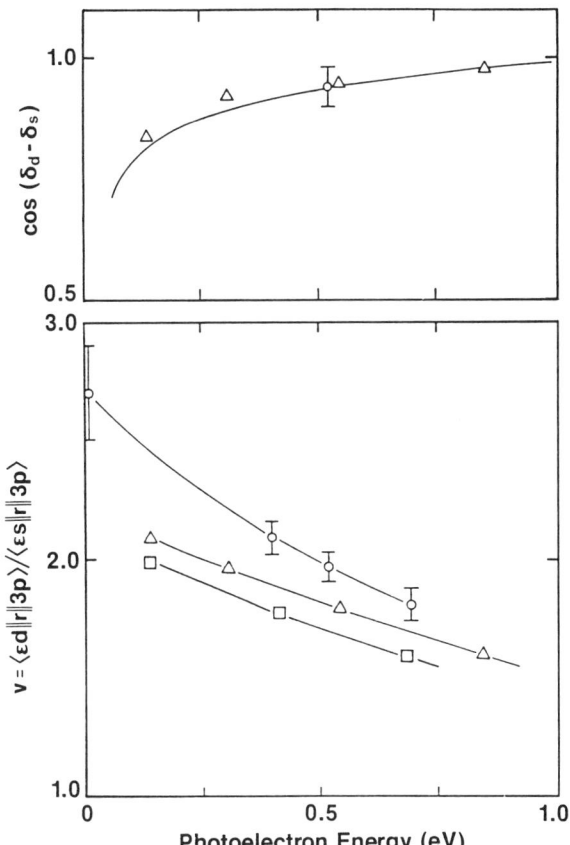

FIG. 8. Top: Comparison of experimental and theoretical results for $\cos(\delta_d-\delta_s)$: ◯, experimental values from Siegel et al. (1983); △ calculation by Chang (1982)—calculated from quantum defects, Siegel et al. (1983). Bottom: comparison of the experimental values of v with theoretical results: ◯, experimental values, are from Siegel et al. (1983); △, calculation of Chang (1982); □, calculation of Luke (1982). The quantity v is a ratio of radial matrix elements. (Figure taken from Siegel et al., 1983.)

that photoionization from excited neon atoms may be treated (to an accuracy of 10%) as a single particle problem.

The results for the ratio, $v, = \langle\varepsilon d|r|3p\rangle/\langle\varepsilon s|r|3p\rangle$, of the radial matrix elements between the $3p$ initial state and εd final state and the $3p$ initial state and the εs final are shown in the lower frame of Fig. 8. The results of the experiment (shown as ◯) are compared with a many body calculation by Chang (1982), (shown as △) and a multiconfiguration close coupling calculation by Luke (1982), (shown as □). The theoretical calculations are about 10% lower than the observations.

3. Argon, Krypton, and Xenon

The work on photoionization from metastable rare gas atoms was continued by Dunning and Stebbings (1974a) in argon and krypton and by Rundel et al. (1975) in xenon. In these atoms the authors focused their attention on the autoionizing resonances that decay to the $np^5\,{}^2P_{1/2}$ continuum. In argon and krypton transitions from metastable $np^5(^2P_{1/2})$ $(n+1)s'\,3p_{2,0}$ to final states $np^5(^2P_{1/2})n'p'[\frac{3}{2}]J=1$ and $np^5(^2P_{1/2})n'f'[\frac{5}{2}]J=3$ were observed. The value for n is 3 for argon and 4 for krypton. The primary focus of this work was to study the spectroscopy of these transitions and the configuration interaction among the Rydberg series converging to the $np^5(^2P_{1/2})$ series limit. Perturbations in both the p' series and f' series were observed, but the level perturbations were much smaller in the case of the f' series. The widths of the autoionizing resonances were comparable to the laser bandwidth. The widths of the p' series resonance were qualitatively larger than those of the f' series as expected from lifetime arguments.

Two photons from a pulsed tunable ArF laser were used by Bokor et al. (1980) to populate the $4p^5(^2P_{3/2})6p$ levels in krypton. The photoionization cross section at $\lambda = 1935$ Å for the $4p^5(^2P_{3/2})6p[\frac{3}{2}]J=2$ was determined to be $3.2(2.0)\times 10^{-19}$ cm^2. This value can be compared to the calculated value of 1.8×10^{-19} cm^{-2}, which was based on a simple quantum defect approximation (Burgess and Seaton, 1960). For these measurements a gas cell was used. The cross section was determined from the laser intensity, the radiative decay rate, the intensity of fluorescence from the excited state, and the collisional self-quenching rate.

The resonances studied in xenon were due to transitions from $5p^5(^2P_{3/2})6s[\frac{3}{2}]J=2$ and $5p^5(^2P_{3/2})6s'[\frac{1}{2}]J=0$ metastable states to autoionizing states:

$$5p^5(^2P_{1/2})n'p'[\tfrac{3}{2}]J=1,2; \quad 5p^5(^2P_{1/2})n'p'[\tfrac{1}{2}]J=1$$

$$5p^5(^2P_{1/2})n'f'[\tfrac{7}{2}]J=3; \quad 5p^5(^2P_{1/2})n'f'[\tfrac{5}{2}]J=2,3.$$

These resonances were broad enough so that the parameters (Fano, 1961) that characterize the resonance shape (q), width (Γ), and energy position (E_r), could be determined. Figure 9 is an example of the resonance profile of two resonances due to transitions $5p^5\,6s\,{}^3P_0 \to 5p^5\,{}^2P_{1/2}\,8p' \leftarrow (\tfrac{3}{2})J=1$ and $5p^5\,6s\,{}^3P_0 \to 5p^5(^2P_{1/2})8p'[\tfrac{1}{2}]J=1$. These profiles were fitted to a Fano profile which describes the interaction of discrete states with one or more ionization continua. The oscillator strength, f, for the transitions studied were in the range $10^{-5} \lesssim f \lesssim 10^{-3}$.

Spectroscopic measurements for high-lying Rydberg and autoionizing levels in xenon was continued by Grandin and Husson (1981). These authors

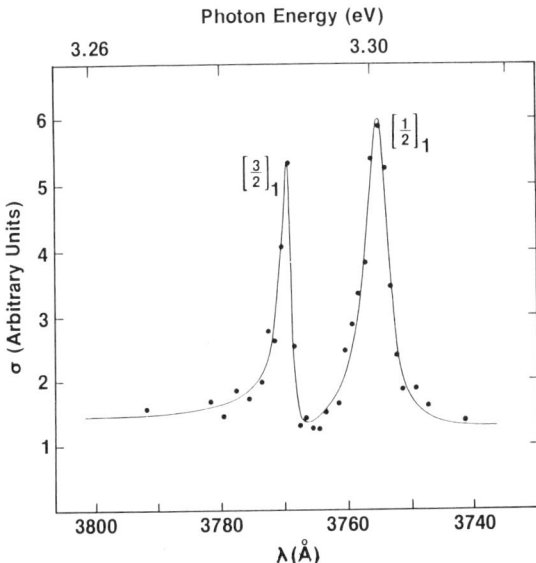

FIG. 9. Measurements of the relative photoionization cross section of autoionizing resonances. Two overlapping resonances are shown due to the transitions $Xe5p^56s\ ^3P_0 \to Xe5p^5\ ^2P_{1/2}8p'[3/2]_1$ and $Xe5p^56s\ ^3P_0 \to Xe5p^5\ ^2P_{1/2}\ 8p'\ [1/2]_1$. Solid dots show the experimental data, the solid line shows the fitted Beutler Fano profile (Fano, 1961). (Figure taken from Rundel et al., 1975.)

used the optogalvanic detection method to study transition to levels $5p^5(^2P_{1/2})n'p'[\frac{1}{2}]J = 1$. Very little perturbation was observed in the $n'f(\frac{3}{2})J = 1$ series because each $nf[\frac{3}{2}]J = 1$ level is situated approximately equidistant from adjoining $p[\frac{1}{2}]J = 1$ levels and the interaction of these levels is approximately equal and opposite.

B. PHOTOIONIZATION OF EXCITED ALKALI ATOMS

1. Lithium

The only photoionization cross section measurements of the $2p$ excited electron of lithium these authors are aware of have been obtained by Roth (1971). Using the recombination radiation method (Sect. II.A.), he obtained a threshold value of 19.7 ± 3.0 Mb (1 Mb $= 10^{-18}$ cm^2) for photoionization of the $2p$ electron assuming an oscillator strength of 0.00417 for the $5s\ ^2S_{1/2}$–$2p\ ^2P_{1/2}$ transition. If we use the more recent oscillator strength value from the computation of Martin and Wiese (1975), of 0.0044 for this transition, Roth's threshold value of the cross section should be adjusted to 20.8 ± 3.0 Mb.

TABLE I

A Comparison of Lithium 2p Excited
State Cross Section:

$$\sigma_{2p}(\lambda) = \sigma_{Th}\left(\frac{\lambda}{\lambda_{Th}}\right)^{\delta}$$

Method	σ_{Th} (10^{-18} cm^2)	Slope δ
Quantum Defect[a]	15.7	2.9
Quantum Defect[b]	16.7	3.3
Self Consistent Field[c]	15.0	3.0
Model Potential[d]	15.2	
Experiment[e]	20.8(3.0)	1.8

[a] Moskvin (1963)
[b] Ya'akobi (1956)
[c] Gezalov and Ivanova (1958)
[d] Aymar et al. (1976)
[e] Roth (1971)

A comparison of the theory with the experiment is shown in Table I. Roth's measurements are consistantly higher than theory throughout the measurement interval (3.4 eV–4.3 eV) and the cross section is proportional to the square of the wavelength. Theory predicts a more hydrogenic behavior with a cross section more nearly proportional to the cube of the wavelength.

Roth carefully calibrated his equipment and analyzed his data to eliminate systematic errors, but neither the threshold value nor the slope are in good agreement with a number of calculations.

2. Sodium

Sodium has been a showcase for demonstrating the techniques that can be used to study excited state ionization. Photoionization from the $3p\,^2P$ excited state in sodium has been observed (Roth, 1969), by studying the recombination radiation emitted by a sodium seeded plasma, as outlined in Sect. II.A. The cross section has also been measured using the pump and ionize techniques described in Sect. II.B. In the latter case, the ions were detected in the experiment of Duong et al. (1978), and the photoelectrons were detected in the experiment of Hanson et al. (1980). The saturation method, (Ambartzumian et al., 1976), described in Sect. II.B.1., was also utilized by Smith et al. (1980) to study photoionization from the $4d\,^2D$ and the $5s\,^2S$ excited state.

Several calculations of photoionization from the sodium $3p$ state have been performed. Aymar et al. (1976) calculated the photoionization cross section for sodium in np states ($n = 3 - 8$) using a single particle central field approximation with a parametric potential. The optimum potential is determined by minimizing the total energy of the lowest state of a chosen symmetry or by minimizing the root mean square deviation between observed and calculated energies of selected levels. These authors predicted a Cooper minimum (Cooper, 1962) for the $3p \to \varepsilon d$ cross section channel to occur at photon energy of 8.84 eV. The $3p \to \varepsilon s$ cross section channel decreases monotonically. Laughlin (1978) also made a calculation to determine one and two photon ionization of sodium using a parameterized potential. Chang (1974) included core polarization in his many body calculation and found good agreement with the parameterized potential calculations (Aymar et al., 1975; Laughlin, 1978) for photoionization of the outer $3s$ electron. Core polarization would be expected to play an important role in the behavior of the cross section because of the penetrating nature of the $3s$ orbital and the strong cancellation in the radial matrix element. This behavior will prevail to some extent for the photoionization of the $3p$ electron. Two calculations based on the quantum defect method, (Burgess and Seaton, 1960, and Moskvin, 1963), are available for photoionization of the $3p$ excited state and another based on the self consistent field approximation (Rudkjobing, 1940) can also be compared with the experiments.

Figure 10 is a summary of the photoionization cross section of the $3p$ electron in sodium. The results of three experiments and four calculations are shown. The total cross section measurements obtained from recombination radiation measurements are shown as +. The best analytical fit to the experimental data is shown ($\sigma_{3p}(hv) = 2.13 \times 10^{-16}(hv)^{-3}$ cm^2) where the photon energy hv is expressed in eV. The results of Duong et al. (1978) for the $3p \to \varepsilon s$ and $3p \to \varepsilon d$ channels are shown as ■ at selected photoenergies for clarity. In this elegant experiment, Duong et al. used circularly polarized radiation from a dye laser to optically pump the $3\,^2P_{3/2}$ state to $M_L = +1$ and $M_F = +3$. The two senses of circularly polarized ionizing radiation from another dye laser produced transitions to continuum states with $M_L = 0$ or $M_L = 2$. Only εd continuum states can have $M_L = 2$, so the signal with one degree of polarization is proportional to $\sigma(3p \to \varepsilon d)$. The signal observed with the other degree of polarization ($M_L = 0$ continuum states populated) is proportional to a linear combination of $\sigma(3p \to \varepsilon s)$ and $\sigma(3p \to \varepsilon d)$. The wavelength range of the dye lasers limited these measurements to $hv < 3.5$ eV. Recently $3p$ photoionization cross section measurements have been remeasured by Presses et al., (1985) to a photon energy of about 5 eV using a dye laser to pump the $^2P_{3/2}$ level and monochromatized synchrotron radiation to ionize the $3p$ level. These results were in good agreement with those obtained by Duong et al. (1978).

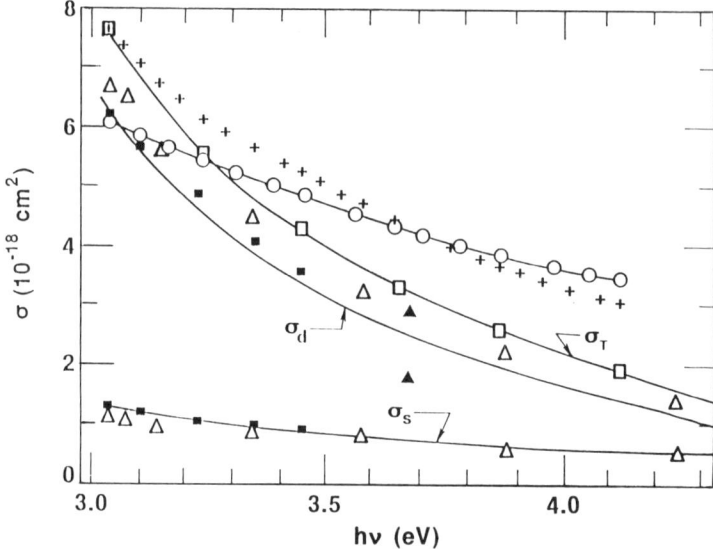

FIG. 10. Photoionization cross section of the excited $3p$ electron in atomic sodium as a function of the photon energy $h\nu$. Three experiments are shown: $+$, Roth (1969); ■, Duong et al. (1978); ▲, Hanson et al. (1980), and four theoretical calculations are shown: solid line, Aymar et al. (1976); ○, Moskvin (1963); △, Laughlin (1978); and □, Rudkjobing (1940). The symbols σ_T, σ_d, and σ_s represent the total cross section, the partial cross section due to final states of d symmetry and the partial cross section due to final states of s symmetry, respectively.

The cross section from the $3p\,^2P_{3/2}$ state has been determined by Hanson et al. (1980) at one photon energy ($h\nu = 3.68$ eV) using a linearly polarized pulsed dye laser to populate the $3p\,^2P_{3/2}$ level and a linearly polarized nitrogen laser to photoionize that level. The angular distribution of the photoelectron was measured. Since the excitation was time-dependent, the coefficients in Eq. 6 were also time dependent. A considerable amount of analysis was necessary before a cross section could be extracted. The cross section derived from this analysis for the $3p \to \varepsilon d$ and $3p \to \varepsilon s$ channels is shown as ▲ in Fig. 10.

The calculations of Aymar et al. (1976) and Laughlin (1978) are shown as a solid line and △ respectively for the $3p \to \varepsilon s$, the $3p \to \varepsilon d$ partial cross section. These authors used a parameterized model potential and have obtained essentially the same results.

The total cross section calculation of Rudkjobing (1940), shown as □, using a self-consistent field with exchange, is almost in exact agreement with that of Aymar et al. The calculation of Moskvin (1963), who used a quantum defect method, is shown as —○— in Fig. 10. This calculation is in good agreement with other calculations at threshold but has a different functional form at photon energies greater than threshold.

The measurements of Duong et al. are in particularly good agreement with the calculation of Aymar et al. and with Laughlin. The total cross section measurements of Roth (1969) are in accord with all the calculated total cross sections (except Moskvin) at the threshold but are about 50% higher than all the calculations (except Moskvin) at $hv = \sim 4.2$ eV.

The value of $\sigma(3p \to \varepsilon d)$ obtained by Hanson et al. (1980) is in good agreement with the calculations of Aymar et al. and Laughlin, but their value for $\sigma(3p \to \varepsilon s)$ is a factor of two higher than the cross section obtained by Laughlin (1978) and by Aymar et al. (1976).

The results presented in Fig. 10 illustrate a number of points. The first is that the model potential used in several calculations gives an apparently reliable representation of the excited $3p$ photoionization cross section. Secondly the recombination radiation method produces good agreement with theory at threshold but yield results considerably higher about 1 eV above threshold. The current use of cw dye lasers, while providing excellent agreement with several calculations, can probe a limited energy range above the ionization threshold. The use of pulsed lasers is less desirable because of the time dependence they introduce in the data. Synchrotron radiation provides the broad photon energy range of excitation to study the dynamics of the photoionization process over a large enough energy range to include such effects such as inter shell coupling and the Cooper minimum.

The use of circularly polarized radiation by Duong et al. (1978) to probe the εs and the εd channels suggests the use of circularly polarized synchrotron radiation to photoionize the laser aligned atom over an energy range much broader than that of present day lasers.

Photoionization from states other than the $3p$ state has been observed. Smith et al. (1980), obtained values of the photoionization cross section from the sodium $4d\,^2D$ and $5s\,^2S$ states using the saturation method and taking the alignment of the states into account. At a photon energy of 1.17 eV, their result of $15.2(1.7) \times 10^{-18}$ cm^2 for the cross section $4d\,^2D \to \varepsilon f$ can be compared with the calculation of Aymar (1976), who obtained a value of 14.9×10^{-18} cm^2 for this cross section. Smith et al. (1980) obtained a value of $1.49 \pm 0.13 \times 10^{-18}$ cm^2 for the cross section $5s\,^2S \to \varepsilon p$, which can be compared with 1.40×10^{-18} cm^2 obtained by these authors from a quantum defect calculation.

The total photoionization cross section for np and ns states show the expected monotonic decrease with increasing photon energy. Msezane and Manson (1984a) calculated the cross sections for the sodium $3d$ electrons within the framework of the Hartree-Fock approximation. In this approximation, the partial cross section for the $3d \to \varepsilon p$ channel is zero at a photon energy above the ionization threshold. However, the total cross section calculated by Msezane and Manson shows no effect of the zero in the $d \to \varepsilon p$ channel because the total cross section is dominated by the $d \to \varepsilon f$ channel. Measurements similar to those by Duong et al. (1978), or by Hanson et al. (1980), are necessary to sort out the behavior of the different channels.

The results presented on sodium show an emerging picture of the near threshold dynamics of the excited state photoionization process which suggests that the cross section is dominated by single particle processes. Experimentally photoionization cross sections have been measured over a limited energy range and the angular asymmetry parameter has been obtained for only a few discrete photon energies and not enough information exists to probe the details of the models used for the calculation.

3. Potassium

The only measurements of the potassium $4p$ photoionization cross section that have been reported (Nygaard et al. 1978) did not use lasers. The $4p\,^2P_{3/2,1/2}$ levels of the potassium atomic beam were pumped by a resonance lamp and the ionization was produced by a mercury-xenon lamp which was monochromatized by a small monochromator. The results of these measurements are compared with two calculations in Fig. 11. The agreement between the experiment and the theory is good at threshold. However, the measurements (shown as dots with error bars in Fig. 11) suggest that the cross section decreases like λ^4. The cross section based on quantum defect calculation, Moskvin (1963), (shown as +) reach a maximum near threshold and decrease less rapidly than the experiment. More recent calculations, by Aymar et al. (1976), (shown as □) based on a parameterized model potential have the same shape as the quantum defect cross section but are 20% to 30% higher. This seems to be a systematic effect since one could make the same observation for similar calculations of the sodium $3p$ photoionization cross section. The results presented in Fig. 11 do not present a definitive case for any one method but seem to cry out for new measurements and raise questions about calculations based on the quantum defect method that should be in good agreement with the model potential calculation, since this is the energy range where a hydrogenic approximation is expected to be valid. Hartree–Fock calculations of the photoionization of the $3d$ electron in sodium and potassium by Msezane and Manson (1984a) indicate that, while the cross section is dominated by the $3d \to \varepsilon f$ ionization channel, the angular asymmetry parameter is very sensitive to the zeros in the $3d \to \varepsilon p$ channel. As yet, no measurements of the angular asymmetry parameter have been made to test this assertion.

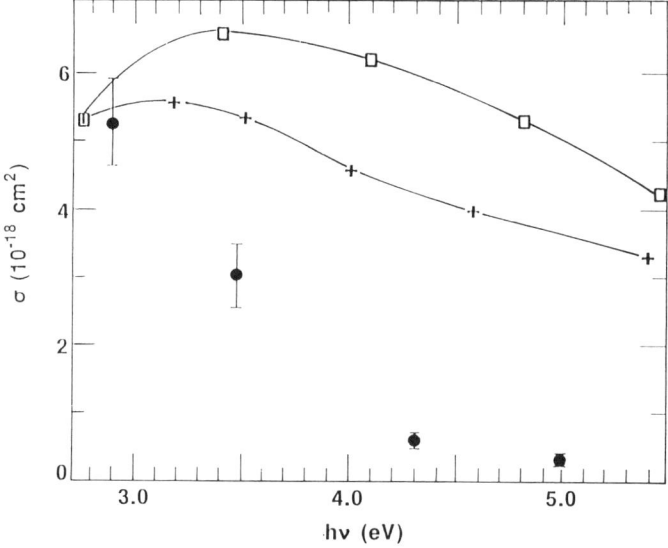

FIG. 11. Photoionization cross section of the excited $4p$ electron in potassium as a function of the photon energy $h\nu$. The experimental data from Nygaard et al. (1978) are shown as points with error bars; two calculations are shown: □, Aymar et al. (1976) and —, Moskvin (1968).

4. Rubidium

Photoionization from the $5p\,^2P$ and from the $6p\,^2P$ excited states of rubidium has been measured. The results are shown in Table II along with calculations based on the quantum defect method. The $5p\,^2P$ state was populated by a resonance lamp and ionized with 4400Å radiation from an argon ion laser. The ions were collected in an ion chamber. The saturation technique (Sect. II.B.1) was used to obtain the cross section at 3471Å and at 6943Å. A ruby laser delivering 0.5J of energy pumped a dye laser which in turn was used to pump the $6p\,^2P$ state of rubidium. The experimental results are in agreement with calculations on the $10\% - 30\%$ level.

5. Cesium

Cesium has received a great deal of attention because it is important in a variety of laboratory (Nygaard et al., 1975) and astrophysical processes (Norcross and Stone, 1966) such as isotope separation and stellar opacity. From a theoretical point of view, cesium is an excellent test (Msezane and Manson, 1984b) case for photoionization because the spin-orbit effect is large and because the core polarization is more important in cesium than in the

TABLE II

Excited State Photoionization of Rubidium

Initial state	σ (10^{-18} cm^2)	λ (Å)	Method
$5p\,^2P$	9.6 ± 0.2	4400	Experiment[a]
	10.5		Quantum defect[b]
	11.5		Quantum defect[c]
$6p\,^2P_{3/2}$	17 ± 4	5943	Experiment[d]
$6p\,^2P_{1/2}$	15 ± 4		Experiment[d]
$6p\,^2P$	10		Quantum defect[d]
$6p\,^2P_{3/2}$	1.0 ± 0.5	3471	Experiment[d]
$6p\,^2P$	2.9		Quantum defect[d]

[a] Klyucharev and Sepman (1975).
[b] Shevelko (1970).
[c] Moskvin (1963).
[d] Ambartzumian et al. (1976).

lighter alkali metals. Cross sections have been obtained for photoionization from the $6p\,^2P_{1/2}$, $5d\,^2D_3$, $6d\,^2D_{3/2}$, $7p\,^2P_{3/2}$, and $7s\,^2S_{1/2}$ states.

Experimental results for the photoionization cross section from the $6p\,^2P$ state have been obtained from early measurements by Mohler and Boecker (1929), Mohler (1929), Mohler (1933), and Agnew and Summers (1965) using radiative recombination (Sect. II.A). A comprehensive set of measurements using crossed beams and non-laser sources was carried out more recently by Nygaard et al. (1975). A cross section measurement was made by Klucharev and Dobrolege (1973) at a single wavelength 4880Å. This point is a refinement of an earlier measurement by Klucharev and Rayazanov (1972a). Two model potentials have been used to calculate (Weisheit, 1972, and Norcross, 1973) the cross section. These calculations are an improvement to the quantum defect method (QDM) developed by Burgess and Seaton (1960) and by Moskvin (1963) and an adjusted form of the QDM Norcross and Stone (1966). The agreement between theory and experiment is good at threshold but the most recent calculations are systematically high at short wavelengths. This was also the case for the other alkalies except sodium, where at least one experiment was in good agreement with a model potential calculation (Aymar et al. 1976) over the rather limited measurement range (see Fig. 10).

Photoionization from the $6d\,^2D$ excited state of cesium is particularly interesting because zeros in the $nd \to \varepsilon f$ dipole transition matrix elements have been predicted by Msezane and Manson (1975). In Fig. 12, where a Hartree Fock (HF) calculation of the cross section is compared with a central

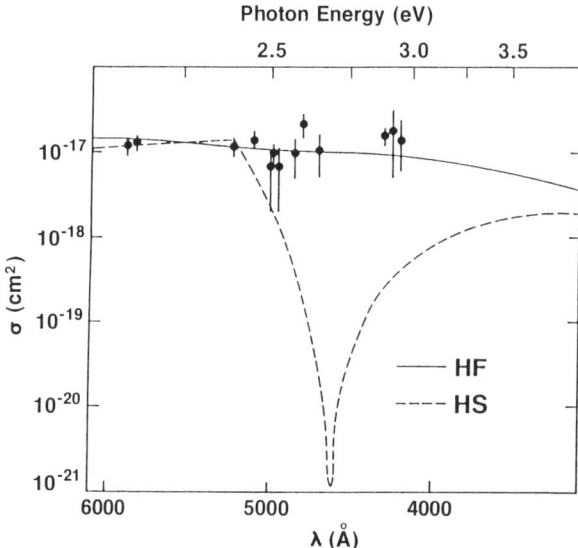

FIG. 12. Photoionization cross section for the excited $6d$ electron in cesium. Experimental points ●, are from Gerwert and Kollath (1983). Solid curve is from Msezane and Manson (1984b), and the dashed curve is from Lahiri (1981). (Figure taken from Msezane and Manson, 1984b.)

field calculation by Lahiri (1981) (dashed line in Fig. 12). The cross section computed using the central field method shows a minimum at 4600Å. The cross section computed by Msezane and Manson (1984b) in the HF approximation also has a minimum, but it is shifted to higher photon energies corresponding to a wavelength of about 900Å. This wavelength is not readily accessible by current laser technology but is easily accessible to users of synchrotron radiation. The experimental results of Gerwert and Kollath (1983) for photoionization from the $6d\,^2D_{3/2}$ excited state are shown as dots with error bars and suggest that the HF approach is correct. The results were obtained from a crossed beam experiment using colinear pulsed dye lasers, one of fixed frequency tuned to resonantly excite the $6d\,^2D_{3/2}$ state and the other of variable frequency to photoionize the excited state.

The apparatus of Gerwert and Kollath (1983) was also used to obtain the relative cross section for photoionization from the $7p\,^2P_{3/2}$ state. As expected from theory (Burgess and Seaton, 1960; Norcross and Stone, 1966) the experimental results decrease monotonically with decreasing wavelength. The measured cross section was put on an absolute basis by normalizing it to the absolute value of the cross section obtained by Heinzmann et al. (1977) using the ionization saturation technique (Sect. II.B.1) at one wavelength. The measurements are about 50% higher than calculations based on either the QDM (Burgess and Seaton, 1960), or the modified QDM (Norcross and Stone, 1966). The other parameters that define photoionization from the $7p\,^2P_{3/2,1/2}$ states were obtained at one wavelength from a measurement by Kaminski et al. (1980) of the spin polarization of the electron ejected by the photoionization process.

A measurement of the photoionization cross section from the $7s\,^2S_{1/2}$ excited state has been reported by Gilbert et al. (1984). This measurement obtained at a wavelength of 5400Å employed a fluorescence technique to measure the density of excited states in an atomic beam of cesium. The cross section obtained is $1.14 \pm 0.10 \times 10^{-19}$ cm^2. The cross section obtained from a central field calculation (Lahiri and Manson, 1984) is a factor of two smaller. A more accurate HF calculation is not available, and so this result is not particularly surprising. The ground state photoionization cross section obtained by Cook et al. (1977) for photoelectrons of the same kinetic energy (0.7 eV) is about 2×10^{-20} cm^2.

C. Photoionization of Excited Alkaline Earths

In the alkaline earths, long lived triplet states are available at energies higher than the ground singlet state and may be used as initial states for excited state photoionization. McIlrath (1969) measured absorption from laser pumped calcium atoms. His experiment was one of the pathfinding experiments in the field to study the photoionization from excited state atoms. In a later paper by Eshvick et al. (1976), two photon non-resonant techniques were used to investigate even parity $4snd$ and $4sns$ resonances that form Rydberg series converging to the $4s\,^2S$ term in ionized calcium. This work laid the ground work for extensive spectroscopic measurements (Eberly and Karczewski, 1977; Eberly and Gallagher, 1979, 1980, and 1981; Eberly et al., 1984; and Kimura, 1985) of high-lying Rydberg levels in a number of elements.

The first measurements of the cross section for photoionization of a selectively excited short-lived atomic singlet state was made by Bradley et al. (1976) in magnesium using two pulsed lasers. By this technique the autoionizing resonance due to the transition between the $3s3p\,^1P$ excited initial state and the $3p^2\,^1S_0$ final state was observed to have a peak cross section of $(8 \pm 4) \times 10^{-16}$ cm^2, a width of 34 meV, and a resonance position of 4.12 eV (3009Å). The authors also determined the ratios of the cross section of the $^1S_0$ to the $^1D_2$ terms of the $3p^2$ configuration to be 14 to 1.

The spectroscopy of the p^2 configuration has also been studied in barium by Wynne and Armstrong (1978) and Camus et al. (1982) by two-photon laser excitation, where it has been the subject of extensive analysis. In particular the $^1S_0$ term of the $6p^2$ configuration interacts strongly with neighboring $5dnd$ $J = 0$ resonances and a four channel multichannel quantuum defect analysis (Aymar et al. 1982), was necessary to sort out the spectrum.

Very recently the p^2 configuration was investigated by Clark et al. (1985) in beryllium using two-photon laser excitation. The energy position of the $2p^2\,^1S_0$ term was expected to be less than the ionization energy but the resonance has been found by Clark et al. to lie at energies greater than the ionization threshold and subject to autoionization. This particular excitation offers the possibility to detect very small quantities of beryllium by the technique of resonance ionization mass spectrometry.

This paper has dealt exclusively with photoionization of excited electrons in a two-step process that were excited by one photon and then ionized by another photon. As mentioned in the introduction a great number of experiments have been conducted to study multiphoton photoionization; while they are beyond the scope of this paper we single out one as an example. The $5p^2\,^3P_2$ level in strontium was populated by two photons from a pulsed laser and then subsequently ionized by a third photon. In this experiment by Feldman and Welge (1982), circularly polarized radiation was used and the angular distribution of the photoelectrons was measured for the $5p^2\,^3P_2$ state ionization. With three photons producing ionization one would expect an additional term in Eq. 6 containing the Legendre polynomial of order six. The authors found this high order term to be small compared to the lower order terms. Detailed calculations of the geometric factors governing the excitation-ionization factors for pulsed laser were not reported. It is very difficult to extract this information because of the pulsed nature of the source and the time dependence of the coefficients in Eq. 6.

Barium has been an especially interesting test bed for the development of an understanding of the correlation between two highly excited electrons. Lasers have been used extensively to populate autoionizing two electron states. A pulsed laser system was used by Bradley et al. (1973) initially to study photoexcitation and photoionization of the $6s6p\,^1P_1$ level. The $6s6p\,^1P_1$ state was resonantly populated and the excited atoms were exposed to the continuum radiated by a laser exposed dye. The absorption spectrum was observed with a spectrograph. In a later experiment by Carlsten et al. (1974), photoionization from the $6s5d\,^3D$ metastable levels was observed in barium vapor. A pulsed laser populated the $6s6p\,^3P_1$ level via the inner combination transition. The ^3P level subsequently decayed to the $6s5d\,^3D$ metastable level which was photoionized by the background continuum from a flash lamp. The absorption spectrum was recorded on a spectrograph and absolute cross sections were obtained from a determination of the density of excited atoms and a measurement of the transmission of irradiated vapor.

The cross section has been measured and has an average value of 20 Mb from the ionization threshold at 2500 Å to about 2030 Å (about 5 eV). Superimposed on the continuum cross section are many doubly-excited autoionizing states.

As techniques developed, very high resolution studies of highly excited states were possible. The spectroscopy of these states was determined by Wynne and Armstrong (1978) and Camus et al. (1983) by using the $6s5d$ $^{1,3}D$ metastable states as a springboard for two photon excitation of autoionizing resonances with the same parity as the ground state that lay between the $6s\,^2S_{1/2}$ and the $5d\,^2D$ limits of singly ionized barium. The final states were composed of the configurations: $5dnl$, $6pnp$, $5dnd$.

A multi-laser experiment by Jopson et al. (1984) and by Bloomfield et al. (1984) has been developed to study autoionizing two electron excitation states in barium of the type $7snd$ and $msnd$. Pulsed lasers were used to study correlation effects in barium where both electrons were promoted to levels with large principle quantum numbers. These experiments are particularly important for the study of correlation effects in two electrons. For two photon transitions of the type $6snd\ ^1D - 7sn'd\,^1D$ there is apparently no shielding of the $7s$ core electron by the outer ns Rydberg electron and the autoionizing rate increases rapidly as a function of m for $msnd$ states but remains constant for $msns$ final states.

Lasers are ideally suited for studies of this type and play an important role in understanding the dynamics of highly excited atoms. However, laser-synchrotron hybrid experiments enable one to probe inner shell excitation and provide a continuum source over a broad energy range.

V. Results from Synchrotron Radiation Ionization of Laser-Excited Atoms

A. GENERAL BACKGROUND

The basic characteristics of laser and synchrotron radiation sources are quite different. An important feature of synchrotron radiation provided by an electron or positron storage ring is its broad spectral range extending from the infra-red up to the x-ray region, typically 1000 eV for a low energy storage ring and more than 10,000 eV for a high energy storage ring. This can be contrasted to the restricted photon energy range of lasers, which is typically 1 eV to 5 eV for cw dye lasers and up to 15 eV for pulsed lasers using different methods for non-linear frequency generation. On the other hand, monochromatized beams of synchrotron radiation are presently less intense and spectrally broader by orders of magnitude than those available from continuously tunable dye lasers. For example, in the VUV region, about 10^{12} photons/sec in a band width equal to 1% of the photon energy is the highest flux routinely available from a toroidal grating monochromator, illuminated by the radiation emitted from a bending magnet of a modern storage ring. A resolution of about 1000 is considered the maximum one can maintain with flux levels adequate for photoelectron spectroscopy. In contrast, with a conventional cw dye laser one can obtain routinely 10^{18} photons/sec in a 20 MHz bandwidth corresponding to a resolution of 10^7. In addition, the time structure of these two sources is quite different. Synchrotron radiation is usually emitted as a 0.1 nsec to 1 nsec pulse, with a typical repetition rate of 1 to 100 MHz. Pulses from a nonsynchronously pumped pulsed laser have a temporal width of the order 100 psec to 1 μsec with a repetition rate of 10 to 1000 Hz. The number of photons in one pulse of synchrotron radiation after monochromatization is 10^5 to 10^6 photons and the absorbing medium must be almost transparent in electron spectrometry to VUV photons. These two conditions make these pulsed lasers a poor source for the experiments described here, because a 1 μs-wide, 10 Hz pulsed laser combined with a 1 ns-wide, 10-MHz pulsed synchrotron radiation source, would shine on only 10 synchrotron pulses, yielding a combined laser-synchrotron duty factor of about 10^{-5}. Thus, at this time cw dye lasers have been the only practical source for these photoionization studies on excited atoms.

B. Results from Continuum Photoabsorption Experiments in Laser–Excited Atoms

Some of the pioneering experiments on excited atoms have been described in Sect. IV. Those experiments dealt primarily with absorption by outer shell electrons. Using the apparatus shown in Fig. 3, McIlrath and Lucatorto (1977) studied absorption by core electrons. They directed a pulsed dye laser tuned to the $1s^2 2s\,^2S \to 1s^2 2p\,^2P$ lithium resonance line into a column of lithium vapor (density = 10^{16} atoms cm^{-3}). The excited vapor column was subsequently excited by a pulsed continuum discharge (Balloffet et al., 1961), to high-lying autoionizing states such as $1s2p^2\,^2S$, ^2D, ^2P, $1s2s3s\,^2S$ and $1s2s3d\,^2D$, having the same parity as the ground state. McIlrath and Lucatorto were able to record the photoabsorption spectrum of excited lithium atoms by delaying the VUV pulse until the maximum density of excited atoms was reached. When the VUV pulse was delayed for a longer time, it was also possible to measure the photoionization of lithium ions. The spectroscopy of the doubly-excited states by stepwise excitations from laser–continuum sources has been documented by Lucatorto and McIlrath (1980), but no oscillator strengths or cross sections for excited state transitions could be determined.

Similar data were obtained in sodium by Lucatorto and McIlrath (1976) and by Sugar *et al.* (1979). An inner shell $2p$ electron was excited to autoionizing final states in sodium atoms where the $3s$ electron was laser excited to a $3p$ orbital. The excited states belong primarily to the configuration $2p^53s3p$. This absorption spectrum is shown in Fig. 13. With the aid of multiconfiguration intermediate coupling calculations, the resonances were assigned to doublet and quartet terms associated with configuration $2p^5(^2P)3s3p(^{1,3}P)$. Corrections to the single configuration energies due to configuration interaction ranged from 0.2 to 1.3 eV. The largest shifts were associated with the $2p^5(^2P)3s3p(^1P)$ coupled states. This work confirmed the identity and energy position of the resonances and was extremely useful for the success of laser-synchrotron experiments. However, once again, it was not possible to obtain the numerical value of the oscillator strengths for these transitions.

Very recently, new photoabsorption data have been produced in the study of laser-excited calcium atoms (Lucatorto *et al.*, 1985).

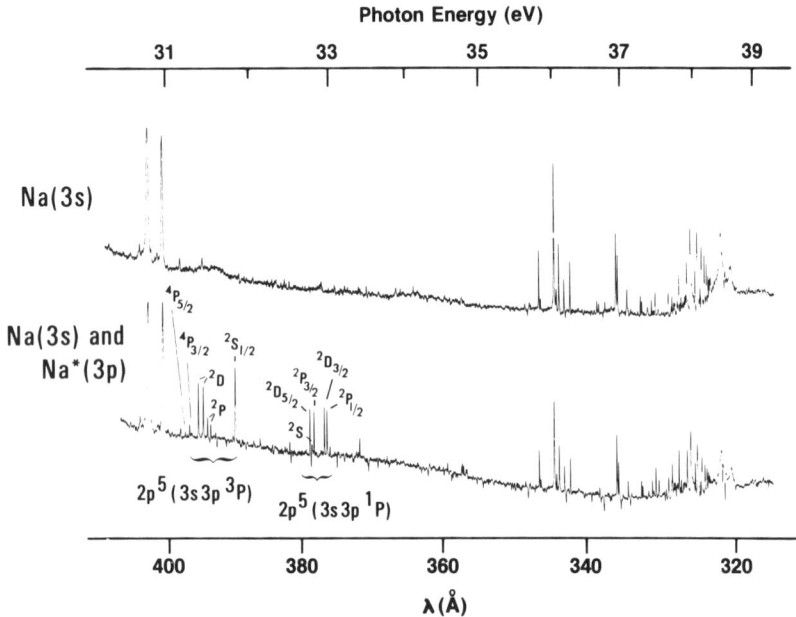

FIG. 13. Inner shell photoabsorption from atomic sodium. Upper trace: autoionizing resonances due to transitions from Na $2p^63s \rightarrow$ Na $2p^53snl$. Lower trace: autoionizing resonances due to transitions from Na $2p^63p \rightarrow$ Na $2p^53s3p$. Final state terms of the Na $2p^53s3p$ configuration are shown in the figure. (Figure taken from Sugar *et al.*, 1979.)

C. Laser-Synchrotron Radiation Combination: The Feasibility Experiment

In these paragraphs we would like to sketch out the various steps that are necessary to ensure the success of a combined laser-synchrotron radiation experiment. To demonstrate the feasibility of a photoemission experiment combining these two sources, a simple system was chosen because of the complexity involved in these measurements. Several experimental techniques have to work simultaneously under optimum conditions:

(i) An atomic beam of density suitable for photoelectron studies must be made. This density has to be high enough to provide a significant electron counting rate and a good signal to noise ratio, but low enough to avoid inelastic scattering of electrons in the medium under investigation.

(ii) The laser must have high efficiency and easy tunability in the energy range of the transition to the first excited state of the atom.

(iii) The monochromatized beam of synchrotron radiation must have the highest flux possible and the best resolution in the photon energy region where it is available.

(iv) It is desirable that the element to be studied will not have too many subshells with binding energies included in the photon energy range of interest. This restriction is desirable to avoid untangling a complicated electron spectrum due to the presence of higher order photons in the x-ray spectrum diffracted by the monochromator (see Fig. 5 for example).

(v) It is important that the photoabsorption spectrum of the system be well known in the ground state as well as in the excited state, because even under powerful laser irradiation, the medium is likely to contain a large percentage of atoms in the ground state.

(vi) Finally, the photoelectron emission spectrum of the ground state atoms is an important piece of information to have, because electron correlations are known to lead to the production of several ionic states following photoionization of an inner-shell electron or an outer-shell electron. Accurate ground state spectra are an important aid in the interpretation of the data obtained from excited atoms.

After evaluating these criteria, we chose sodium as the best candidate. An atomic beam of 10^{12} atoms cm^{-3} can be produced at relatively low temperatures, between 500K and 600K. The ground state configuration of sodium is $1s^2 2s^2 2p^6 3s\,^2S$. Not too many subshells are present and the binding energies of the electrons are conveniently located. The 3s, 2p, and 2s electrons have a binding energy of 5.14 eV, 38 eV, and 71 eV respectively. The energy of the resonant transition $3s\,^2S_{1/2} \rightarrow 3p\,^2P_{3/2}$ is 2.11 eV, or 5890Å, which can be efficiently pumped by a laser using Rhodamin 6G. The photoabsorption spectrum of sodium in the ground state was measured with synchrotron radiation. (Wolff et al., 1972) and is shown in Fig. 14. Only the relative shape of the photoabsorption spectrum was measured at that time and normalized to theoretical calculations. Later, the absolute photoabsorption cross section was measured in the vapor phase (Codling et al., 1977). From the spectrum shown in Fig. 14, it is possible to choose the spectral ranges $48\,\text{eV} < h\nu \leq 66$ and $h\nu \geq 71\,\text{eV}$ as those especially suitable for direct photoionization into the continuum. We note that between 33 eV and 48 eV, and between 66 eV and 71 eV there are numerous absorption lines due to inner-shell excitation of $2p$ and $2s$ electrons. These discrete transitions make the aforementioned energy ranges less suitable for a straight forward interpretation of the photoionization spectrum. The $2p$ photoelectron spectrum of sodium had also been extensively investigated (Krummacher et al., 1982) by using the same electron spectrometer that is used in the laser-synchrotron experiments. These measurements in ground state sodium served to fully characterize the photoionization process in the $2p$ subshell. Figure 15 presents an example of such a spectrum. Photoionization of sodium in the first closed shell ($2p$ subshell) leaves the residual positive ion with two open shells whose coupling produces a multiplet structure corresponding to the $1s^2 2s^2 2p^5 3s\,^{1,3}P$ final ionic states in Na$^+$ (binding energy of 38.0 eV and 38.4 eV, respectively). However, the ionization process may involve excitation of the $3s$ electron as well. Apart from the peak corresponding to this single $2p$ ionization, there are other peaks in Fig. 15, corresponding to two-electron transitions in which the residual positive ion is left in some excited states. Shake-excitation, leading to final states like $2p^5 4s$ (peak d in Fig. 15), as well as conjugate shake-up processes leading to final states such as $2p^5 3p$ (peak e in Fig. 15) or $2p^5 3d$ (merged in peak d) are possible. These excited ionic states result from relaxation effects and final state electron correlations. Their photon energy dependence has been measured (Krummacher et al., 1982) and has been recently reproduced by Hartree-Fock calculations (Craig and Larkins, 1985). Of special interest for our purpose is the group of ionic levels belonging to the $2p^5 3p$ configuration. Six fine structure levels of this configuration have been tabulated. Five of them are within a 0.5 eV band and they have an average binding energy close to 42.1 eV. The relative intensity of this group of satellites was measured and found to have an approximately constant value of about 8% for photon energies between 75 eV and 80 eV, relative to the $2p^5 3s\,^{1,3}P$ ionic states. Peak f, at a kinetic energy 2.10 eV lower than the kinetic energy of the main $2p^5 3s\,^{1,3}P$ ionic states, is due to the relatively high density in the atomic beam. This peak is a result of the $2p$ subshell photoelectrons inelastically scattered and causing $3s \rightarrow 3p$ excitation in another sodium atom.

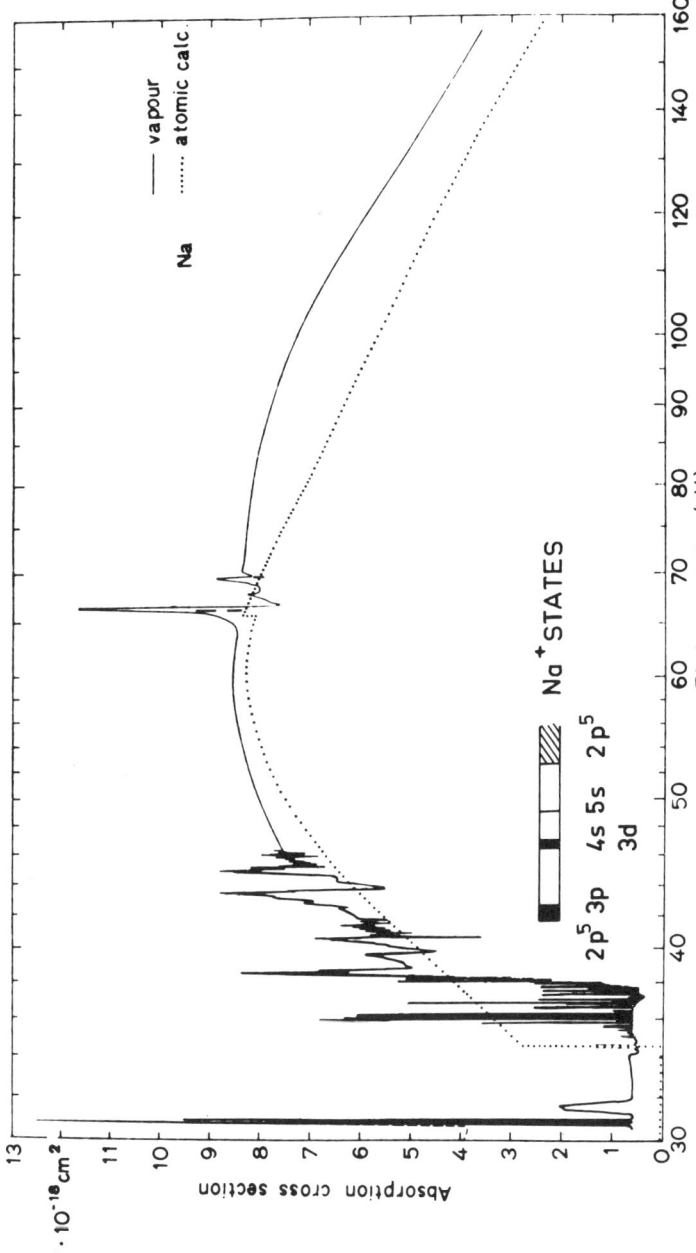

FIG. 14. Spectral dependence of the absorption cross section of atomic sodium (solid line) in the energy range 30-150 eV. The data are scaled to the absolute cross section calculated by McGuire (dotted line, ...). The binding energies of 2p and 2s electrons are 38.0 eV and 71.0 eV respectively. (Figure taken from Wolff et al., 1972.)

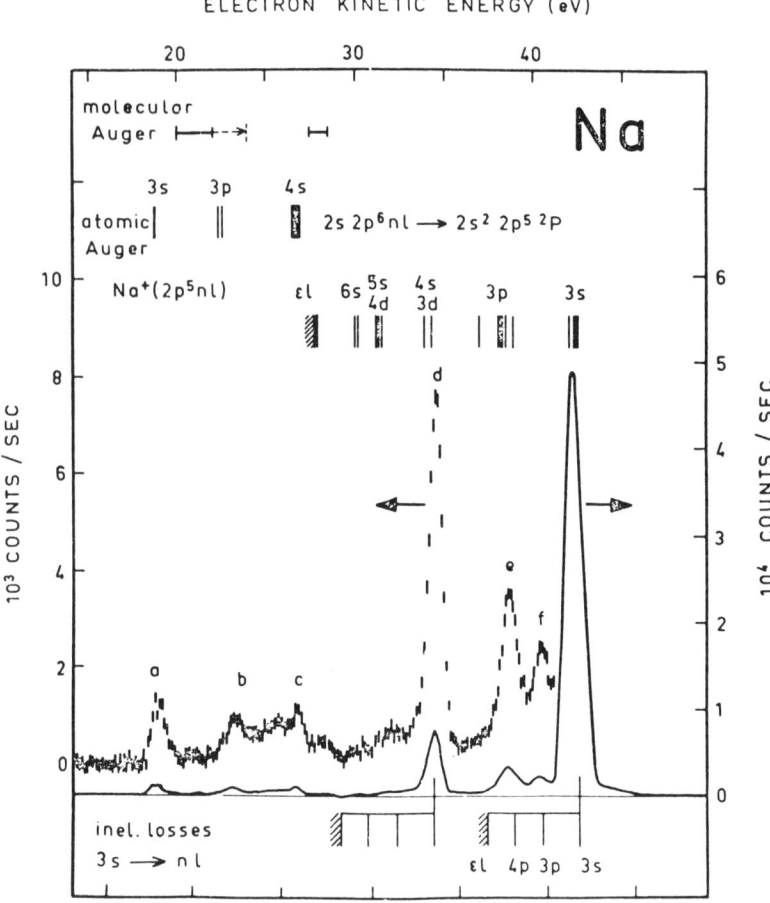

FIG. 15. Ejected-electron spectrum of sodium vapor observed at 80 eV photon energy with a bandpass of 0.9 eV. The energy positions of several spectral features are indicated: molecular Auger transitions, atomic Auger transitions, atomic photoionization with $2p^5nl$ final ionic states. Inelastic loss peaks due to $3s \to np$ excitations in neutral sodium atoms are indicated below the spectrum for the two most intense peaks. (Figure taken from Krummacher et al., 1982.)

The first observation of a photoelectron line produced by ionization of a laser-excited atom with synchrotron radiation was achieved during the series of experiments in 1981 (Bizau et al., 1981 and 1982). Figure 16 shows a spectrum obtained at that time to test the feasibility of the method. In the upper frame, part of the photoelectron spectrum of sodium atoms in the

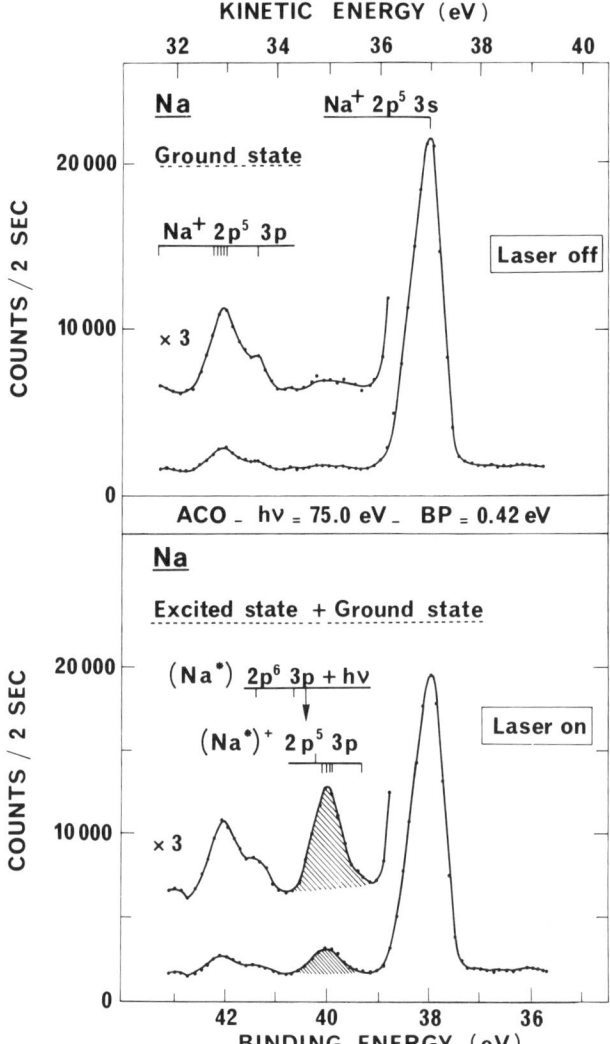

FIG. 16. Photoelectron spectra of atomic sodium. Upper frame: a 2p subshell spectrum of atomic sodium obtained with the laser off. The large peak is due to the direct photoionization of a 2p electron. The small peaks are due to ionization of a 2p electron and excitation of the 3s electron to a 3p orbital. Lower frame: the 2p subshell photoelectron spectrum obtained with the laser on. In addition to the laser-off spectrum, a new peak appears (hatched area) due to 2p ionization of the laser excited sodium atoms. (Figure taken from Bizau et al., 1981 and 1982.)

ground state is observed at a photon energy of 75 eV, which is a structureless region of the ground state photoabsorption spectrum. We will focus on the ionic states with 38 eV binding energy and the first group of satellites corresponding to the ion being left in the $2p^53p$ excited state. These photolines are especially interesting because they correspond to the same final state configuration as the one which can be reached via the ionization of a $2p$ electron in the $2p^63p$ excited atom.

One can note a significant difference between the spectrum in Fig. 16 and the corresponding part of the spectrum in Fig. 15. The peaks e and f in Fig. 15, due to inelastic scattering, are nearly absent in Fig. 16. This is because ten times more flux was available for the new experiment. The density of sodium atoms in the ground state could be reduced and the monochromator could be operated at somewhat higher resolution. The laser was tuned to the $3s\,^2S_{1/2}(F=2) \to 3p\,^2P_{3/2}(F'=3)$ transition of atomic sodium at 2.10 eV and directed into the source volume of the CMA, as shown in Fig. 4. The spectrum in the lower frame of Fig. 16 was taken with the laser and synchrotron beam on. A new electron line appears at about 40 eV binding energy, corresponding to photoionization of a $2p$ electron in the laser-excited $2p^63p$ atom. Because of the broad bandpass of the monochromator, it is not possible to distinguish between the various terms of the $2p^53p$ configuration. Each term belonging to this configuration has a binding energy relative to the $2p^63p$ excited state, which is equal to the excitation energy of each of the terms produced from the ground state of sodium atoms minus the energy delivered by the laser photon. Thus, 40 eV is the mean binding energy of a $2p$ electron in the different $2p^53p$ states and the observation of the additional line is a clear signature of the presence of atoms in the excited state $2p^63p$ in the vapor. The high relative intensity of this feature makes it possible to study the photoionization of the excited sodium atoms.

The binding energy of core electrons in the excited atom is expected to be larger because the mean radius of the electronic density is larger for the $3p$ orbital than for the $3s$ orbital. The more distant $3p$ electron screens the $2p$ inner electrons less and they are imbedded in a more attractive Coulomb potential making them more tightly bound.

In experiments carried out with higher resolution (Wuilleumier, 1982b), the group of lines centered around 42.1 eV binding energy (corresponding to terms of $2p^53p$ ionic states produced by ground state photoionization) had a different intensity distribution than the group centered at around 40 eV which correspond to terms of $2p^53p$ ionic states produced by excited state photoionization. One concludes that the various final states belonging to the $2p^53p$ configuration are not populated in the same way via the one-photon-two-electron excitation route and via the two-photon excitation route. Systematic studies at higher resolution and under varying conditions for the density of atoms in the ground state and for the laser excitation would be of great interest to understand this different behavior.

Calculations by Chang and Kim (1982a, 1982b) predict that the partial cross section for ionization of a $2p$ electron is nearly the same, independent of the excitation of the valence electron. Using these results, the relative population of atoms in the laser-excited state was determined from the ratio of integrated areas under the $2p$ photoelectron peaks at 40 eV (excited state) and at 38 eV (ground state). In Fig. 16, this ratio is close to 10%. In the feasibility experiments, it was not possible to obtain a higher population. In later experiments, the laser was stabilized by locking it to the fluorescence from a collimated auxiliary sodium beam, and higher atomic densities in the source volume of the CMA could be achieved by efficient liquid nitrogen trapping. Under these operating conditions, it was possible to pump up to 25% of the atoms into the $3p$ state.

The second feature detected during this first phase of the experiments was the observation of electrons produced by resonant photoemission of the excited $3p$ electrons. The search for direct photoionization into the continuum of the $3p$ excited electron was not successful because of the very low cross section (about 0.05 Mb) in the photon energy range available in this experiment. However, we were able to observe the electrons emitted in the autoionization of doubly excited states due to transitions of the type $2p^63p \rightarrow 2p^53s3p$ (Bizau et al., 1982; Wuilleumier, 1982a, 1982b) by setting the VUV monochromator to the excitation energy of the $2p^63p \rightarrow 2p^53s3p$ transitions in laser-excited sodium (Sugar et al., 1979, and also Fig. 13). These two observations provided a demonstration that a two-photon experiment was possible, involving the simultaneous absorption of a photon from the laser and a photon from the synchrotron.

Since our experiments, two other experiments have been performed at other synchrotron radiation facilities to test the feasibility of an experimental program similar to ours at other storage rings. In one measurement, the VUV photoelectron spectra of laser-excited barium atoms have been successfully investigated by Nunneman et al. (1985) and in the other, the near UV photoionization of sodium vapor was investigated by Preses et al. (1985). Nunneman et al. (1985) used an apparatus similar to ours to study the excitation of electrons in the $5p$ and $6s$ shells in barium vapor in the 20 eV photon energy range. Preses et al. (1985) studied the behavior of the cross section for photoionization of the excited $3p$ electrons in sodium by measuring the total ion signal produced by photoionization over a limited energy range of 2 eV above the $3p$ ionization threshold.

Following our feasibility tests, we have made quantitative measurements of photoionization cross sections at LURE. Some of the results of these experiments have already been presented by Wuilleumier (1984) and by

Bizau et al. (1985a). In the following paragraphs, we will describe three of these experiments carried out between 1982 and 1985.

D. RESONANT $3p$ CROSS SECTION IN LASER-EXCITED SODIUM ATOMS: DETERMINATION OF OSCILLATOR STRENGTHS FOR INNER-SHELL EXCITATIONS

In the first of these new experiments (Bizau et al., 1985a), we have measured oscillator strengths for transitions, in atomic sodium, involving even parity autoionizing levels. The transitions between a laser-excited initial $2p^63p$ configuration and autoionizing levels of $2p^53s3p$ configuration were systematically investigated. Measurements of the oscillator strength for inner shell transitions to autoionizing levels in sodium are of special interest because the quartet terms of the core-excited levels are metastable and are possible upper states for an XUV laser (Holmgren et al., 1984).

The general two-photon excitation and decay scheme is the following:

$$2p^63s + hv(\text{Laser}) \rightarrow 2p^63p$$

$$2p^63p + hv(\text{SR}) \rightarrow 2p^53s3p$$

$$2p^53s3p \rightarrow 2p^6 + e^-$$

We used the same experimental set-up as in the feasibility experiment but improved the method of the laser excitation and sodium beam production, as already mentioned. We used the measurements of Sugar et al. (1979) to select the photon energy range covering the excitation energy of a particular transition.

A typical photoelectron spectrum is shown in Fig. 17. In the upper panel a spectrum is shown with the laser off and the monochromator set at a photon energy hv_1 equal to 32.73 eV, close to the excitation energy of the $2p^5(^2P)3s3p(^1P)^2D_{5/2}$ state. Photons of twice the first order photon energy (63.46 eV) are also transmitted through the monochromator and are the only photons able to photoionize the $2p$ shell electrons observed as peaks 1 and 3 in the figure. This spectrum is quite similar to the spectrum of Fig. 16 (upper frame) except that it is produced by second order photons. With the laser switched on and tuned to the $3s\,^2S_{1/2}(F=2) \rightarrow 3p\,^2P_{3/2}(F'=3)$ transition (middle panel in Fig. 17) new electron peaks, noted 2 and 4, appear. Peak 2 is due to the photoionization of $2p$ subshell electrons with the $3s$ electron laser-excited to a $3p$ orbital. It is the same peak as already observed at 40 eV binding energy in the feasibility experiments (see Fig. 16), but now its intensity is about three times the intensity of the satellite peaks $2p^53p$ produced directly in the photoionization of the ground state atoms, around

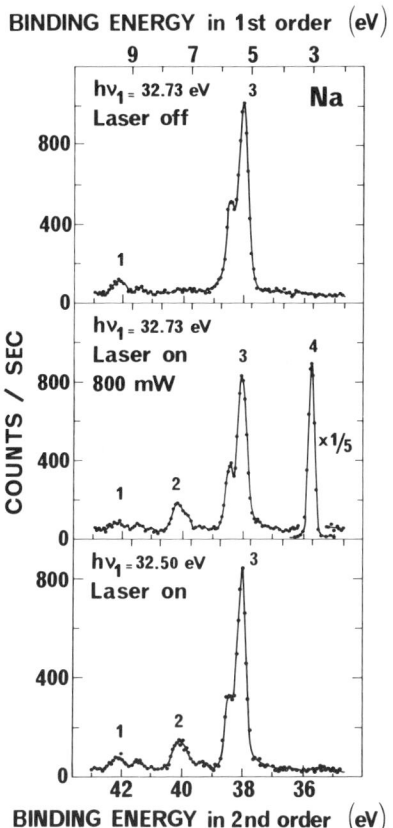

FIG. 17. Top: A photoelectron spectrum of sodium taken at $hv_1 = 32.73$ eV with the laser turned off. Peaks 1 and 3 are due to ionization of the $2p$ shell electron by photons of energy equal to twice hv_1 or 63.46 eV. The appropriate binding energy (E_B) scale is at the bottom of the figure and is computed from the kinetic energy (E_{kin}) according to $E_B = 2\ hv_1 - E_{kin}$. Middle: A spectrum taken at $hv_1 = 32.3$ eV with the laser turned on. Peak 2 is due to photoionization of the $2p$ subshell electron with the $3s$ electron excited to a $3p$ orbital leaving an ionic state $2p^53p$. Peak 4 is due to autoionization of the $2p^5(^2P)\ 3s3p\ ^1P\ ^2D$ doubly excited state. The appropriate E_B scale for peak A is at the top of the figure and is computed from E_B according to $E_{kin} = hv_1 - E_{kin}$. Bottom: A spectrum taken off resonance at $hv_1 = 32.50$ eV with the laser on. Peak 4 disappears, but the peak due to the $2p$ ionization of the laser-excited atom is still present. (Figure taken from Bizau et al., 1985a.)

42 eV binding energy. This means that, under the improved experimental conditions, the relative density of excited atoms was increased to a maximum of 25% of the total atomic density. The maximum value achievable is 31% when the polarization of the laser and the hyperfine structure of the $3s\ ^2S_{1/2}$ and $3p\ ^2P_{3/2}$ states is taken into account (Hertel and Stoll, 1974, Fischer and Hertel, 1982).

In Fig. 17, peak 4 is due to autoionization from the discrete state resonantly excited by hv_1, which is the final state of the transition

$$2p^63p\,^2P + hv_1 \to 2p^5(^2P)3s3p(^1P)^2D_{5/2}.$$

The evidence of this decay is manifested by the presence of the intense peak at the binding energy of 3.03 eV (the binding energy of a $3p$ electron). When the VUV monochromator is tuned off resonance (lower panel of Fig. 17) at 32.50 eV, peak 4 disappears, because, even with the lower background achieved in the new series of experiments, the nonresonant $3p$ photoionization cross section is too weak at this photon energy, but the peaks 1, 2, and 3, due to photoionization of $2p$ shell electrons by second order radiation remain.

With a higher electron counting rate from atoms in the excited state produced by more efficient laser excitation, it was possible to test the prediction by Chang and Kim (1982a, 1982b) that, at the same final state energy, the inner shell photoionization cross section of a sodium atom with an excited outer electron is the same as the photoionization cross section of the atom in the ground state. The photoionization cross sections calculated by Chang and Kim (1982a, 1982b) are plotted in Fig. 18. The calculated photoionization cross section for the $2p \to \varepsilon d$ transition is shown for sodium in different initial states. The black triangles, the crosses, and the open circles are the values of the cross section for the atoms in the ground $3s$ state, in the $3p$ excited state, and the $3d$ excited state respectively. In the photon energy range of our measurements (60–66 eV) the predicted equality of the cross section was checked. If the ground state cross section and the excited state cross sections are equal, the integrated area under the photolines 2 ($2p$ photoionization in the excited $3p$ state) should equal the change in the peak 3 intensity with the laser on and off, after the spectrum is corrected for the transmission of the CMA. These quantities were equal within the overall experimental accuracy of 15%. More systematic measurements covering a broader photon energy range are possible, and would be desirable to make a more definitive comparison between the experimental and the theoretical cross sections.

From the information contained in Fig. 17, the oscillator strength for excitation of an autoionizing resonance could be measured. The area N_{hv} of the photoelectron peak 2 for $2p$ ionization of the excited atoms is proportional to the product of the excited state density, n_{3p}, and the photoionization cross section, $\sigma(hv)$

$$N_{hv} = KI_{hv}E_{hv}n_{3p}\sigma(hv), \tag{8}$$

where K is the spectrometer constant, I_{hv} is the photon flux integrated over the monochromator bandpass ΔE at energy hv and E_{hv} is the kinetic energy of

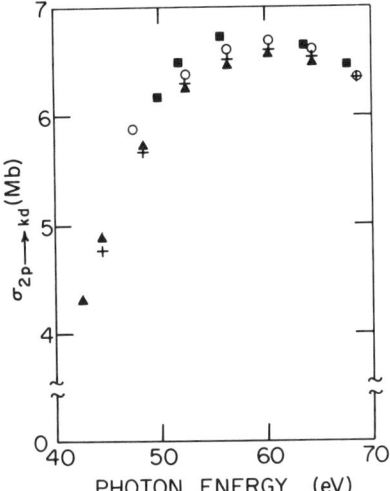

FIG. 18. The calculated photoionization cross section for the $2p \to kd$ transition for sodium initially in the ground state, $\triangle$, the excited state with valence electron in $3p$ orbit, $+$, the excited state with valence elecron in the $3d$ orbit, $\bigcirc$. (Figure taken from Chang and Kim, 1982a.)

the photoelectron. Similarly, the integrated area N_R of the electron peak (4 in Fig. 17) for the autoionizing resonance is given by:

$$N_R = 110 K' \left(\frac{I_R}{\Delta E_R}\right) E_R n_{3p} \int_{\Delta E_R} \frac{df_R}{d\varepsilon} d\varepsilon, \qquad (9)$$

where the symbols have the same meaning as before and are taken at a photon energy corresponding to the resonance energy. The quantity K' is the product of the spectrometer constant and the branching ratio for autoionization. The cross section and the monochromator bandpass are expressed in units of 10^{-18} cm^2 and eV respectively, and the quantity 110 is a scaling factor converting cm^2, and eV to oscillator strength.

This equation is valid only if the direct photoionization process is negligible compared to the resonant process, which is true for the sodium resonances. Under this condition there is negligible interference between the resonance channel and continuum channel. The resulting line shape will have a Lorentzian profile (Fano, 1961). In our experiment the resonance profile shape is masked because the width of the resonance is smaller than the monochromator bandpass by about one order of magnitude.

Because of this large bandpass, the integral over the oscillator strength density $df_R/d\varepsilon$ can be replaced by the oscillator strength of the resonance f_R. Solving Eq. (8) and Eq. (9) for f_R yields:

$$f_R = 0.0091 \frac{K}{K'} \frac{I_{hv}}{I_R} \frac{E_{hv}}{E_R} \frac{N_R}{N_{hv}} \sigma(hv) \Delta E_R \qquad (10)$$

These equations can be written because the sodium beam is almost transparent for the VUV photons.

To check the validity of Eq. (10), we first verified that the ratio between the intensity of the autoionization line (peak 4, proportional to N_R) and the intensity of the 2p photoelectron in the excited atom (peak 2, proportional to N_{hv}), remains constant, for a constant bandpass of the monochromator. This ratio was studied as a function of the laser power, the density of sodium atoms in the ground state, and the flux of synchrotron radiation. Then the linear dependence of f_R with ΔE_R was checked by fixing the VUV monochromator to a photon energy of 31.78 eV. This photon energy is the excitation energy of an isolated resonant transition to the $2p^5(^2P)3s3p(^3P)^2S_{1/2}$ final state, shown in Fig. 13. This transition was the only one that could be excited within the bandpass of the monochromator set at this photon energy. It was possible then to change the bandpass of the monochromator without changing the number of resonances simultaneously excited by the VUV photon beam, as it would have been the case at other photon energies, where several resonances would overlap the monochromator bandpass. The electron spectrum measured at 31.78 eV is shown in Fig. 19. It is very similar to the upper and middle frame of Fig. 17, except that the intensity ratio between peak 4 and peak 2 is quite different because of the different value of the oscillator strength for excitation of this single transition. It is also interesting to note that the kinetic energy scale has been shifted on purpose in the lower frame of Fig. 19 to illustrate the large value of the kinetic energy shift due to the creation of a plasma potential in the medium when the laser is on. This phenomenon will be discussed in detail in Sect. VI. The existence of this potential is due to the plasma produced by collisional ionization between excited atoms. It can be as high as 1 eV, adds to the contact potential of the spectrometer, and contributes to the retardation of the electrons in the field-free space of the CMA. The presence of the 3s photoline is also of interest because of the discrepancy between the measured 3s photoionization cross section by Hudson and Carter (1968) and calculated cross section. This weak photoelectron line due to 3s photoionization of Na atoms in the ground state occurs at a binding energy of 5.14 eV.

At this photon energy, the bandpass of the monochromator was systematically varied, and the electron spectrum was recorded for the values shown in Fig. 20. The peaks noted here: A^*, P, and P^* correspond respectively to peaks 4, 3, and 2 in Fig. 17. When the bandpass is narrowed, an improvement in spectral resolution can be seen in Fig. 20 by observing that peak P splits into two peaks. Peak P is a doublet corresponding to $2p^53s\,^3P$ and ^1P final ionic states. At the smallest bandpass (the second order bandpass $BP_2 = 0.16$ eV) almost complete resolution of the doublet is achieved. As the entrance and the exit slits were closed, the absolute intensity of all the spectral lines

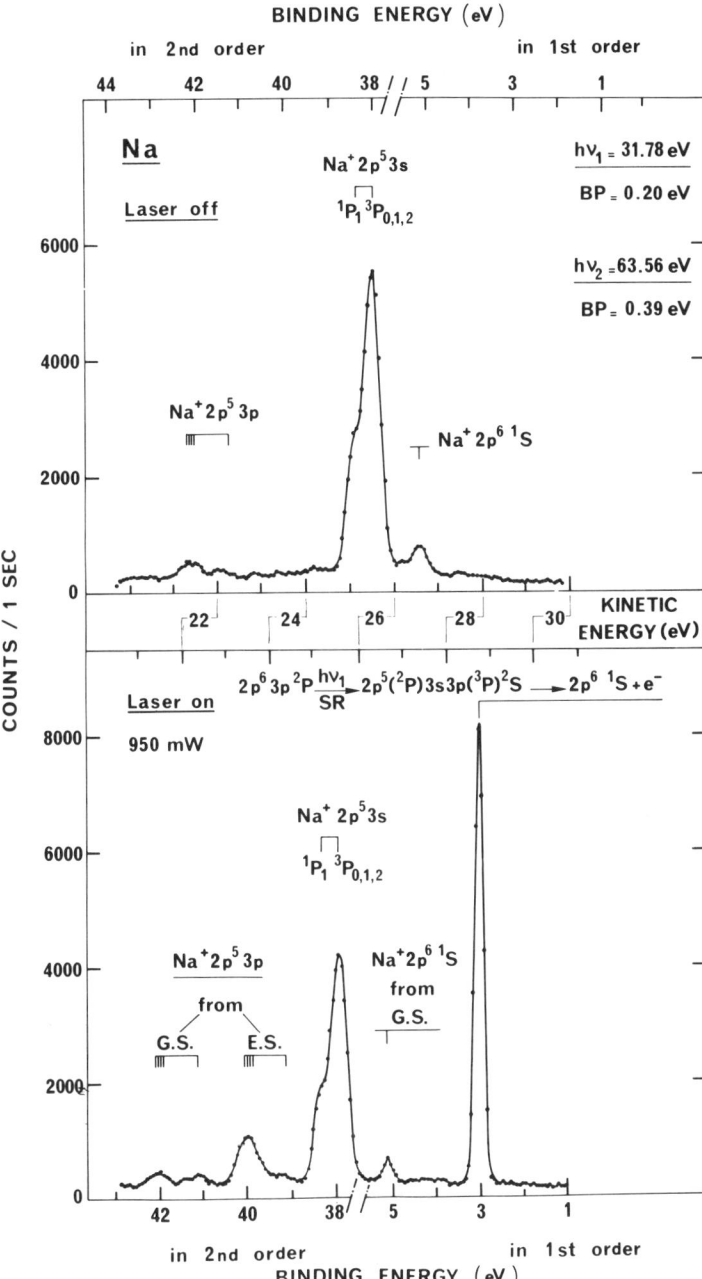

FIG. 19. Spectra of electrons ejected from the atoms in the ground state (top, laser off) and from a mixed medium partly formed of atoms in the $2p^6 3p\ ^2P_{3/2}$ state (bottom, laser on) by 31.78 eV and 63.56 eV photons simultaneously transmitted by the monochromator. See text and Fig. 17 for further explanation. (Figure taken from Wuilleumier, 1984.)

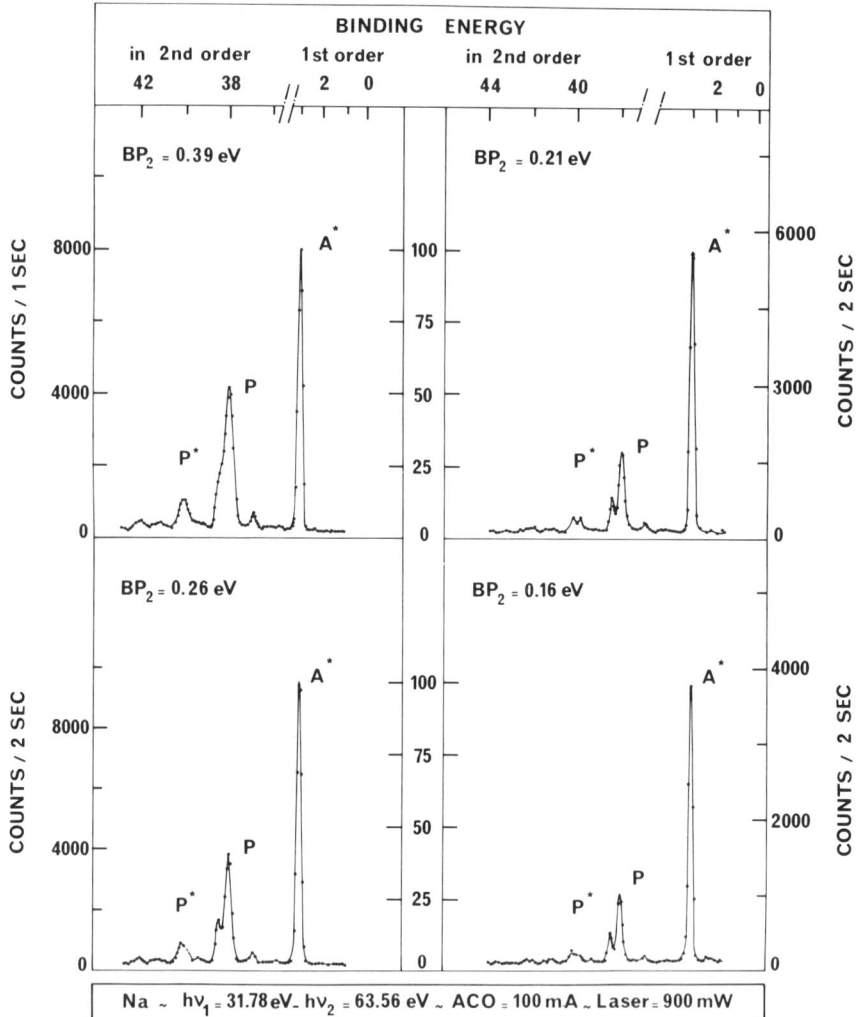

FIG. 20. Spectra of electrons ejected from Na atoms, with the laser on, obtained using the same photon energies as the spectra of Fig. 19 with four different values of the monochromator bandpass. The top left spectrum is identical to the spectrum in Fig. 19. A^*, P, and P^* corresponds to peaks 4, 3, and 2 in Fig. 17 respectively. BP_2 is the bandpass of the monochromator for photons diffracted in the second order. The corresponding bandpass in the first order are 0.18 eV, 0.11 eV, 0.076 eV, and 0.059 eV, respectively.

decreased and the relative intensity of the peak P and P* compared to peak A* changed. The variation of the ratio between the integrated area under the peaks A* and P*, each corresponding to electrons ejected from the laser-excited atoms, is shown in the upper frame of Fig. 21 as a function of the width of the entrance and exit slits. The kinetic energy and the width of peak A* as shown in Fig. 20 remain unaltered, as expected, since only one

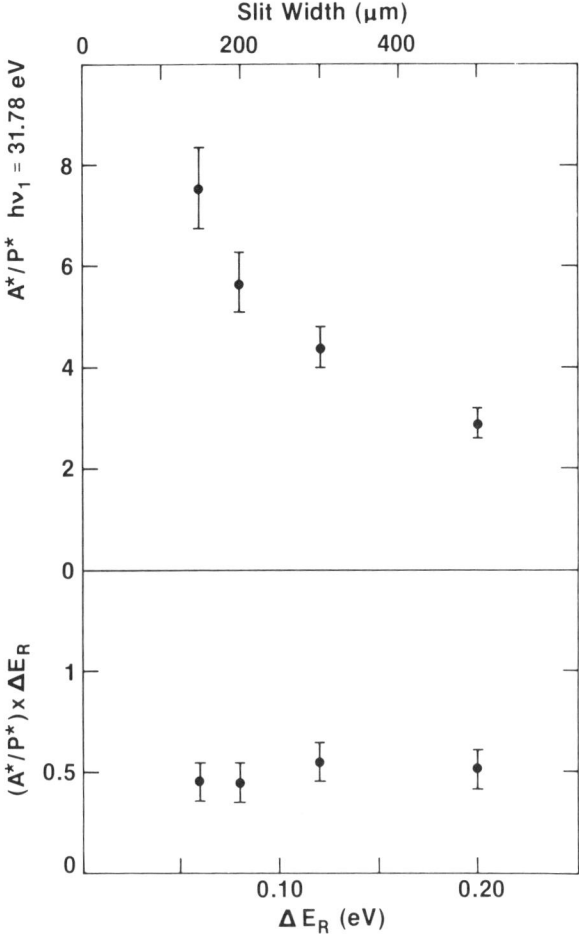

FIG. 21.. Top: Variation of the ratio between the integrated area under the autoionization line (A* in Fig. 20) and under the 2p photoionization line (P* in Fig. 20) as a function of the widths of entrance and exit slits of the monochromator (upper scale) and bandpass in first order (lower scale). Bottom: Variation of the same ratio after multiplication by the monochromator bandpass. This quantity is constant within the experimental accuracy and is proportional to the oscillator strength of the resonance at 31.78 eV.

resonance is excited and, even for the smallest bandpass, the natural width of the doubly excited state is by far narrower than the monochromator bandpass. The observed width of peak A^* is dominated by the width of the electron spectrometer function whose width is 0.9% of the kinetic energy of the electrons. This width is much larger than the width of the resonance, and is independent of the bandpass of the monochromator.

In the lower frame of Fig. 21 the product of the ratio of areas and ΔE_R is plotted as a function of ΔE_R the monochromator bandwidth. If Eq. (10) is valid, the product must be a constant. The product shown in the lower frame of Fig. 21 is constant within our experimental uncertainty of 15–20%.

All the quantities in Eq. (10) could be measured in the experiment or were known from other data. The kinetic energies, E_{hv} and E_R, are measured from the electron spectrometer setting. The contact potential and the laser plasma potential can be deduced from the known photon energy and from the binding energies of $2p$ and $3p$ electrons. The quantities N_{hv}/N_R and ΔE_R are accurately determined by measuring the intensity of photoelectrons emitted from rare gases with known photoionization cross sections, such as neon (Wuilleumier and Krause, 1979). The photoionization cross section $\sigma(hv)$ is equal to the $2p$ photoionization cross section $\sigma_{2p}(hv)$ for ground state atoms. This assertion was tested as described in earlier paragraphs. The quantity σ_{2p} is determined by proportioning the total photoabsorption cross section for atomic sodium (Codling et al., 1977) using the branching ratios determined in the photo-emission experiments on sodium in the ground state (Krummacher et al., 1982). The branching ratio between autoionization and radiation is

TABLE III

Oscillator Strengths for Autoionizing Resonances in Laser Excited Sodium
$2p^6 3p\,^2P \rightarrow 2p^5 3s 3p$

Classification	hv^a	hv^b	f_R^c
$2p^5(^2P)3s3p(^3P)^2S_{1/2}$	31.78	31.77	0.022(4)
$2p^5(^2P)3s3p(^3P)^2D_{5/2}$	31.40	31.40 ⎫	0.087(12)
$^2D_{3/2}$	31.34	31.34 ⎭	
$2p^5(^2P)3s3p(^3P)^4P_{5/2}$	31.19	31.19	< 0.01
$2p^5(^2P)3s3p(^1P)^2D_{5/2}$	32.68	32.69	0.063(11)
$^2D_{3/2}$	32.85	32.87	0.045(3)
			$\Sigma = 0.22(4)$

[a] Present measurement of resonance energy.
[b] Resonance energy from Sugar et al. (1979).
[c] The number in parenthesis is the estimated probable error of the measurement.

assumed to be equal to one. This assumption is valid for strict LS coupling, which applies to low Z elements. The oscillator strength for inner-shell excitation to $2s^22p^5(^2P)3s3p(^3P)^2S_{1/2}$ was thus obtained and is listed in Table III.

The situation is more complicated for the other terms of the $2p^53s3p$ configuration. When the $2p$ electron is excited to the $3s$ orbital, the parent configuration $3s3p$ couples through the exchange interaction to form $^{1,3}P$ terms around 31.40 eV and 32.70 eV. These two terms in turn couple with the $2p^5\,^2P_{3/2,1/2}$ core to form terms of the type $^{2,4}D$, $^{2,4}P$, and $^{2,4}S$ and form the two groups of excited states shown in Fig. 13. The group of resonances associated with the $2p^5(^2P)3s3p(^1P)$ parent configuration was scanned to produce the excitation curve shown in Fig. 22. Spectra were recorded every 0.03 eV with a bandpass of 0.10 eV, and the excitation function is equal to the ratio of the area under peaks 4 and 2, shown in Fig. 17. The widths of the dashed curve in Fig. 22 were adjusted for a best fit to the data. The solid curve is the sum of the dashed curves, which have the mathematical form of the instrument function and are normalized to the peak counting rates. This fit was made assuming the ^2P terms decay radiatively and the observed intensity is due to just the autoionization of the ^2D terms. The width of each

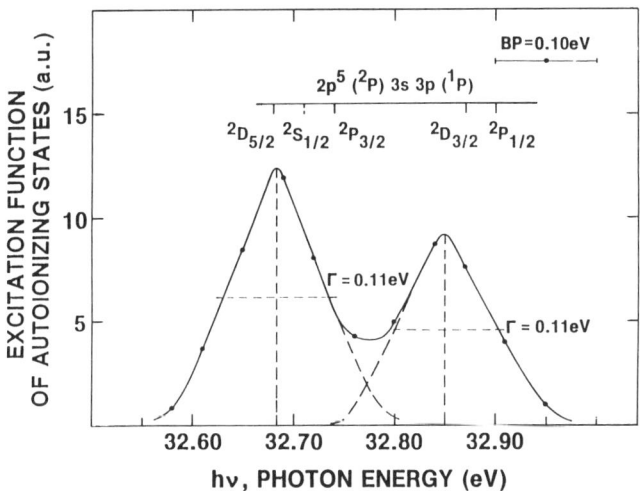

FIG. 22. Excitation function of autoionizing lines as a function of photon energy for a monochromator bandpass (BP in Figure) of 0.10 eV in the energy range of the $2p^5(^2P)3s3p$ (4P) excited states. The scale of the ordinate has arbitrary units (a.u.). The configuration and terms of the autoionizing resonances in the spectral range shown in the figures are indicated according to Sugar et al. (1979). The energy position of the $2p^5(^2P)3s3p(^1P)^2D_{3/2}$ term differs by 0.02 eV from Sugar et al. (1979). The width of the structure, $\Gamma = 0.11$ eV, includes the resonance width plus the monochromator bandpass. (Figure taken from Bizau et al., 1985a.)

component is slightly larger than the bandpass. The even parity ^2P terms cannot autoionize because autoionization requires the parity of the initial and final state to be the same, and the only final state available $2p^6(^1S_0)\varepsilon p\,^2P$ has odd parity. The ^2S term can autoionize but seems to be too weak to be observed in these measurements.

Our interpretation of this excitation curve is supported by the variation of the kinetic energy of the autoionization line versus the photon energy shown in Fig. 23. When the direct photoionization process is negligible compared to the resonant photoionization process, the kinetic energy of the electrons from the autoionizing state will be constant as long as one resonance and only one resonance is included in the bandpass of the monochromator. This is observed in Fig. 23. When the photon energy is varied between 32.60 eV and 32.72 eV, the kinetic energy of the autoionization electrons is constant and is equal to the energy of the $2p^5(^2P)3s3p(^1P)^2D_{5/2}$ transition (32.68 eV) minus the binding energy of the $3p$ electron (3.03 eV). When the bandpass of photon energies transmitted by the monochromator overlap with both ^2D resonances, the kinetic energy of the electrons changes with photon energy until it reaches a second plateau corresponding to the excitation of the $^2D_{3/2}$ resonance. The $^2S_{1/2}$ resonance does not seem to have any more influence on the kinetic energy of the electrons than it had on the excitation curve of the autoionization states.

The spectrum is further complicated when several resonances of comparable intensities are excited in the finite bandwidth of the monochromator, as it is the case around 31.40 eV. A similar excitation curve, Fig. 24, was made in this energy region for the terms having $2p^5(^2P)3s3p(^3P)$ coupling with two different values for the monochromator bandpass. In this case, most of the strength was in the $^2D_{5/2,3/2}$ levels with some indication of the presence of the $^4P_{3/2}$ level. Unfortunately, the best resolution of the monochromator is not sufficient to resolve the fine structure of the ^2D states. Excitation functions taken with two values of the bandpass, peak at the same photon energy of 31.40 eV, which is the excitation energy of the $^2D_{5/2}$ state.

In strict LS coupling, the ^2P levels would not autionize. Their presence implies mixing with ^2D levels that do autoionize. The quartet manifold has been studied recently (Holmgren *et al.*, 1984) in a series of experiments that used a discharge to populate highly excited states and a laser to pump them to a radiating state.

The value of the oscillator strength for each transition, whose oscillator strength was measured, is summarized in Table III. The sum of the oscillator strengths for all the measured transitions of the $2p^63p \rightarrow 2p^53s3p$ array is 0.22(4). While calculations for the oscillator strengths of the individual transitions of this array have not been published, several calculations have been made (Kastner *et al.*, 1967; Shorer, 1979) of the oscillator strength of the

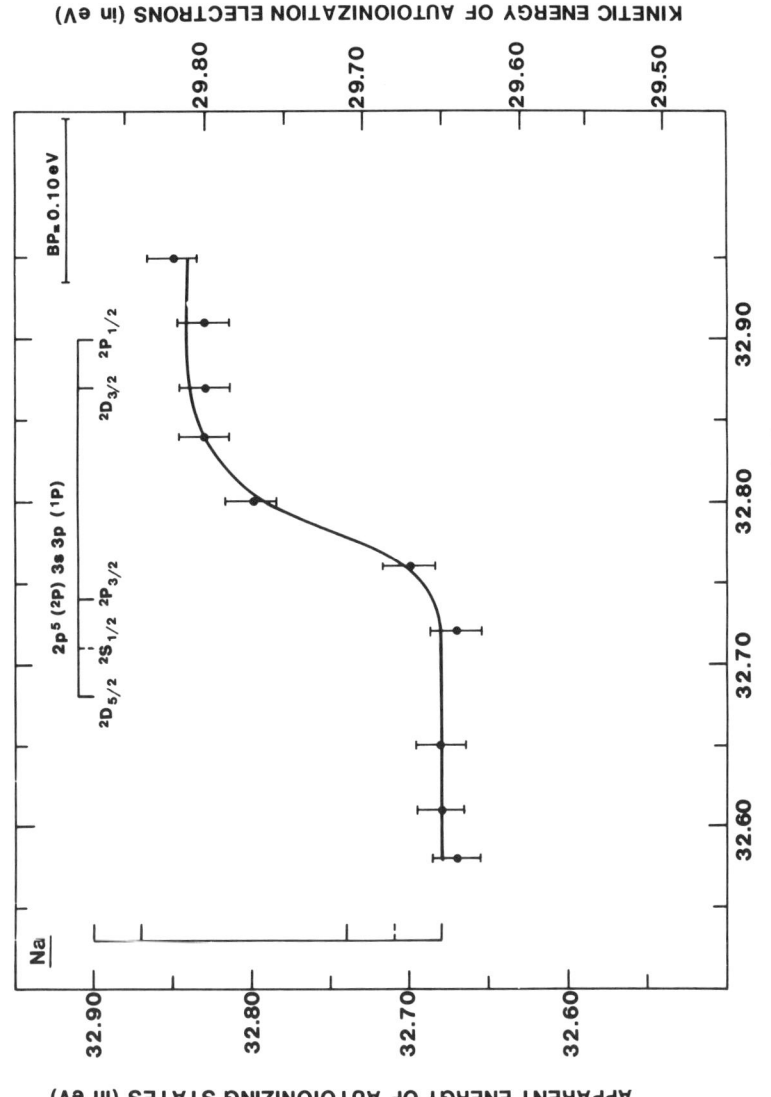

FIG. 23. Apparent energy of the autoionization line as a function of the photon energy. The vertical scale on the left was obtained by shifting the scale of the measured kinetic energy on the right by 3.03 eV, the binding energy of the $3p$ electron. The observation of a plateau means that there is only one resonance with significant oscillator strength included in the bandpass of the monochromator *and* that the direct photoionization process is negligible compared to the resonant process. The energy of the resonances are obtained from the energy position of the plateau.

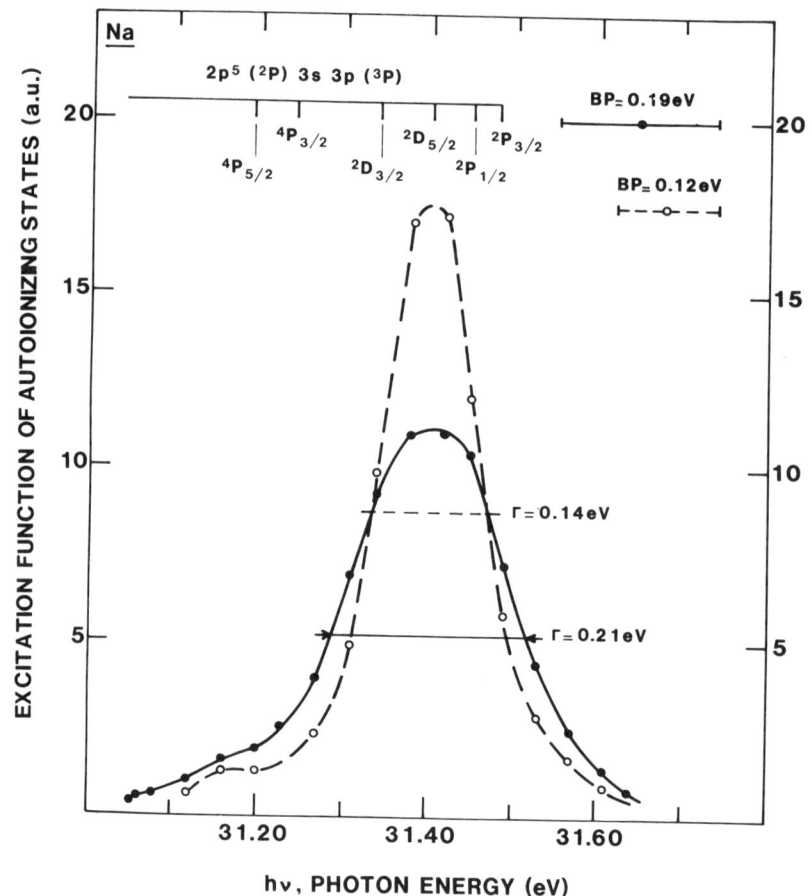

FIG. 24. Excitation function of autoionizing resonances, as a function of photon energy in the range of the $2p^5(^2P)\,3s3p\,(^3P)$ excited states, for two different values of the monochromator bandpass (BP = 0.12 eV and 0.19 eV). Configuration and terms were marked as for Fig. 22. The autoionization of the ^4P term appears below 31.20 eV in both curves. (Figure taken from Wuilleumier, 1984.)

transition $2p^6 \rightarrow 2p^5 3s$ for the isoelectronic sequence of the Ne-like ions. The results were: 0.17 for NeI, 0.20 for NaII, 0.21 for AlIV and SiV. The value we obtain for the oscillator strength of the entire transition array is a lower bound, but the close correspondence between our measurements and the calculations suggests that LS coupling dominates and that the excited $3p$ electron acts chiefly as a spectator.

This successful experiment demonstrated that quantitative information can be obtained from stepwise laser-VUV two-photon absorption. We think these new results open up challenging new opportunities to investigate atomic systems.

E. INNER SHELL AND OUTER SHELL PHOTOIONIZATION IN LASER-EXCITED ATOMIC BARIUM

1. VUV Photoelectron Spectroscopy of Laser-Excited Barium

After sodium we felt barium was the best candidate for continued exploration of the potentialities of this new method. Barium meets most of the criteria that have been listed in Sect. V.A, to successfully achieve an experiment combining synchrotron and laser radiation. Barium also provides an excellent opportunity for studying electron correlations in excited states because both outermost 6s electrons are easily excited. However, the photoionization spectrum of barium is far more complex because many subshells can be ionized in the VUV photon energy region, making the use of electron spectroscopy compulsory. The complexity of the ground state spectra was known from earlier photoabsorption measurements by Connerade *et al.*, (1979) and Connerade and Martin (1983). VUV photoemission was studied by Hotop and Mahr (1975) and by Rosenberg *et al.*, (1979). Stepwise laser ionization of barium has been briefly described in Sect. IV.C. All these experiments shows the important role configuration interaction and electron correlation play in the correct description of the excitation spectra.

For our experiments, the laser was tuned to the $6s^{2\,1}S_0 \to 6s6p\,^1P_1$ transition at 2.24 eV photon energy (5535Å). The laser was able to deliver up to 600 mW of power. The first VUV photoelectron spectra of laser excited barium were obtained in 1982 (Wuilleumier 1982b; Bizau *et al.*, 1983).

In Fig. 5 of Sect. II we have already presented a photoelectron spectrum of barium atoms in the ground state to demonstrate the need of electron spectrometry for this type of experiment. Here we would like to describe these spectra in detail. The upper panel of Fig. 5 (Bizau *et al.*, 1985) is a spectrum obtained with the laser off and the monochromator set at a photon energy hv_1 equal to 45.0 eV. Photons of this energy are able to photoionize the 5p electrons (binding energies 22.72 eV and 24.76 eV, corresponding to $5p^5 6s^{2\,2}P_{3/2}$ and $^2P_{1/2}$ final ionic states, respectively) and the 6s electrons (binding energy = 5.21 eV, final ionic state $5p^6 6s\,^2S_{1/2}$). Photons of twice the first order photon energy (90.0 eV) are also transmitted through the monochromator and photoionize the 5s electrons (binding energy 38.0 eV, the final ionic state is $5s5p^6 6s^{2\,2}S_{1/2}$). Photons of three times the first-order photon energy (135.0 eV) photoionize the 4d electrons (binding energies 98.4 eV and

101.0 eV, $4d^95s^25p^66s^2\ ^2D_{5/2}$ and $^2D_{3/2}$ final states, respectively). Thus, photoionization processes in the 4d, 5s, 5p, and 6s subshells are simultaneously observable in this spectrum. With the laser switched on (lower panel) new photoelectron peaks, noted 2, 4, and 6, appear with higher kinetic energies than electrons corresponding to photoionization of the 4d, 5s, and 5p subshells in ground state atoms. In addition, the relative intensity and shape of the peaks noted 1 and 3 are distinctly modified. A new photoelectron peak, noted 5, also appears at a binding energy of 3.8 eV, which is that of a 5d electron in the $5p^66s5d\ ^1D$ excited state. No new peak is observed at the binding energy of a 6p electron in the $5p^66s6p\ ^1P$ excited state (2.97 eV). This was the case at each photon energy between 22 eV and 130 eV. The absence of the 6p photoline in our spectra is due likely to be a very low population of this state rather than a low value of the 6p photoionization cross section.

The binding energies of inner-shell electrons with one of the 6s electrons being excited to a 5d orbital have been calculated. The calculated binding energies are all shifted by 1.5 eV to 2 eV, to lower values, because the atomic valence charge is compressed following 6s − 6d transfer (Wendin, 1985). While exciting the outer electron of sodium decreased the screening of the core by the outer electron, exciting the 6s electron to a 5d orbital increases the screening. This change decreases the influence of the nuclear charge and lowers the binding energy, in accordance with our observations. Conversely, the average radius of the 6p electron density is larger than the radius of the 6s electron, but the difference is smaller than for the 5d electron density (Wendin, 1985). Therefore, one would not expect a significant increase in the binding energies for inner-shell electrons with one 6s electron in a 6p orbital. The shift has been estimated to be of the order of a few tenths of an eV (Wendin, 1985). Despite this small shift, one would observe a significant broadening in the spectrum arising from photoionization of inner-shell electrons in ground state atoms, if a sizable percentage of the excited atoms were in the $6p\ ^1P$ state. Broadened lines are not observed in the spectrum of barium shown in the upper panel of Fig. 25 which was taken at a quite different photon energy than that shown in Fig. 5 (Wuilleumier, 1984). At this photon energy, 92.06 eV, only first order photons are diffracted by the grating. Because the photon energy is lower than the binding energy of the 4d electrons, the only photoelectrons observable are ejected from the 5s, 5p, and 6s subshells in the ground state. The spectral purity of the photons makes the photoelectron spectrum simpler to interpret. When the laser is on (lower panel), one notes that the intensity of the 5d photoline is now comparable to the intensity of the 6s photoline from the ground state. The 5p photoelectron spectrum is complex but at kinetic energies corresponding to 5s photoionization, there are only two 5s photolines, one from the atom in the ground state, the other one from the atoms in the 5d excited state. No increase in the full

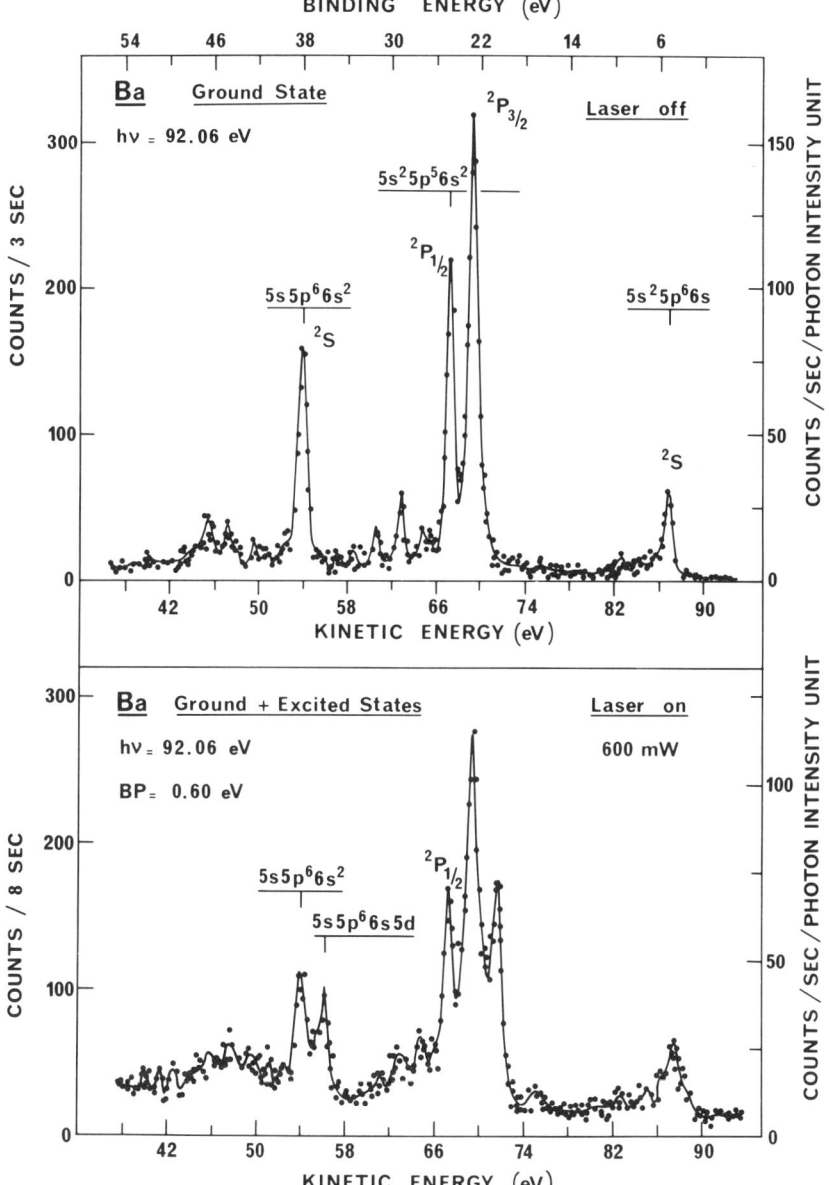

FIG. 25. Spectra of photoelectrons ejected from barium atoms in the ground state (upper frame) and in the $6s5d$ $^{1,3}D$ excited states (lower frame) by 92 eV photons. (Figure taken from Wuilleumier, 1984.)

width at half maximum (FWHM) of the 5s photoline from the ground state atoms is detectable, which puts an upper limit of a few percent on the number of atoms that are promoted to the $6p\,^1P$ excited state under our steady state conditions. This low population density can be easily understood when one considers the short lifetime (8.4 ns) of the $6s6p\,^1P$ state excited by the laser. This excited state decays radiatively to the ground state and to the $6s5d\,^1D$ state, which is a metastable state with a lifetime in the millisecond range. The ground state population is transferred to the $6s5d\,^1D$ state through the $6s6p\,^1P$ state.

In the experiments made at higher resolution, it was observed that the $6s5d\,^3D$ is also populated with a branching ratio, $^1D:{}^3D$ close to 2:1. The $6s5d\,^3D$ state is most likely populated by the decay of the $6s6p\,^3P$ state formed by collisional transfer from the $6s5d\,^1D$ state.

Using the information in the preceeding paragraphs, one can easily interpret the laser excited spectrum at 45 eV (Fig. 5) or at any other photon energy (Fig. 25). Peaks 1 and 3 in Fig. 5 are due to final ionic states produced by photoionization of ground and excited state atoms. Peaks labeled 2, 4, and 6 are due respectively to the photoionization of electrons from the $5p$, $4d$, and $6s$ subshells of barium atoms in the excited $5p^6 6s5d\,^{1,3}D$ metastable states. Since there are three open shells in the positive ion produced by inner shell photoionization of $6s5d\,^{1,3}D$ excited atoms, a large number of final ionic states results from the coupling of these open shells. The fine structure of the peaks is not resolved in the experiment.

2. Measurement of the 5d Photoionization Cross Section

From spectra similar to the ones shown in Fig. 5 and Fig. 25, the relative density of atoms transferred into the excited states was obtained from the reduction in intensity of either the $^2P_{1/2}$ or $^2D_{3/2}$ photolines when the laser was on. In these series of experiments, up to 50% of the atoms were brought into excited states. The relative variation of the 5d photoionization cross section was obtained from the normalized integrated area under the 5d photoline. At each photon energy, the photoelectrons from 5p and 5d subshells were simultaneously recorded. The absolute scale of the cross section was established by a least squares adjustment of the relative 5p to local density-based random phase approximation (LDRPA) calculations (Wendin, 1985) over the entire photon energy range covered in our experiment.

The variation of the 5d photoionization cross section as a function of the photon energy (Bizau et al., 1985b, 1985c, 1986) is shown in Fig. 26 between 15 eV and 40 eV and in Fig. 27 between 40 and 140 eV. In Fig. 26, the results of three different calculations are shown. The one-electron calculations using

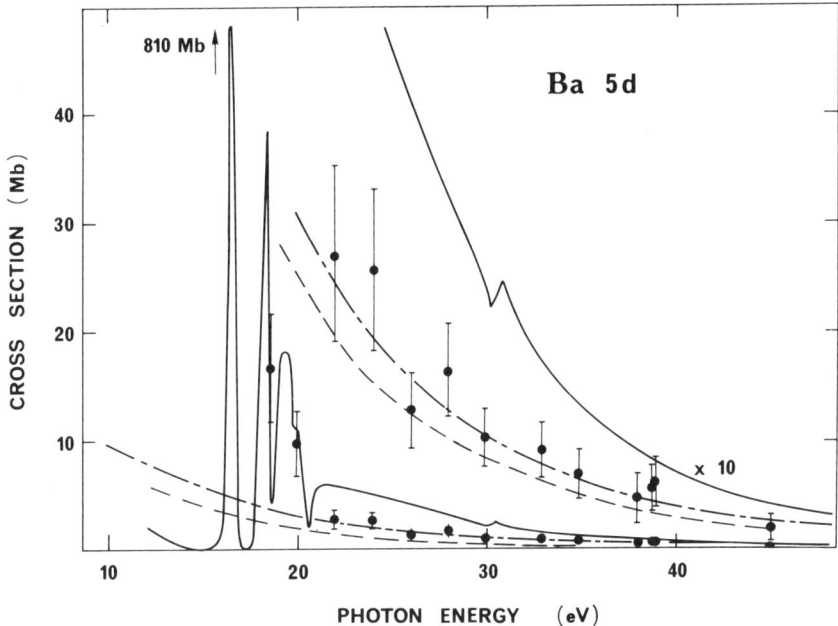

FIG. 26. Variation of the 5d photoionization cross section in Ba as a function of the photon energy from 10 eV to 40 eV. One electron theoretical results are from Hartree-Fock-Slater calculations (—·— Theodosiou, 1983) and from LDA calculations by Wendin (——— in Bizau et al., 1986). LDRPA calculations, taking into account electron correlations (—) are from Wendin (in Bizau et al., 1986). (Figure taken from Bizau et al., 1986.)

the Hartree-Fock-Slater model (Theodosiou, 1985) or the local density approximation (Wendin, 1985) reproduce the general nonresonant behavior of the cross section rather well, although such an excellent agreement should be considered as somewhat fortuitous. A third one-electron calculation (Salzmann, 1983) not shown in the figure is also in agreement with the experiment. The LDRPA calculation (Wendin, 1985), which includes some of the electron correlation effects, is appropriate in this photon energy range because it couples the 5p shell resonances that occur below 20 eV to the 5d continuum channel. One other measurement of the 5d cross section has been made by Carlsten et al. (1974). They obtained a value of about 20 Mb between 4 eV and 5 eV using a laser to photoionize the laser-excited atoms.

At higher photon energies, in Fig. 27, one observes a strong resonance enhancement of the 5d cross section when the 4d ionization channels open up. Similar resonance effects had already been seen in the photoionization of xenon atoms (West et al., 1975, and Adam et al., 1978) and in solid barium (Hecht and Lindau, 1984) in the ground state. However, these results are the

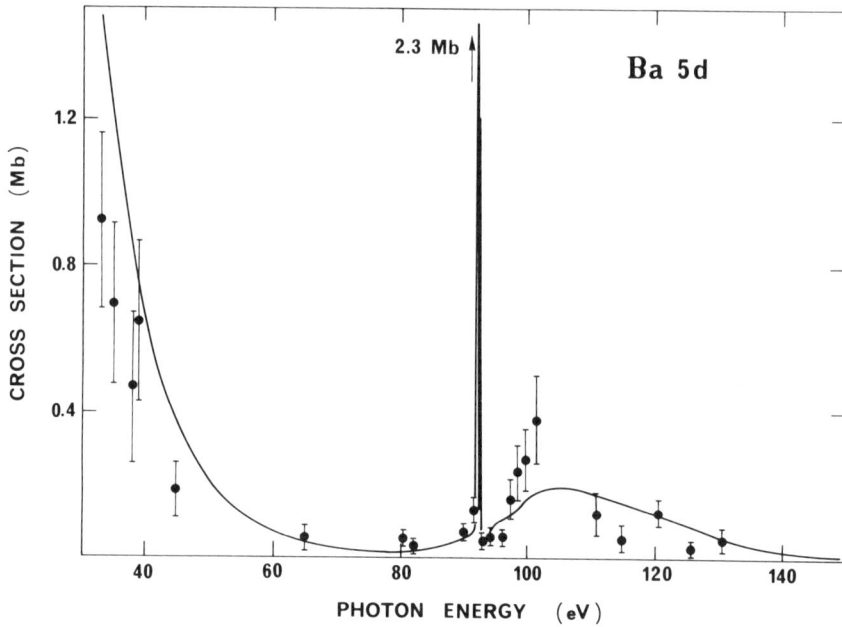

FIG. 27. Variation of the 5d cross section above 40 eV photon energy. The 4d thresholds in the excited atoms are around 96 eV and 98 eV. Only the LDRPA calculations from Wendin (—, in Bizau et al., 1985b, 1986) are able to reproduce the resonance enhancement of the cross section with a maximum above 100 eV when the 4d ionization channels open up. (Figure taken from Bizau et al., 1986.)

first demonstration that such effects do exist in the photoionization of an excited atom. They can be reproduced theoretically in the LDRPA model only by including intershell interactions between the 4d and 5d electrons.

A resonant enhancement of the 5d cross section, around 92 eV, due to the interference with discrete excited states of the $4d^9 5s^2 5p^6\, 6s5dnl$ configuration is observed in the experiment and is reproduced by the LDRPA calculations. The variation of the 5d cross section in this energy region is shown on an expanded scale in Fig. 28. In order to get good agreement with experiments around 92 eV, the Auger lifetime of the 4d-hole must be included in the calculation. The dashed portion of the curve in Fig. 28 shows how the Auger lifetime broadens the $4d \to 6p$ resonance and reduces its strength in the 5d ionization channel.

These results are the first experimental determination of a photoionization cross section for an atom in an excited state over a broad photon energy range, extending up to 40 times the binding energy and including several inner-shell ionization thresholds.

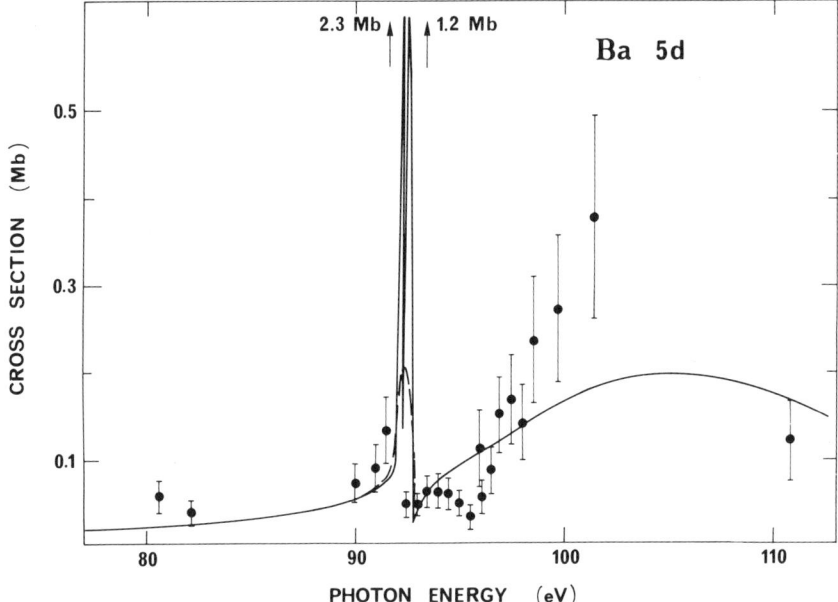

FIG. 28. The variation of the 5d photoionization cross section enlarged on a magnified scale between 80 eV and 110 eV. In addition to the resonance enhancement in the continuum, one also observed the resonant enhancement of the 5d cross section around 92 eV, due to the interference with discrete excited states of the $4d^9 5s^2 5p^6\ 6s5dnl$-type. The agreement between experimental results and the calculation, around 92 eV, is good only when the Auger lifetime of the 4d hole is taken into account (— · —, Wendin, 1985).

3. 6s-5d Resonant Photoemission in Laser-Excited Barium Atoms

Photoabsorption (Connerade *et al.*, 1979) and photoemission measurements (Rosenberg *et al.*, 1979; Kobrin *et al.*, 1984) of ground state barium atoms in the energy range of the 5p excitations (17 eV to 23 eV) revealed strong electron correlation phenomena due to the mixing of the 6s, 5d, and 6p orbitals. The spectra are dominated by the coupling of the inner-shell excited states to different ionization continua. They show up in the photoelectron lines corresponding to different final ionic states. The spectrum in this photon energy range would be expected to be even more complicated in laser-excited barium. Since inner-shell excitations of autoionizing states in sodium have been successfully investigated (Bizau *et al.*, 1985a), it seemed natural to apply this method to the study of laser-excited barium. Experiments of this type are continuing at two laboratories (Nunneman *et al.*, 1985, Cubaynes *et al.*, 1985). In this section we would like to give two examples of preliminary results to illustrate the richness of these new observations.

As for the observation of direct inner-shell photoionization into the continuum (Wuilleumier, 1982 and Bizau et al., 1983), and the measurement of the 5d photoionization cross section (Bizau et al., 1985b and 1986), the laser was tuned to the $6s^2 \to 6s6p\,^1P$ resonance, but most of the excited atoms were in the $5p^66s5d\,^{1,3}D$ excited states.

Figure 29 is an electron spectrum obtained at 18.24 eV photon energy (Cubaynes et al., 1985, 1986a). This energy corresponds to a low intensity resonance in the 5p photoabsorption spectrum of ground state barium (Connerade et al., 1979). This resonance has not been identified in the latest interpretation of this spectrum (Connerade and Martin, 1983), but the intermediate inner-shell excited state is likely to belong to a $5p^55dnln'l'$ configuration. In the upper panel, the spectrum of barium atoms in the ground state shows a strong signal in the $5p^65d$ ionic channel, corresponding to the autoionization of this discrete excited state. The $5p^66s$ continuum channel is very weak at this photon energy, showing that there is little autoionization to the ground state of the ion. The rest of the spectrum is almost featureless which implies that the branching ratio is close to unity in favor of autoionization to the $5p^65d$ channel. The direct photoionization process is also very weak to other channels at this photon energy. When the laser is on (lower panel of Fig. 29), the electron peaks due to resonant outer shell ionization of barium atoms in the ground state are reduced by about 50%. In addition, strong signals appear now at the binding energies of a 6s and of a 5d electron in the $5p^66s5d\,^{1,3}D$ excited states. At this photon energy in the laser-excited atoms, there is likely a resonant excitation of a 5p electron to a doubly excited state of the type $5p^56s5dnl$ which decays to both of the $5p^66s$ and $5p^65d$ continuua with comparable intensities. As in the photoionization of barium atoms in the ground state, no other electron peaks appear at this photon energy with a significant intensity, which implies that other continuum channels do not play an important role in the photoionization process.

The second example, taken at 19.71 eV and shown in Fig. 30 (Cubaynes et al., 1985, 1986a) looks quite different. This photon energy coincides with the excitation energy of one member of Rydberg series observed by Connerade et al. (1979), belonging to a mixture of $5p^56s^2nl$, $5p^56s5dnl$ and $5p^55d^2nl$ configurations. The ground state spectrum, (upper panel of Fig. 30) shows the resonant enhancement of a number of satellite lines corresponding to excited ionic states of Ba^+: $5p^65d$, $5p^66p$, and $5p^67s$. When the laser is on (lower panel of Fig. 30), the electron lines from atoms in the ground state practically disappear. The most intense one ($5p^66s$ final state of Ba^+) is reduced in intensity by more that 90%, which corresponds to the percentage of atoms being transferred to excited states. The spectrum shows, from the relative intensity of $5p^66s$ and $5p^65d$ photolines, that the intermediate ^1D state is

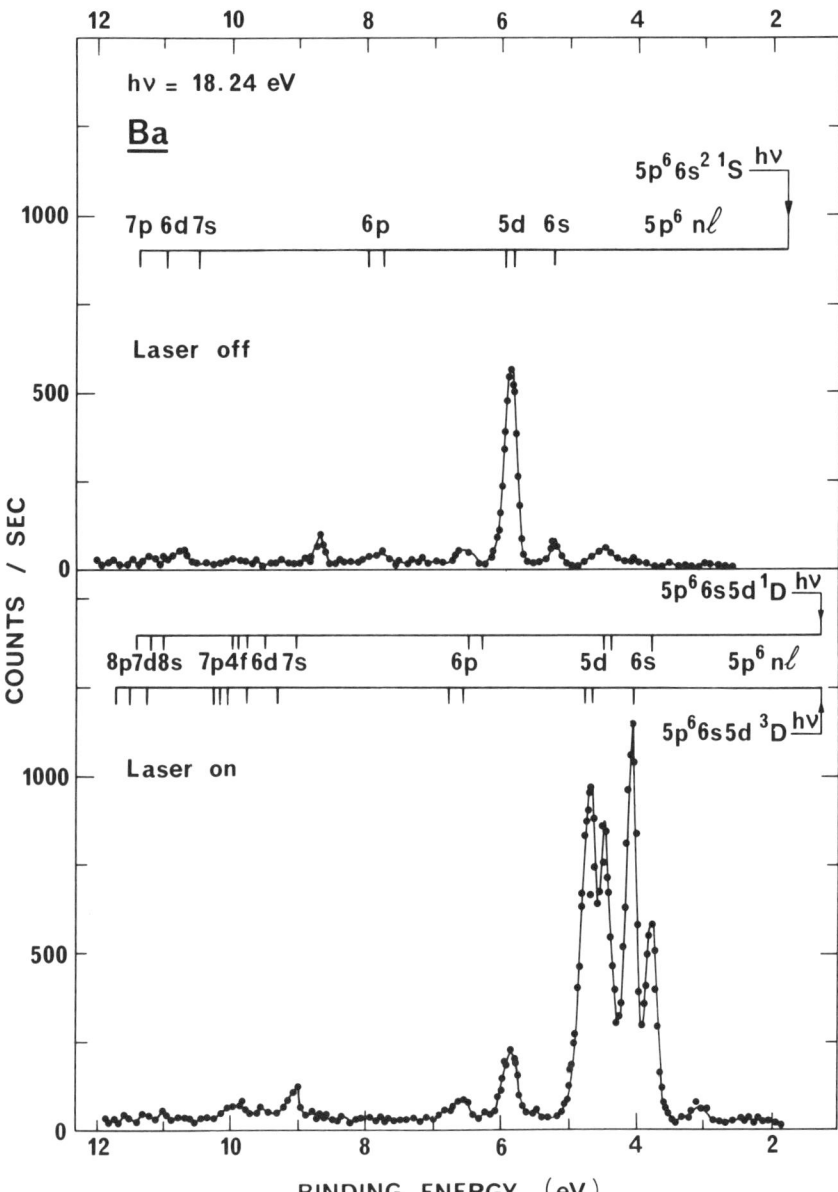

FIG. 29. Electron spectra of atomic Ba taken at $h\nu = 18.24$ eV with the pumping laser off (upper part, atoms in the ground state) and with the laser on (lower part, about 50% of the atoms are in the ground state and 50% are in the $6s5d^{1,3}D$ states). The bandpass of the monochromator was 0.13 eV. See text for detailed explanations. (Figure taken from Cubaynes et al., 1985.)

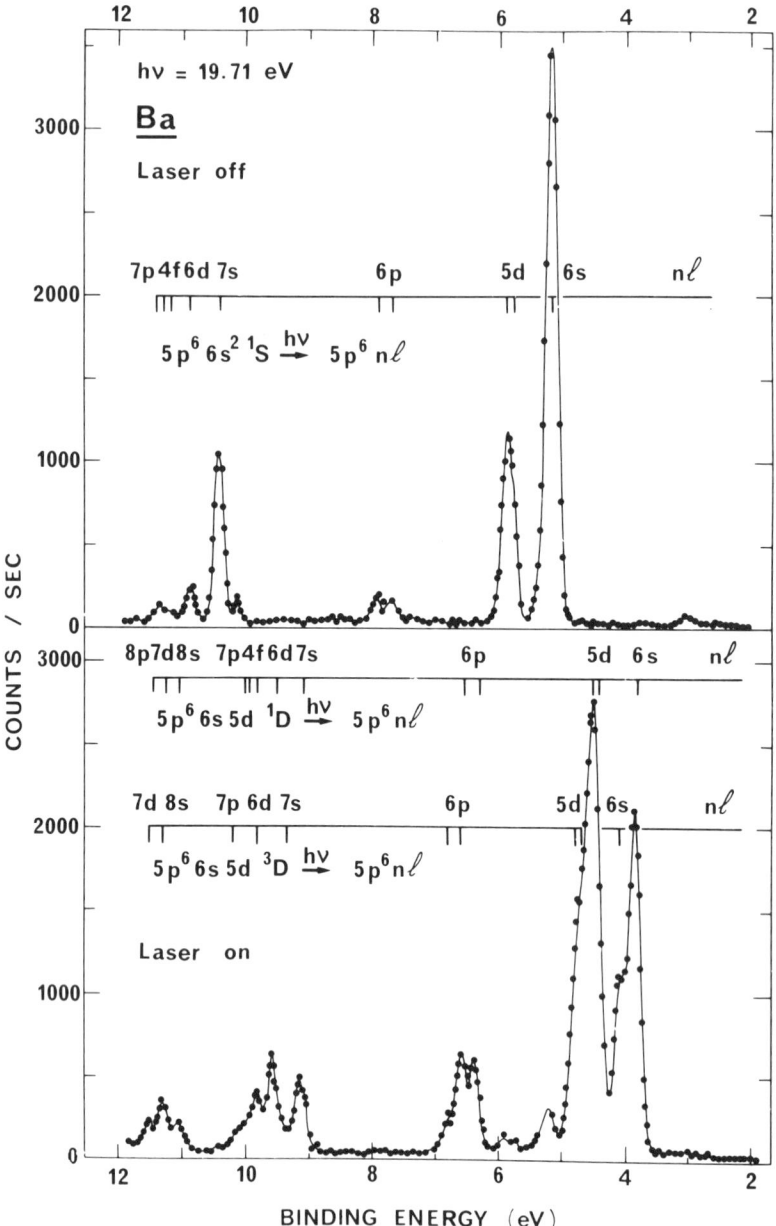

FIG. 30. Electron spectra of atomic Ba taken at a photon energy $h\nu = 19.71$ eV. Upper frame: Ba atoms are in the $5p^66s^2\ {}^1S$ ground state. Lower frame: more than 90% of the atoms have been transferred to the $5p^66s5d\ {}^{1,3}D$ excited states; one observed here, for the first time, an almost pure photoelectron spectrum from the excited states, including a large number of satellites resonantly enhanced by autoionization. (Figure taken from Cubaynes et al., 1985.)

preferentially excited and decays to a number of Ba^+ ionic states. The intensity of several satellite lines produced in autoionization of the ^1D excited state are strongly enhanced. These correspond to the ionic configurations: $5p^66p$, $5p^67s$, $5p^66d$, $5p^67p$, $5p^68s$, $5p^67d$, and $5p^68p$. This is the first observation of an almost pure resonant photoemission spectrum from an atom in an excited state.

By continuously scanning the photon energy range from 17 eV to 22 eV, the excitation functions for autoionization of ^1D and ^3D excited states into several channels have been measured and the oscillator strengths of a few strong excitation transitions have been determined, using the same method developed for sodium (Bizau et al., 1985a). In this energy range, autoionization of intermediate states formed by $5p$ excitation in the $5p^66s6p\,^1P$ laser excited atoms have been observed, but the maximum intensity measured in this channel shows that at most 4% of the atoms were in the ^1P state for our experimental conditions of laser excitation. This observation is at variance with that of Nunneman et al. (1985), who report to excite 20% of the ground state atoms to $6s6p\,^1P$.

In barium there are a large number of discrete intermediate states and continuum channels available for dipole transitions. There is also extensive initial and final state configuration mixing. This highly complex system will require extensive theoretical insight for a detailed interpretation of the $5p$ excitation spectrum. The next generation of experiments must be carried out with higher resolution to better understand the complex interactions that occur in barium. The present experiments have, nevertheless, provided a wealth of new information and demonstrated the applicability of this method.

VI. Collisional Ionization of Laser-Excited Atoms

This review deals primarily with the photoionization processes rather than collision processes. However, we want to at least consider the basic aspects of collisional ionization. In this section devoted to collisions, we shall describe the model case of sodium vapor, irradiated under well-defined conditions by cw monochromatic radiation.

A. General Background

During the last few years, considerable attention has been paid to the phenomenon of ionization of a dense atomic vapor irradiated by intense laser radiation tuned to the resonance line. Lucatorto and McIlrath (1976, 1980) and McIlrath and Lucatorto (1977) first demonstrated nearly total ionization, on a time scale of less than 1 µs, of sodium and lithium vapors of densities $\sim 10^{16}$ cm^{-3}. The vapors were irradiated by pulsed lasers of intensity 10^6 watts cm^{-2} resonant to the first electronic transition. Similar effects have been observed in other experiments on sodium vapor by Stacewicz (1980), Stacewicz and Krasinski (1981), Roussel et al., (1980), Carré et al., (1981a, 1981b), and Krebs and Schearer (1981), and in barium, calcium, and strontium vapors by Lucatorto and McIlrath (1980), Skinner (1980), Bachor and Koch (1980, 1981), Jahreiss and Huber (1983), Huber and Jahreiss (1985), Bréchignac and Cahuzac (1982), and Bréchignac et al. (1985). This efficient and rapid ionization phenomenon, occurring at relatively moderate laser intensities, cannot be accounted for in terms of multiphoton ionization processes. The explanation, originally proposed by Measures (1977), is based on energy transfer via superelastic collisions between an electron and the excited atom. The phenomenon can lead to important applications involving the coupling of laser energy into a plasma.

More precisely, Measures and Cardinal (1981) have developed a physical model of laser ionization based on resonance saturation (LIBORS) in which the phenomenon is separated into successive stages. Seed electrons are first created in the medium through collisional ionization and photoionization processes. These electrons gain energy in superelastic collisions with the laser-excited atoms. They are then able to directly ionize the atoms by electron impact. This process gives rise to an exponential growth of the electron density which finally leads to the ionization burnout.

In our experiments, we have made a detailed study of the first stage of ionization of the medium. In the total ionization experiments of Lucatorto and McIlrath (1976, 1980) and McIlrath and Lucatorto (1977), the atom density ($\gtrsim 10^{15}$ cm^{-3}) is high enough so that the mean free path, λ, of the electrons is much smaller than a typical dimension, l, of the medium. Under such conditions, the electron seeding processes take place within a very short time (a few ns) and they are hidden by the avalanche phenomenon. In contrast, our experiments limit the atom density to a relatively low value ($\lesssim 10^{13}$ cm^{-3}) such that $\lambda > l$. Under these conditions, the individual processes can be observed separately. The use of electron spectrometry has given us the opportunity to unambiguously identify the various electron seeding mechanisms. Moreover, it has allowed us to observe directly the superelastic collisions in a laser-ionized medium, verifying Measures' hypothesis. (Measures, 1970, 1977).

Our first experiments (LeGouét et al., 1982; Carré et al., 1984, 1985, 1986a) have been concerned with the study of sodium vapor. Laser pumping of the resonance line $3s$–$3p$ converts a large fraction of the atoms to the resonance level $3p$. We have observed that, under low intensity ($\lesssim 10$W.cm^{-2}) irradiation with a cw laser, the electron seeding processes are purely collisional.

These processes are associative ionization between two Na(3p) atoms, and Penning ionization of Na atoms in high-lying levels, nl, in collisions with Na(3p) atoms. The nl levels are populated in energy-pooling collision of two Na(3p) atoms (Allegrini et al., 1976, 1985). At intensities greater than 10^4 W cm^{-2} provided by pulsed lasers, photoionization of highly excited nl levels becomes more efficient than Penning ionization (Carré et al., 1985, 1986a, 1986b). At larger intensities, laser-assisted collisional processes could also contribute to the production of primary electrons, according to some authors (von Hellfeld et al., 1978; Weiner and Polak–Dingeles, 1981; Weiner, 1985). We have also observed hot electrons of up to 6.3 eV energy, resulting from 1, 2, or 3 superelastic collisions of low energy seed electrons with Na(3p) atoms (LeGouët et al., 1982; Carré et al., 1984, 1985, 1986a).

We have also undertaken experiments on barium vapor. In the presence of laser radiation tuned to the $6s^2 \to 6s6p$ resonance line, most of the atoms are transferred into the low-lying metastable levels $6s5d\,^{1,3}D$, populated by radiative decay of the $6s6p\,^1P$ resonance level. Here, the observed seed electrons essentially arise from Penning ionization of high-lying nl levels. More energetic electrons are produced in successive superelastic collisions involving mainly the metastable $5d\,^1D$ and ^3D levels (Carré et al., 1986b; Cubaynes et al., 1986, 1987).

B. Experimental Conditions

In the experiments performed at the synchrotron radiation source ACO, collision studies have been developed simultaneously with the photoionization studies, using the same set up shown in Fig. 4 and described in Sect. II.B.2. In a typical spectrum of the electrons emitted from the laser-excited vapor (see Fig. 31), those with low kinetic energy (typically 0–6 eV) are produced in collisional ionization processes, while the higher kinetic energy electrons are produced by photoionization and autoionization processes induced by synchrotron radiation. In cases where photoionization peaks occur at low energy, the collisional phenomena have to be subtracted to extract the photoionization data. Conversely, synchrotron radiation is a useful tool in the collision experiments. It was used for the measurement of the absolute and relative population densities of the various excited levels existing in the vapor in the presence of the laser radiation. Since the vapor is optically thin with respect to synchrotron radiation, this probe did not perturb the collisional medium under study. Besides, suitable photoionization peaks have served to calibrate the electron energy scale in the low kinetic energy range. Such an accurate reference was needed because the apparent kinetic energy depends on the laser power and the presence of the metal vapor.

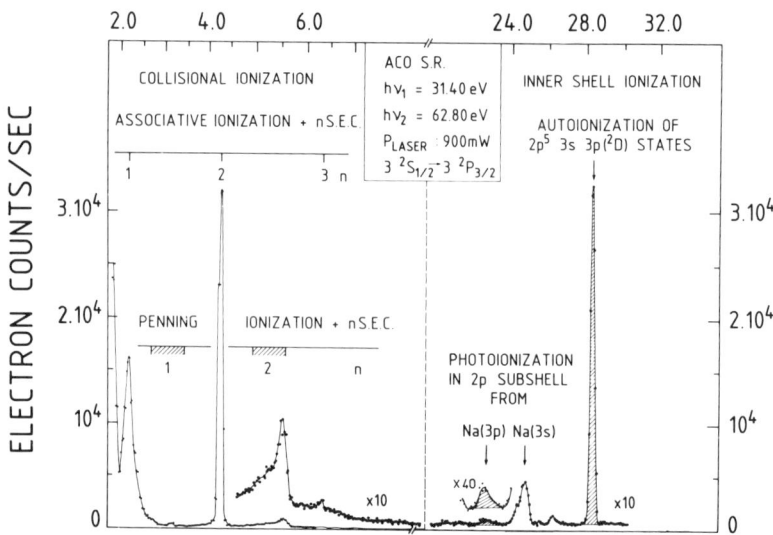

FIG. 31. Electron energy spectrum from Na atoms simultaneously laser-excited to the $3p\ ^2P_{3/2}$ state and irradiated with monochromatized synchrotron radiation from the ACO storage ring (photons of 31.40 eV and 62.80 eV). Right frame: high energy electrons come from photoionization and autoionization processed produced by VUV photons. The spectrum is very similar to the spectrum in Fig. 19, lower panel. Left frame: low energy electrons have been produced in ionizing collisions and are subsequently heated by superelastic collisions, as explained in Section VI.C. The autoionization line in the right frame at a kinetic energy of 28.37 eV serves to calibrate the absolute scale of the kinetic energies and the density of atoms in the $3p$ excited state. (Figure taken from Wuilleumier, 1984.)

We would like to emphasize here a number of characteristic experimental features, which differ from most of the previous studies on collisional ionization of laser-excited metal vapors.

1. Atomic Beam Collision Medium

One of these specific features is the use of a single atomic beam as the collision medium. A single beam (de Jong and van der Valk, 1979) or crossed beams (Weiner and Polak–Dingels, 1981) have already been used in a few ionizing collision experiments. Although the single beam configuration is much simpler than the crossed beam, it offers similar advantages for the definition of the geometrical parameters. Single beam experiments are well suited for studying effects depending on the direction of the collision axis with respect to an external field or the polarization vector of a laser field (Kircz *et al.*, 1982). As discussed by Baylis (1977), the relative velocity distribution of colliding atoms within an effusive thermal beam is narrower than in a gas cell at the same temperature. The calculation for a perfectly collimated beam indicates that the rms velocity is reduced by a factor between two and three over those in a cell. The reduction in the rms velocity corresponds to a temperature, T, much lower in a beam than in a cell, such that $T_{\text{beam}}/T_{\text{cell}} = \frac{1}{7}$.

In our experiments, the divergence of the effusive beam is rather large. The actual behavior of the beam is intermediate between that of a vapor cell and a well collimated beam at the same temperature, and it should be described by an effective temperature such that, $T_{\text{beam}} < T_{\text{eff}} < T_{\text{cell}}$. The divergence angle, θ, for the active atoms, which is roughly the ratio of the characteristic dimension of the source volume of the electron spectrometer to the distance from the oven, is $\theta \sim 0.2$ radians. In the sodium experiments, the oven temperature was varied typically in the range 500–600K, corresponding to measured densities of the beam in the active volume ranging between $5 \times 10^{10}\,\text{cm}^{-3}$ and $5 \times 10^{12}\,\text{cm}^{-3}$.

2. Laser Excitation

In the experiments reported hereafter, we used a cw single-mode dye laser irradiating the atomic beam at right angles. In the absence of Doppler broadening, such a narrow-band laser can be locked on a specific atomic transition, e.g. the most intense transition $3s\,^2S_{1/2}F = 2 \to 3p\,^2P_{3/2}F' = 3$ of the sodium resonance line, or the $6s^2\,^1S_0 \to 6s6p\,^1P_1$ transition of the most abundant isotope in barium.

By modulating the laser frequency at a low rate and synchronously detecting the fluorescence from an auxiliary atomic beam (Fig. 4), the laser is long-term stabilized to the relevant transition and maintains a constant excited-state population in the collisional medium throughout the experiment. The absorption linewidth due to the residual Doppler broadening is of the order of $\theta \Delta v_D$, where Δv_D is the Doppler-linewidth in a vapor cell at the same temperature. The amplitude of the laser frequency modulation is adjusted to fit this Doppler-broadened linewidth, of typically 100 MHz. This ensures that all atoms in the active volume can be excited, independently of the direction of their trajectories.

In the case of sodium, when the single-mode dye laser is tuned to the $3s\,^2S_{1/2}F = 2 \to 3p\,^2P_{3/2}F' = 3$ hyperfine component of the D_2 line, no optical pumping to the other hyperfine level $F = 1$ of the ground state can take place. When the transition is saturated, one expects to transfer half of the $F = 2$ population (i.e. $\frac{5}{16}$ of the total population) to the $3p$ resonance level under steady-state conditions. The actual situation is more complicated. At the highest cw laser intensities that we used ($\sim 10\,\text{W}\,\text{cm}^{-2}$), the dynamic Stark effect can mix the $F' = 3$ level with the $F' = 2$ neighboring level of the excited state, which allows spontaneous emission to the $F = 1$ level of the ground state (Grove et al., 1977). Because of the Doppler shift, the $F' = 2$ level may also be excited in atoms with oblique trajectories. These two effects, which introduce a leak in the closed $F = 2 \rightarrow F' = 3$ two-level system, decrease the efficiency of the laser excitation. On the other hand, this efficiency is increased by radiation trapping (Holstein, 1947) which becomes significant at atom densities greater than 10^{11} cm^{-3} (Garver et al., 1982; Fischer and Hertel, 1982). As discussed in Sect. V, the experimental observation is that up to 25% of the Na atoms could be converted to the $3p$ level by laser pumping. The single-mode dye laser delivered a power of up to 1 W at the D_2 wavelength (5890 Å), corresponding to an intensity of up to 10 W.cm^{-2}. Crossed polarizers were used to change the laser power without displacing the laser beam, so that the collision signal could be studied as a function of excited-state density, n_{3p}.

3. Electron Spectrometry

Many experiments involving ionizing collisions in metal vapors have used ion detection, sometimes in conjunction with mass spectrometry. Electrons have been involved only in measurements of electron temperatures (Stacewicz, 1980, and Roussel et al., 1980) or total electron yields (Stacewicz and Krasinski, 1981). Electron spectrometry is a tool that permits one to observe individual ionization channels of the atoms and collisional energy transfers between the electrons and the atoms. Besides our studies on metal vapors, electron spectrometry has also been used for studying collisional ionization in laser-excited rare gas systems by Ganz et al. (1982, 1983, and 1984) and by Bussett et al. (1985). Optical methods such as VUV absorption (Lucatorto and McIlrath 1976, 1980, and McIlrath and Lucatorto, 1977), hook interferometry (Skinner, 1980; Bachor and Koch, 1980, 1981; Jahreiss and Huber, 1983; Huber and Jahreiss, 1985), laser-induced fluorescence (Krebs and Schearer, 1981 and Bréchignac and Cahuzac, 1982) or laser-induced absorption (Bréchignac et al., 1985) can provide complementary, time-resolved information on the individual level populations of the neutral and ionized species present in the laser-irradiated medium.

Our electron spectrometer is shown in Fig. 4 and described in Sect. II.B.2. The polarization vector of the laser radiation is set parallel to the CMA axis. In future experiments, it would be interesting to study the angular distribution of the electrons produced in the collision as a function of the laser polarization to measure the effect of the alignment or orientation of the excited colliding atoms (Kircz et al., 1982 and Nienhuis, 1982).

In the collision experiments, the use of the CMA raises specific problems connected with the measurement of electron kinetic energies between 0 and

5 eV. The transmission factor of the spectrometer rises from zero for electrons with zero kinetic energy to a plateau for the kinetic energies greater than about 5 eV. Moreover, repulsive potentials prevent very low energy electrons to diffuse freely out of the collisional sample. These are a contact potential ($V_c \lesssim 2V$), which develops on the walls of the collision chamber, and a plasma potential, which develops inside the collisional medium itself (electron density $\sim 10^6$ cm^{-3}). The plasma potential has a magnitude of V_p, between 0.1 and 0.8 eV depending on the laser intensity. Conical accelerating grids are placed around the interaction volume to extract the electrons with kinetic energies lower than about 3 eV.

4. Synchrotron Radiation

Since the apparatus is on line with the ACO storage ring, the problems connected with the electron spectrometer can be solved with the use of synchrotron radiation. We calibrate the transmission factor of the CMA with or without the accelerating grids, according to the experiment, by using photoelectron lines of known intensity produced by the photoionization of rare gases by synchrotron radiation. In order to determine the absolute energies of the electron peaks, despite the apparent energy shifts due to the repulsive potentials, we use reference lines of known energy produced by photoionization of the metal itself, or subsequent decay processes (Auger lines), or by photoionization of an auxiliary gas (Xe) injected in the vacuum chamber. This method is advantageous because it provides a real time energy calibration.

Other crucial parameters in the collision experiments are the absolute and relative densities of ground state and excited state atoms. These are also obtained in real time, without appreciable perturbation of the collisional medium, by photoionization of the metal vapor with synchrotron radiation. The photoionization signals are recorded simultaneously with the collision signals in a higher energy window of the electron spectrometer. By comparing the intensities of the photoionization signals of atoms in the ground state and in the excited state, one can deduce the excited state fractional population. This is illustrated in Fig. 16 for $2p$ inner shell photoionization of Na. In the case where several excited states are present in the laser-irradiated vapor, the method also yields the relative excited state populations. Moreover, as we discussed for sodium and barium in Section V, the wavelength of the monochromatized synchrotron radiation can be adjusted to coincide with the excitation energy of autoionizing resonances in the excited atom as shown for example in Fig. 31. These resonances produce photoelectron lines that are much more intense than nonresonant photoionization lines, and provide more sensitive measurements of the excited state density as examplified in

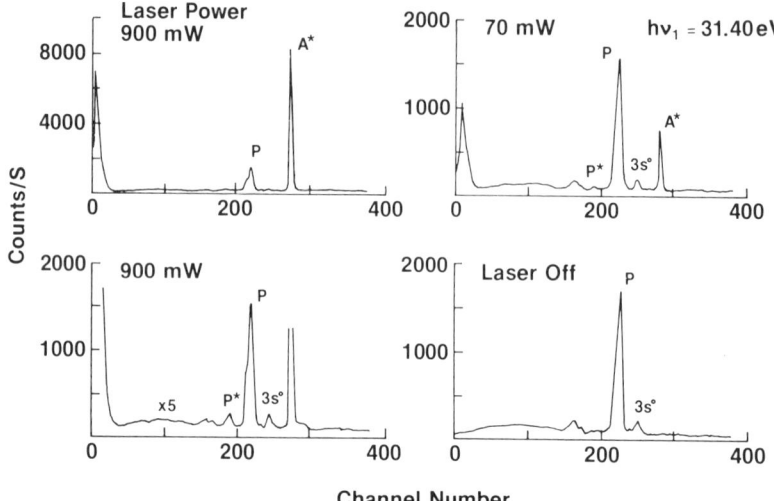

FIG. 32. Variation, as a function of the laser power, of the electron energy spectrum from sodium atoms shown in Figure 31. The density of excited atoms at a laser power of 70 mW is about one order of magnitude lower than in the spectrum at 900 mW. At 70 mW, the photoelectron line P^* arising from $2p$ shell ionization in the sodium atoms laser excited to the $3p$ state, is barely visible, while the autoionizing line A^* is still intense and can be used for calibration. The $3s$ photoline is due to $3s$-photoionization of sodium atoms in the ground state by photons of 31.40 eV energy. The left lower part of the figure is a magnification (fractor 5) of the left upper part.

Fig. 32. For example, with the use of the strongest autoionization resonance in sodium at 31.40 eV photon energy (Fig. 32), it is possible to measure a relative population of $3p$ excited sodium atoms of less than 1% with an acuracy of 15–20%.

C. THE SHOW CASE OF SODIUM

1. Electron Spectra

Figure 33 shows a typical energy spectrum of the electrons ejected out of the laser-excited sodium beam for an oven temperature $T = 580$K. The left-hand part (a) of the spectrum was obtained with the use of the accelerating grids. It essentially displays the spectrum of seed electrons produced by collisional ionization. The intense peak, at near zero kinetic energy, arises from associative ionization between two Na($3p$) atoms:

$$\text{Na}(3p) + \text{Na}(3p) \rightarrow \text{Na}_2^+ + e \quad (11)$$

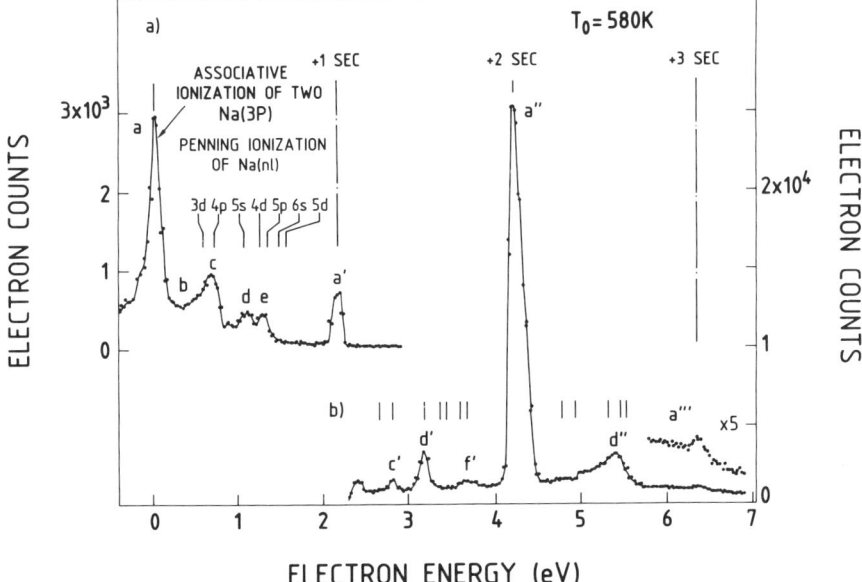

FIG. 33. Energy spectra of the electrons emitted from a sodium atomic beam (oven temperature $T = 580$ K) irradiated by cw laser radiation tuned to the resonance line. Spectrum (a) was obtained with the use of extracting grids. Spectrum (b) was produced by the electrons escaping freely out of the irradiated volume. The spectra are not corrected for the transmission function of the electron spectrometer and the grids, i.e., the intensities in spectra (a) and (b) are not directly comparable. Peak (a) arises from associative ionization, peaks b, c, d, and e from Penning ionization. Peaks labelled with p primes result from the effect of p superelastic collisions. (Figure taken from Carré et al., 1984.)

The series of peaks: b, c, d, e, between 0.5 and 1.5 eV correspond to Penning ionization of different high-lying excited levels of nl quantum numbers in collisions with Na($3p$) atoms:

$$\text{Na}(nl) + \text{Na}(3p) \rightarrow \text{Na}^+ + \text{Na}(3s) + e(2.1 \text{ eV} - E_{nl}), \qquad (12)$$

where E_{nl} is the binding energy of the nl level.

The right-hand (b) of the spectrum was obtained without the grids. It shows the higher energy electrons which have been heated in superelastic collisions with $3p$ excited atoms:

$$\text{Na}(3p) + e(E) \rightarrow \text{Na}(3s) + e(E + 2.1 \text{ eV}). \qquad (13)$$

This part of the spectrum approximately reproduces the seed electron spectrum (range 0–2.1 eV), translated by the effect of one superelastic collision (2.1–4.2 eV) or two superelastic collisions (4.2–6.3 eV). The associative ionization peak, a, at about zero kinetic energy gives rise to the peaks, a

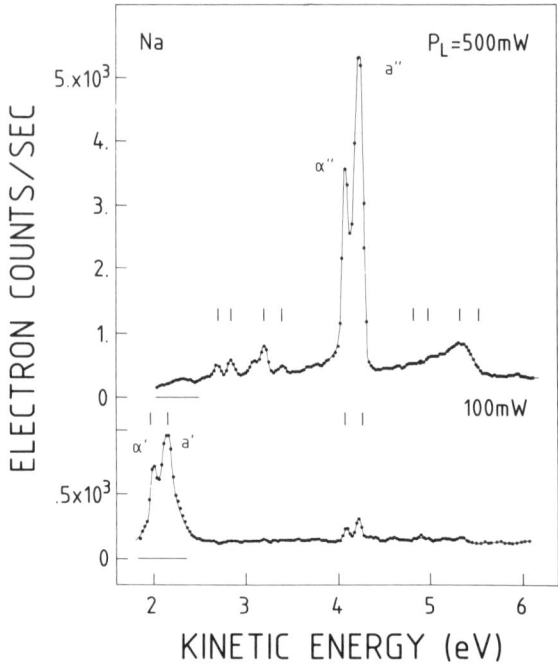

FIG. 34. Energy spectra of the electrons diffusing freely out of the laser-irradiated volume of the sodium atomic beam, at the oven temperature $T = 520$ K, for two different laser powers P_L. The figure shows the effect of the plasma potential (increasing with P_L), which repels the low energy electrons. The associative ionization signal splits into two peaks with relative intensities independent of P_L and the number of superelastic collisions.

with p primes, after p superelastic collisions. The peak a' at 2.1 eV is observed in both spectra (a) and (b). Because of the higher transmission of the CMA at this energy, the peak a'' at 4.2 eV appears more intense. The peak a''', resulting from associative ionization followed by three successive superelastic collisions, is still visible at 6.3 eV. The same labeling convention b, c, d, e with p primes is used to label the peaks produced by Penning ionization followed by p superelastic collisions ($p = 1, 2$). Because broadening occurs with each superelastic collision, the Penning ionization peaks are not resolved after two consecutive superelastic collisions.

Figure 34 displays the apparent energy shift depending on the radiation intensity, which we discussed in Sect. VI.B.3. This effect arises from the increase in the plasma potential, V_p, for increasing laser intensities, i.e. for increasing rate of ionization. The figure shows spectra obtained without extracting grids at $T = 520$K, for two different laser powers. The repulsive potentials $V_c + V_p$ (V_c contact potential) allow only the electrons with a high enough kinetic energy to escape out from the active volume. At a laser power of 100 mW, we observe the associative ionization + 1 superelastic collision signal a' at 2.1 eV. At a power of 500 mW, this signal does not appear anymore, because V_p increases. However, the excited state density n_{3p} is much greater and the associative ionization + two superelastic collisions signal a'' is more intense. Note that at this relatively low temperature of 520K, a double-peaked structure appears in the associative ionization signals in Fig. 34. Possible origins of this structure will be discussed in the following paragraph.

2. Physical processes

a. Associative ionization. When the total energy of the two-atom colliding system is larger than the ionization threshold of a single atom, Penning ionization producing atomic ions can occur. If this condition is not fulfilled, ionizing collisions may still occur in the form of associative ionization giving rise to molecular ions. In the case of two Na(3p) atoms, the total electronic excitation energy (4.22 eV) is less than the atomic sodium ionization threshold (5.14 eV), but just greater than the bottom of the potential energy well of the ground electronic state $^2\Sigma_g$ of the Na_2^+ molecular ion. These conditions are shown with the potential energy curves in Fig. 35. Associative ionization of two Na(3p) atoms can thus take place, and it gives rise to Na_2^+ ions and low energy electrons.

One has, in fact, to consider the various forms of energy which may be exchanged in the associative ionization process in sodium:

$$Na(3p) + Na(3p) + W_{kin} \rightarrow Na_2^+(^2\Sigma_g, v) + e(E_v), \tag{14}$$

where E_v is the kinetic energy of the electrons emitted when the Na_2^+ ion is formed in the vibrational level, v, of the ground state $^2\Sigma_g$, and W_{kin} the mean relative kinetic energy of the colliding atoms in the beam. From the value of W_{kin} (W_{kin} = 50–65 meV for T_{eff} = 400–500K), the known values of the binding energy (Leutwyler et al., 1981) and the vibrational spacing (Carlson et al., 1981) of the Na_2^+ molecule, one can deduce that electrons are emitted with a kinetic energy between 0 and 100 meV, leaving the Na_2^+ ion in vibrational states lying between $v = 3$ and $v = 10$. This proposition is consistent with the spectral feature, a, shown in Fig. 33 located at an energy 0.05 eV and having a width of about 50 meV. The double-peaked structure observed in Fig. 34 may be connected with the production of different sets of vibrational levels of Na_2^+, according to the collisional entrance channel of the quasi-molecule Na_2. Another possible explanation is that negative ions Na^- are formed in the excited state which lies just below the 3p level of the neutral atom (Andrick et al., 1972). This negative ion would dissociate emitting an electron with kinetic energy slightly less than 2.1 eV.

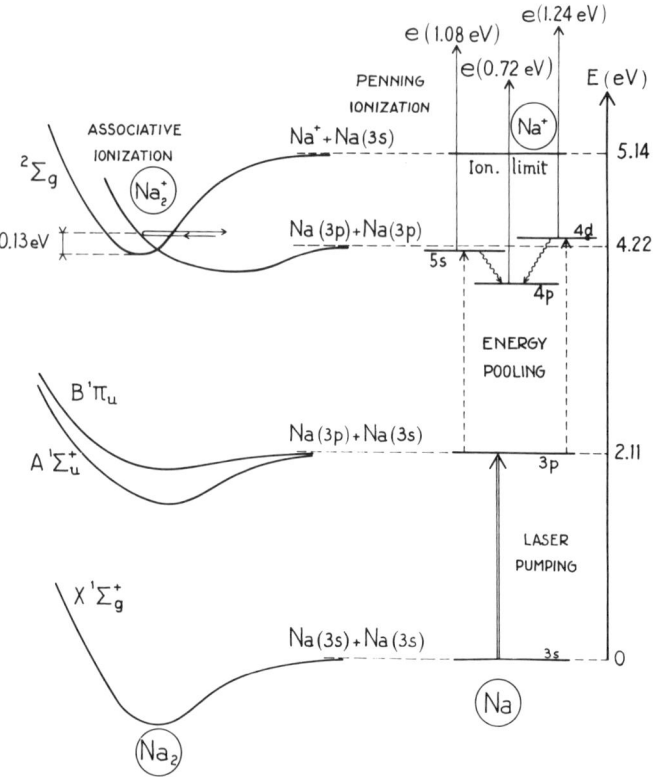

FIG. 35. Energy levels of the sodium atom Na and of the molecule Na$_2$ involved in the collisional processes which take place in the sodium vapor when the atom is laser–excited to the 3p state.

In order to gain more insight into the fine structure of the associative ionization process, additional experimental and theoretical effort will be necessary. New experiments are being planned that use atomic beams with well defined velocity distributions to decrease the width of the relative kinetic energy distribution in the initial state. A better defined initial state kinetic energy distribution will enable one to resolve the vibrational distribution in the final state. Supersonic beams have been used by Wang et al. (1985) to measure associative ionization cross sections as a function of velocity. Laser induced fluorescence can also be used to probe the vibrational distribution in the $^2\Sigma_g^+$ state of Na$_2^+$. Preliminary calculations by Valance and Nguyen (1981) and by Montagnani et al. (1983) have predicted that a crossing point exists at short internuclear distance between the $^1\Sigma_g^+$ excited state of Na$_2$ and the $^2\Sigma_g^+$ ground state of Na$_2^+$. However, other molecular states correlated to

Na(3p) + Na(3p) can be formed in the collision. More complete calculations providing the various potential curves of the system have been recently published by Henriet et al. (1985). These authors have found four crossings between the curves correlated to separated sodium atoms in the 3p state and the Na_2^+ ground state.

Associative ionization between two alkali atoms excited to the first resonance level has also been studied in cesium (Klucharev and Ryazanov, 1972b), rubidium (Borodin et al., 1975), and potassium (Klucharev et al., 1977a, 1977b). The numerous studies of the Na(3p) + Na(3p) system, in a vapor cell (Bearman and Leventhal, 1978; Kushawaha and Leventhal, 1980, 1982a) or an effusive atomic beam (de Jong and van der Valk, 1979) have led to disparate cross section values ranging over two orders of magnitude. Differences in the experimental conditions such as the density of the medium, polarization and spatial distribution of the laser radiation, fluorescence radiation trapping, and detection techniques have been suggested as causes of the discrepancies. Our electron spectrometry measurement (Carré et al., 1984) of the associative ionization cross section in sodium is 4×10^{-17} cm^2 at $T = 520$K, and agrees reasonably well with the most recent ion detection measurements by Huennekens and Gallagher (1983a, 1983b) in a vapor cell, or by Bonanno et al. (1983) in crossed beams. The dependence of the associative ionization rate on the polarization properties of the laser radiation has been demonstrated experimentally by Kircz et al. (1982) and investigated theoretically by Nienhuis (1982).

For high radiation intensities $\gtrsim 10^5$ W cm^{-2}, the associative ionization cross section between two 3p sodium atoms is enhanced, according to Weiner and Polak-Dingels (1981). This would be due to a laser-assisted associative ionization process,

$$Na(3p) + Na(3p) + hv_L \rightarrow Na_2^+ + e, \qquad (15)$$

arising from radiation-induced couplings in the quasi-molecule formed during the collision (Bellum and George, 1978). Such a radiatively aided process can occur in systems where purely collisional associative ionization is not energetically permitted, such as Li(2p) + Li(2p) (von Hellfeld et al., 1978).

Associative ionization between a highly excited alkali atom and one in the ground state, of the type

$$Na(nl) + Na(3s) \rightarrow Na_2^+ + e, \qquad (16)$$

has also been investigated. The process Na(np) + Na(3s) is allowed for $n \geq 5$. Crossed-beam measurements for $5 \leq n \leq 15$, by Boulmer et al. (1983), show that the cross section reaches a maximum of more than 10^{-14} cm^2 near $n = 11$. Similar observations have been made in cell experiments for Cs(np) by Devdariani et al. (1978) and for Rb(np) by Klucharev et al. (1980).

Associative ionization involving ns or nd levels and the ground state have been studied in Na by Kushawaha and Leventhal (1982a) and in Rb by Chéret et al. (1981, 1982a).

b. Penning ionization. In the process described by Eq. (12), the transfer of the resonance energy of a $3p$ excited atom can ionize an atom in a highly excited nl level of binding energy $E_{nl} < 2.1$ eV (Fig. 35). The energies of the peaks b, c, d, e in Fig. 33 allow one to interpret them as resulting from Penning ionization of the nl levels $3d$, $4p$, $5s$, and $4d/4f$, respectively. Figure 36 shows on a larger scale the Penning ionization peaks b', c', d', e' observed after one superelastic collision, which are better resolved at the lower temperature $T = 520$ K.

It is to be noted that electrons with the same kinetic energy, 2.1 eV-E_{nl}, would be produced by photoionization of the nl level by the laser radiation tuned to the resonance line:

$$\text{Na}(nl) + hv_L \to \text{Na}^+ + e(2.1 \text{ eV} - E_{nl}). \tag{17}$$

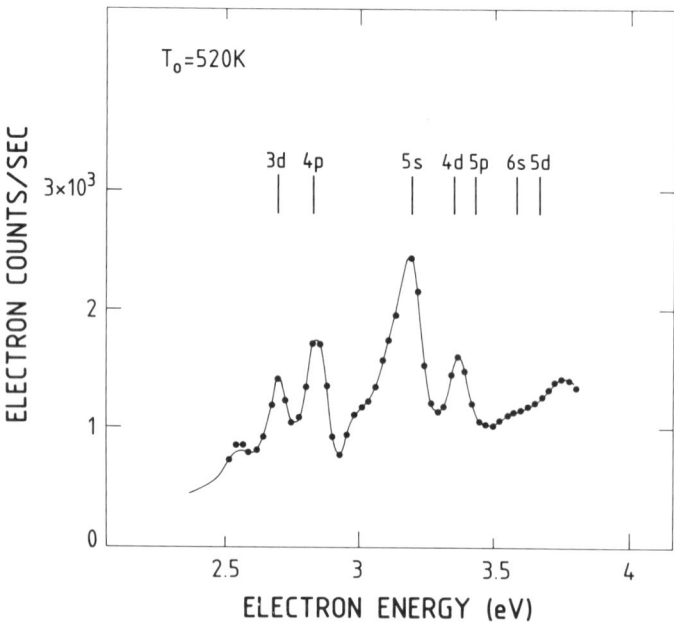

FIG. 36. Partial electron spectrum from excited sodium atoms showing the electrons produced by Penning ionization followed by one superelastic collision, at the oven temperature $T = 520$ K.

However, an analysis based on the calculated photoionization cross sections by Aymar et al. (1976) suggests that, with the intensity available ($\leqslant 10$ W.cm^{-2}) from the cw dye laser, the probability of the photoionization process (17) is typically 100 times less than the probability of the collisional processes of Eq. (12). At the higher intensities provided by pulsed dye lasers, the situation would be reversed. On the same set-up with a flash lamp pumped dye laser of intensity 10^5 Wm^{-2}, we verified that the electron peaks observed at the same energies were produced mainly by photoionization (except for the 5s level, which has a very low photoionization cross section) (Carré et al., 1985, 1986a). Another possible collisional process, which involves the same partners as process (12) but would produce electrons of different energy, is the associative ionization process:

$$\text{Na}(nl) + \text{Na}(3p) \rightarrow \text{Na}_2^+ + e. \qquad (18)$$

However, this process has a lower gross section than Penning ionization (Chéret et al., 1982a, 1982b) and should produce broader electron spectral features. It does not contribute appreciably to our electron spectra.

Our electron spectrometry measurements of Penning ionization of highly excited levels in sodium yield large cross sections, in the range $10^{-13} - 10^{-12}$ cm^2. For example our measured cross section for Penning ionization of the 5s level at $T = 520$K was 1.1×10^{-12} cm^2 (Carré et al., 1984). These values are about one order of magnitude larger than geometrical cross sections. By contrast with associative ionization, Penning ionization occurs at large internuclear distances of the colliding atoms and involves long range dipole-dipole interactions. Mass resolved ion detection studies by Chéret et al. (1982a, 1982b, 1985) of Rb(nl) + Rb($5p$) collisions, for series of ns and nd high lying levels, have given similarly large values of Penning ionization cross sections. Penning ionization cross sections were observed to be about 100 times larger than associative ionization cross sections for the same systems.

We have shown by our electron spectrometry experiments that Penning ionization of high lying excited levels as well as associative ionization play an important role as an electron seeding mechanism in laser-excited sodium vapor. Since Penning ionization has a larger cross section but involves less populated levels, it has a comparable efficiency with associative ionization. This result will be especially valuable in the interpretation of the spectra of other metal vapors like barium (Cubaynes et al., 1986, 1987), in which associative ionization does not play a major role.

c. Energy pooling collisions. The highly excited nl levels involved in Penning ionization lie around twice the energy of the $3p$ resonance level (dashed line at 4.22 eV in Fig. 35) or below. Levels such as $4d$ or $5s$ are populated via energy pooling collisions between two Na($3p$) laser excited atoms:

$$\text{Na}(3p) + \text{Na}(3p) \rightarrow \text{Na}(nl) + \text{Na}(3s) + \Delta E_{nl} \qquad (19)$$

where ΔE_{nl} is the electronic excitation energy mismatch of the reaction. For the $4d$ and $5s$ states, ΔE_{nl} is -78 meV and $+88$ meV, respectively. In the case of $4d$ level, the energy defect is of the order of the thermal energy and can be covered by the kinetic energy of the colliding atoms. Process (19) can take place via dipole-dipole couplings at crossings occuring at large internuclear distances (~ 20 a.u.) between molecular potential curves of Na$_2$ correlated to $3p + 3p$ and $nl + 3s$, respectively (Kowalczyk, 1979). The lower $4p$ or $3d$ levels are populated via radiative decays from the above collisionally populated $4d$, $5s$ states.

Energy pooling collisions between two alkali atoms excited in the resonance level have been studied in cesium (Klucharev and Lazarenko, 1972), potassium (Allegrini et al., 1982), rubidium (Barbier and Chéret, 1983), and in mixtures of two elements (Allegrini et al., 1981). Theoretical calculations based on the construction of adiabatic molecular terms at large internuclear distances have been performed for systems with a small energy defect, such as: Cs($6p$) + Na($6p$) $\rightarrow$ Cs($6d$), (Borodin and Komarov, 1974); Na($3p$) + Na($3p$) $\rightarrow$ Na($4d$), (Kowalczyk, 1979, 1984); Rb($5p$) + Rb($5p$) $\rightarrow$ Rb($5d$), (Barbier and Chéret, 1983). The bulk of experimental studies has been carried on sodium vapor, using cw laser excitation of the $3p$ level and spectral analysis of the fluorescence from the high lying levels (Allegrini et al., 1976, 1983, and 1985; Bearman and Leventhal, 1978; Kushawaha and Leventhal, 1980, 1982b). Other sodium experiments involving pulsed laser excitation allowed the study of time-dependent effects (Kopystynska and Kowalczyk, 1978; Krebs and Schearer, 1981; Hunnekens and Gallagher, 1983). The most recent measurements of Allegrini et al. (1983) and Huennekens and Gallagher (1983) yield energy pooling cross sections with a value of about 2×10^{-15} cm^2 for the production of the $4d$ and $5s$ levels in Na($3p$) + Na($3p$) collisions. This value is consistent with the one we can deduce indirectly from our electron spectrometry measurements of the subsequent Penning ionization. Under our experimental conditions of relatively low atom density and cw irradiation, this corresponds to typical populations of the highly excited nl levels that are a factor of 10^{-4} less than the population of atoms in the $3p$ state.

d. Superelastic collisions. The superelastic collision process described by Eq. (13) is the symmetrical process of the more heavily investigated electron impact excitation process:

$$\text{Na}(3s) + e(E) \rightarrow \text{Na}(3p) + e(E - 2.1 \text{ eV}). \qquad (20)$$

The cross section of superelastic collisions, $\sigma_{SEC}(E)$, can thus be evaluated from experimental values of the electron impact excitation cross section of the sodium resonance line (Enemark and Gallagher, 1972). Calculations of the electron scattering cross sections from the excited states of sodium are also available (Moores et al., 1974). In our range of interest, σ_{SEC} varies from about 10^{-14} cm^2 at $E = 0$ to 10^{-15} cm^2 at $E = 5$ eV.

The superelastic collision process has been observed in crossed electron beam laser-excited sodium beam studies by Hertel and Stoll (1974). Similar studies have been performed on barium by Register et al. (1978). Our experiments provide the first observation of the process in a laser-excited collisional medium. They confirm the original hypothesis of Measures (1977) and validate the theoretical model of Measures et al. (1979), and Measures and Cardinal (1981), in which superelastic collisions play a central role in the kinetics of efficient laser ionization of a metal vapor.

The observation of up to three successive superelastic collisions (Fig. 31) may seem unusual for a medium with our low atomic density. However, the low energy electrons are trapped in the medium for a time period corresponding to many collisions, because of the plasma potential and the contact potential.

This analysis is supported by Fig. 37, which shows the observed cubic dependence of the associative ionization signal after two superelastic collisions (the 4.2 eV peak in Fig. 33) as a function of the excited state density

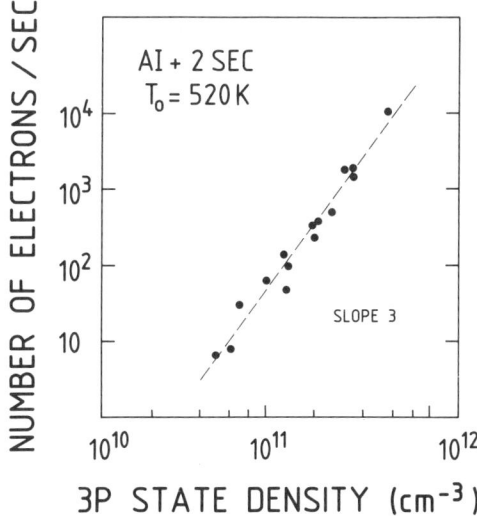

FIG. 37. Variation of the signal produced by electrons with 4.2 eV kinetic energy as a function of the density n_{3p} of excited sodium atoms. These electrons are produced by associative ionization followed by two superelastic collisions. (Figure taken from Carré et al., 1984.)

n_{3p}. Since the signal results from one ionizing collision of two $3p$ atoms followed by two collisions of an electron with a $3p$ atom, one would expect at first glance a fourth power dependence on n_{3p}. In fact, because of the quasi-infinite diffusion time for the seed electrons, the number of electrons which undergo one superelastic collision varies as the associative ionization rate, i.e. as $(n_{3p})^2$. Then the number of electrons which undergo two superelastic collisions varies only as $(n_{3p})^3$.

VII. Conclusion

The results described in the various sections of this chapter illustrate the level presently reached in combined laser-synchrotron radiation experiments for ionization studies in excited atoms, using the flux emitted from the bending magnet of a storage ring. Much more progress is still possible by optimizing the various parameters involved in these experiments. In particular, improvements in electron collection can be made and another geometrical arrangement can be chosen, which would allow angular distribution measurements to be made, as a function of the laser polarization producing alignment of the excited atoms. In the excitation range where the cw dye lasers are tunable nowadays, the main limitations come from the low intensity of synchrotron radiation. New opportunities are possible in the future, when SR intensity can be greatly enhanced through the insertion of undulators in the straight section of a storage ring (Billardon et al., 1983).

In the new positron storage ring, Super ACO, being built in Orsay six sections will be available for insertion devices. The flux output, from the undulators at the 800 MeV, 500 mA facility, is conservatively expected to be one or two orders of magnitude greater than is presently available from a bending magnet at ACO. For example, 10^8 photons/1% bandwidth are expected to be emitted in each pulse from a 30 pole undulator mounted on a storage ring such as Super ACO. Either the undulator beam or the laser beam could be used to prepare an excited atom. The other radiation could be used to probe the excited state. Ultimately, the most attractive excitation-probe mode would be to mode-lock a laser to the synchrotron pulse train (Koch, 1982). Such a system would provide experimentalists with a powerful tool to study the dynamics of the excited states. A greater flux level will make the development of a new group of experiments attractive. Some possible candidates are:

(a) measurements involving transition metals and rare earths, which are difficult to vaporize;

(b) development of crossed beam experiments to study collisional ionization between excited atoms;
(c) the study of the angular distribution of electrons from collisional and photoionization processes;
(d) time history of excitation processes.

There are so many exciting opportunities. We all eagerly await the improved sources that will soon be available.

ACKNOWLEDGEMENTS

We would like to thank Ms. Julie Robertson for her technical assistance in the preparation of this manuscript. We are also indebted to the technical staff at LURE and thank them for providing the synchrotron radiation for the experiments described in Sect. V and Sect. VI. We would like to thank our colleagues who have contributed to the new research combining laser and synchrotron radiation. We appreciate the support from NATO grant No. 1735 and the continuing support from J. C. Lehmann, Y. Petroff, and P. Jaeglé.

REFERENCES

Adam, M. Y., Wuilleumier, F., Sandner, N., Krummacher, S., Schmidt, V., and Mehlhorn, W. (1978). *Jap. J. Appl. Phys.* **17-2**, 170.
Agnew, L. and Summers, C. (1965). Proceedings of the Seventh International Conference on Ionized Gases, Belgrade, p. 574 (unpublished).
Allegrini, M., Alzetta, G., Kopystynska, A., Mori, L., and Orriols, G. (1976). *Opt. Comm.* **19**, 96.
Allegrini, M., Bicchi, P., and Gozzini, S. (1981). *Opt. Comm.*, **36**, 449.
Allegrini, M., Gozzini, S., Longo, I., Savino, P., and Bicchi, P., (1982). *Nuovo Cimento D* **1**, 49.
Allegrini, M., Bicchi, P., and Moi, L. (1983). *Phys. Rev. A* **28**, 1338.
Allegrini, M., Gabbanim, G., and Moi, L. (1985). *J. Physique* **46**, C1-61.
Aller, L. B. (1984). "Physics of Thermal Gaseous Nebulae: Physical Processes in Gaseous Nebulae" (Dordrecht, Holland, Boston, MA).
Ambartzumian, R. V., Furzikov, N. P., Letokhov, V. S., and Puretsky, A. A. (1976). *Appl. Phys.* **9**, 355.
Amusia, M. Ya., Cherepkov, N. A., and Chernysheva, L. V. (1971). *Sov. Phys. JETP* **33**, 90.
Andrick, D., Eyg, M., and Hofmann, M. (1972). *J. Phys. B* **5**, L15.
Arnous, E., Klansfeld, S., and Wane, S. (1973). *Phys. Rev. A* **7**, 1559.
Aymar, M., Luc-Koenig, E., and Combet-Farnoux, F. (1976). *J. Phys. B* **9**, 1279.
Aymar, M., and Crance, M. (1980). *J. Phys. B* **13**, 2527.
Aymar, M., Camus, P., and El Hidey, A. (1982). *J. Phys. B* **15**, 1759.
Bachor, H. A. and Koch, M. (1980). *J. Phys. B* **13**, 1369.
Bachor, H. A. and Koch, M. (1981). *J. Phys. B* **14**, 2793.

Balloffet, G., Romand, J., and Vodar, B. (1961). *S. R. Acad. Sci.* **252**, 4139.
Barbier, L. and Chéret, M. (1983). *J. Phys. B.* **16**, 3213.
Baylis, W. E. (1977). *Can. J. Phys.* **55**, 1924.
Bearman, G. M. and Leventhal, J. J. (1978). *Phys. Rev. Lett.* **41**, 1227.
Bellum, J. C. and George, T. F. (1978). *J. Chem. Phys.* **68**, 134.
Beutler, H. (1935). *Z. Phys.* **93**, 177.
Billardon, M., Deacon, D. A. G., Elleaume, P., Ortega, J. M., Robinson, K. E., Bazin, G. Bergher, M., Madey, J. M. J., Petroff, Y., and Velghe, M. (1983). *J. Physique* **44**, C1.
Bizau, J. M., LeGouét, J. L., Ederer, D. L., Koch, P., Wuilleumier, F., Picqué, J. L., and Dhez, P. (1981). XII.ICPEAC, Abstracts of post-deadline papers, (S. Datz, ed.) p. 1.
Bizau, J. M., Wuilleumier, F., Dhez, P., Ederer, D. L., Picque, J. L., LeGouet, J. L., and Koch, P. (1982). "Laser Techniques for the Extreme Ultraviolet Spectroscopy." (T. J. McIlrath, R. Freeman, eds.) American Institute of Physics Conference Proceedings No. 90 (Am. Inst. of Phys., New York) p. 331.
Bizau, J. M., Carré, B., Dhez, P., Ederer, D. L., Gérard, P., Keller, J. C., Koch, P., LeGouët, J. L., Picqué, J. L., Wendin, G., and Wuilleumier, F. (1983). XIII ICPEAC, Abstracts of contributed papers (J. Eichler, W. Fristch, I. V. Hertel, N. Stolterfoht, and U. Wille, eds.), p. 27.
Bizau, J. M., Wuilleumier, F., Ederer, D. L., Keller, J. C., LeGouët, J. L., Picqué, J. L., Carré, B., and Koch, P. M. (1985a). *Phys. Rev. Lett.* **55**, 1281.
Bizau, J. M., Cubaynes, D., Gérard, P., Wuilleumier, F., Keller, J. C., LeGoüet, J. L., Picqué, J. L., Carré, B., Ederer, D. L., and Wendin, G. (1985b). XIV ICPEAC, Abstracts of contributed papers, (M. J. Coggiola, D. L. Huestis, and R. P. Saxon, eds.), p. 8.
Bizau, J. M., Cubaynes, D., Gerard, P., Wuilleumier, F., Keller, J. C., LeGouët, Picqué, J. L., Carré, B., Ederer, D., and Wendin, G., (1985d). LURE Activity Report, p. 29.
Bizau, J. M., Cubaynes, D., Gerard, P., Wuilleumier, F., Picqué, J. L., Ederer, D., and Wendin, G. (1986). *Phys. Rev. Lett.* **57**, 306.
Bloomfield, L. A., Freeman, R. R., Cooke, W. E., and Bokor, J. (1984). *Phys. Rev. Lett.* **53**, 2234.
Bokor, J., Zavelovich, J., and Rhodes, C. K. (1980). *Phys. Rev. A* **21**, 1453.
Bonanno, R., Boulmer, J., and Weiner, J. (1983). *Phys. Rev. A* **28**, 604.
Borodin, V. M., and Komarov, I. V. (1974). *Opt. Spect.* **36**, 145.
Borodin, V. M., Klucharev, A. N., Sepman, V. Yu. (1975). *Opt. Spet.* **39**, 231.
Boulmer, J., Bonanno, R., and Weiner, J. (1983). *J. Phys. B* **16**, 3015.
Bradley, D. J. (1970). *Phys. Bulletin*, **21**, 116.
Bradley, D. J., Eward, P., Nicholas, J. V., and Shaw, J. R. D. (1973). *J. Phys. B* **6**, 1594.
Bradley, D. J., Dugan, C. H., Ewart, P., and Purdie, A. F. (1976). *Phys. Rev. A* **13**, 1416.
Bréchignac, C. and Cahuzac, Ph. (1982). *Opt. Comm.* **43**, 270.
Bréchignac, C., Cahuzac, Ph., and Débarre, A. (1985). *Phys. Rev. A* **31**, 2950.
Burgess, A. and Seaton, M. J. (1960). *Mon. Notic. Astron. Soc.* **120**, 121.
Burke, P. G. and Taylor, K. (1975). *J. Phys. B* **8**, 2620.
Bussert, W., Bregel, T., Ganz, J., Harth, K., Siegel, A., Ruf, M. W., Hotop, H., and Morgner, H. (1985). *J. Physique* **46**, C1-199 and refs. therein.
Camus, P., Dieulin, M., and El Hindy, A. (1982). *Phys. Rev. A* **26**, 379.
Camus, P., Dieulin, M., El Hindy, A., and Aymar, M. (1983). *Physica Scripta* **27**, 125.
Carlson, N. W., Taylor, A. J., Jones, K. M., and Schawlow, A. L. (1981). *Phys. Rev. A* **24**, 822.
Carlsten, J. L., McIlrath, T. J., and Parkinson, W. H. (1974). *J. Phys. B* **7**, L244.
Caro, R. G., Wang, J. C., Young, J. F., and Harris, S. E. (1984). "Laser Techniques in the Extreme Ultraviolet; AIP Conference Proc. 119." S. E. Harris and T. B. Lucatorto, eds.), p. 417, Am. Inst. of Physics, New York.

Carré, B., Roussel, F., Breger, P., and Spiess, G. (1981a). *J. Phys. B* **14**, 4277.
Carré, B., Roussel, F., Breger, P., and Spiess, G. (1981b). *J. Phys. B* **14**, 4289.
Carré, B., Spiess, G., Bizau, J. M., Dohz, P., Gérard, P., Wuilleumier, F., Keller, J. C., LeGouët, J. L., Picqué, J. L., Ederer, D. L., and Koch, P. M. (1984). *Opt. Comm.* **52**, 29.
Carré, B., Bizau, J. M., Cubaynes, D., Dhez, P., Ederer, D. L., Gérard, P., Keller, J. C., Koch, P. M., LeGouët, J. L., Picqué, J. L., Roussel, F., Spiess, G., and Wuilleumier, F. (1985). *J. Physique*, **46**, C1-163.
Carré, B., Bizau, J. M., Breger, P., Cuyaynes, D., Gérard, P., Picqué, J. L., Roussel, F., Spiess, G., and Wuilleumier, F. (1986a). In "Electronic and Atomic Collisions" (D. C. Lorents, W. E. Meyerhof, J. R. Peterson, eds.), Elsevier, Amsterdam, p. 493.
Carré, B., Roussel, F., Spiess, G., Bizau, J. M., Gérard, P., and Wuilleumier, F. (1986b). "Atoms, Molecules, and Clusters." *Z. Physik D.* **1**, 79.
Carroll, P. K. and Kennedy, E. T. (1977). *Phys. Rev. Lett.* **38**, 1068.
Carroll, P. K., Kennedy, E. T., Sullivan, G. O. (1978). *Opt. Lett.* **2**, 72.
Chang, T. N. (1974). *J. Phys. B* **8**, 743.
Chang, T. N. (1982). *J. Phys. B* **15**, L81.
Chang, T. N. and Kim, Y. S. (1982a). "X-Ray and Atomic Inner Shell Physics." (B. Crasemann, ed.), American Institute of Physics Conference Proceedings No. 94, p. 633, Am. Inst. of Physics, New York.
Chang, T. N. and Kim, Y. S. (1982b). *J. Phys. B* **15**, L835.
Chang, T. N. and Poe, R. T. (1975). *Phys. Rev. A* **11**, 191.
Chéret, M., and Barbier, L. (1985). *J. Physique*, **46**, C1-193.
Chéret, M., Spielfiedel, A., Durand, R., and Deloche, R. (1981). *J. Phys. B* **14**, 3953.
Chéret, M., Barbier, L., Lindinger, W., and Dloche, R. (1982b). *J. Phys. B* **15**, 3463.
Chéret, M., Barbier, L., Lindinger, W., and Deloche, R. (1982b). *Chem. Phys. Lett.* **88**, 229.
Clark, C. W., Fassett, J. D., Lucatorto, T. B., Moore, L. J., and Smith, W. (1985). *J. Opt. Soc. Am.* **8**, 2, 891.
Codling, K. (1979). "Topics in Current Physics: Synchrotron Radiation Technique and Application." p. 231 and refs., Springer-Verlag, Berlin.
Codling, K., Hamley, J. R., and West, J. (1977). *J. Phys. B* **10**, 2797.
Connerade, J. P. (1978). *Contemp. Phys.* **19**, 414 and *J. Phys. B* **11**, L318.
Connerade, J. P., Mansfield, M. W. D., Newson, G. H., Tracey, D. H., Baig, M. A., and Thimm, K. (1979). *Phil. Trans. R. Soc. A* **290**, 327.
Connerade, J. P. and Martin, M. A. P. (1983). *J. Phys. B* **16**, L577.
Cook, T. B., Dunning, F. B., Foltz, G. W., and Stebbings, R. F. (1977). *Phys. Rev. A* **15**, 1526.
Cooper, J., and Zare, R. N. (1969). "Lectures in Theoretical Physics." *Atomic Coll. Proc.* **11C**, 317.
Cooper, J. W. (1962). *Phys. Rev.* **128**, 681.
Craig, B. I. and Larkins, F. P. (1985). *J. Phys. B* **18**, 3369.
Cromer, C. L., Bridges, J. M., Roberts, J. R., and Lucatorto, T. B. (1985). *Appl. Opt.* **24**, 2996.
Cubaynes, D., Bizau, J. M., Gérard, P., Wuilleumier, F., and Picqué, J. L., (1985) LURE Activity Report, p. 30.
Cubaynes, D., Bizau, J. M., Gérard, P., Wuilleumier, F. J., Picqué, J. L., Carré, B., Roussel, F. and Spiess, G. (1986a), in: "Abstracts of VUV8 (P. O. Nilsson, ed.), Chalmers Institute of Technology, Göteborg, vol. III, p. 658.
Cubaynes, D., Bizau, J. M., Gérard, P., Wuilleumier, F., Carré, B., and Picqué, J. L. (1986b). *Annales de Physique* **11**, 197.
Cubaynes, D., Bizau, J. M., Wuilleumier, F. J., Carré, B., Roussel F., Spiess, G. and Picqué, J. L. (1987). *Europhys. Lett.* **4**, 549.
Dalgarno, A., Doyle, H., and Oppenheimer, M. (1972). *Phys. Rev. Lett.* **29**, 1051.

Devdariani, A. Z., Klucharev, A. N., Lazarenko, A. V., and Sheverev, V. A. (1978). *Sov. Tech. Phys. Lett.* **4**, 408.
Duong, H. T., Pinard, J., and Vialle, J. L. (1978). *J. Phys. B* **11**, 797.
Duncanson, J. A., Stand, M. P., Lindgard, A., and Berry, R. S. (1976). *Phys. Rev. Lett.* **37**, 987.
Dunning, F. B. and Stebbings, R. F. (1974a). *Phys. Rev. A* **9**, 2378.
Dunning, F. B. and Stebbings, R. F. (1974b). *Phys. Rev. Lett.* **32**, 1286.
Eberly, J. H. and Karczewski, B. (1970). *Multiphoton Bibliography 1970-1* University of Rochester and Data Management Research Project Lawrence Livermore Laboratory (UCRL-13728).
Eberly, J. H., Gallagher, J. W., Beaty, E. C. (1979). *Multiphoton Bibliography, 1978* NBS LB-92 suppl. 1 (National Bureau of Stnds., Gaithersburg, MD).
Eberly, J. H. and Gallagher, J. W. (1981). *Multiphoton Bibliography 1980*, NBS LP Suppl. 3 (National Bureau of Standards, Gaithersburg, MD).
Eberly, J. H., Piltch, N. D., and Gallagher, J. W. (1982). *Multiphoton Bibliography* 1981, NBS LP-92 Suppl. 4 (National Bureau of Stnds., Gaithersburg, MD).
Enemark, E. A. and Gallagher, A. (1972). *Phys. Rev. A* **6**, 192.
Eshevick, P., Armstrong, J. A., Dreyfus, R. W., and Wynne, J. J. (1976). *Phys. Rev. Lett.* **36**, 1296.
Esteva, J. M. and Mehlman, G. (1974). *Astrophys. J.* **193**, 474.
Fano, U. (1935). *Nuovo Cimento* **12**, 56.
Fano, U. (1953). *Phys. Rev.* **90**, 577.
Fano, U. (1957). *Rev. Mod. Phys.* **29**, 74.
Fano, U. (1961). *Phys. Rev.* **124**, 1866.
Fischer, A. and Hertel, I. V. (1982). *Z. Phys. A* **304**, 103.
Feldmann, D. and Welge, K. H. (1982). *J. Phys. B* **15**, 1651.
Ganz, J., Lewandowski, B., Seigel, A., Bussert, W., Waibel, H., Ruff, M. W., Hotop, H. (1982), *J. Phys. B* **15**, L485.
Ganz, J., Siegel, A., Bussert, W., Harth, K., Ruf, M. W., Hotop, H., Geiger, J., and Fink, M. (1983). *J. Phys. B* **16**, L569.
Ganz, J., Raab, M., Hotop, H., and Geiger, J. ([1984). *Phys. Rev. Lett.* **53**, 1547.
Garver, W. P., Pierce, M. R., and Leventhal, J. J. (1982). *J. Chem. Phys.* **77**, 1201.
Gerwert, K. and Kollath, K. J. (1983). *J. Pl,ys. B* **16**, L217.
Gezalov, Kh. B. and Ivanova, A. V. (1968). *High Temp.* **6**, 4100.
Gilbert, S. L., Noecker, M. C., and Wieman, C. E. (1984). *Phys. Rev. A* **29**, 3150.
Grandin, J. P. and Husson, X. (1981). *J. Phys. B* **14**, 433.
Grove, R. E., Wu, F. Y., and Ezekiel, S. (1977). *Phys. Rev. A.*, **15**, 227.
Hanson, J. C., Duncanson Jr., A., Chien, R. L., and Berry, R. S. (1980). *Phys. Rev. A* **21**, 222.
Hecnz, M. H. and Lindau, I. (1984). *Phys. Rev. Lett.* **47**, 821.
Heckenkamp, C., Schafers, F., Schonhense, G., and Heinzmann, U. (1984). *Phys. Rev. Lett.* **52**, 421.
Heinzmann, U. (1980). *Appl. Opt.* **19**, 4087.
Heinzmann, U., Schinkowski, D., and Zeman, H. D. (1977). *Appl. Phys.* **12**, 113.
Hellemuth, T., Leuchs, G., Smith, S. J., and Walther, H. (1981). "Lasers and Applications," Vol. **26**, (W. O. N. Guimars, C.-T. Lin, and A. Mooradian, eds.), p. 194–203. Springer, Berlin-Heidelberg.
Hellfeld, A. V., Caddick, J., and Weiner, J. (1978). *Phys. Rev. Lett.* **40**, 1369.
Henriet, A., Masnou-Seeuws, F., Le Sech, C. (1985). *Chem. Phys. Lett.* **118**, 507.
Hertel, I. V. and Stoll, W. (1974). *J. Phys. B* **7**, 583.
Hill, Jr., W. T., Cheng, K. T., Johnson, W. R., Lucatorto, T. B., McIlrath, T. J., and Sugar, J. (1982). *Phys. Rev. Lett.* **49**, 1631.

Holmgren, D. E., Walter, D. J., King, D. A., and Harris, S. E. (1984). Laser Techniques in the Extreme Ultraviolet, Ed: S. E. Harris and T. B. Lucatorto, AIP Conference Proc. No. 119 (Am. Inst. of Physics, New York), p. 157.
Holstein, T. (1947). *Phys. Rev.* **72**, 1272.
Hotop, H. and Mahr, D. (1975). *J. Phys. B* **8**, L301.
Huber, M. C. E. and Jahreiss, L. (1985). *J. Physique*, **46**, C1-215.
Hudson, R. D., and Carter, V. L. (1968). *J. Opt. Soc. Am.* **58**, 430.
Huennekens, J. and Gallagher, A. (1983a). *Phys. Rev. A* **28**, 1276.
Huennekens, J. and Gallagher, A. (1983b). *Phys. Rev. A* **27**, 771.
Hurst, G. S., Nayfeh, M. H., and Young, J. P. (1977). *Phys. Rev. A* **15**, 2283.
Hurst, G. S., Payne, M. G., Kramer, S. D. and Young , J. P. (1979). *Rev. Mod. Phys.* **51**, 767.
Jahreiss, L. and Huber, M. C. E. (1983). *Phys. Rev. A* **28**, 3382.
Jacobs, V. (1971). *Phys. Rev. A* **4**, 939.
Johnson, W. R., Cheng, K. T., Huang, K. N., and LeDourneuf, M. (1980). *Phys. Rev.* **22**, 989.
Johnson, W. R. and LeDourneuf, M. (1980). *J. Phys. B* **13**, L13.
de Jong, A. and van der Valk, F. (1979). *J. Phys. B* **12**, L561.
Jopson, R. M., Freeman, R. R., Cooke, W. E., and Bokor, J. (1984). *Phys. Rev.* **29**, 3154.
Kaminski, M., Kessler, J., Kollath, K. J. (1980). *Phys. Rev. Lett.* **45**, 1161.
Kastner, O., Omidvar, K., Underwood, J. H. (1967). *Astrophysics J.*, **148**, 269.
Kelley, H. P. and Simons, R. L. (1973). *Phys. Rev. Lett.* **30**, 529.
Kimura, K. (1985). Advances in Chemical Physics, **60**, (K. P. Lawley, ed.), p. 161-200. John Wiley and Sons, New York.
Kircz, J. G., Morgenstern, R., and Nienhuis, G. (1982). *Phys. Rev. Lett.* **48**, 610.
Klar, H. and Kleinpoppen, H. (1982). *J. Phys. B* **15**, 933.
Klucharev, A. N. and Lazarenko, A. V. (1972). *Opt. Spect.* **32**, 576.
Klucharev, A. N. and Ryazanov, N. S. (1972a). *Opt. Spect.* **32**, 686.
Klucharev, A. N. and Ryazanov, N. S. (1972b). *Opt. Spect.* **33**, 230.
Klucharev, A. N. and Dobrolege, B. V. D. (1973). Abstracts of Papers, VIII International Conference on the Physics of Electronic and Atomic Collisions, Belgrade, (B. C. Cobic and M. V. Karepa, eds.), p. 553-554. Institute of Physics, Belgrade, Yugoslavia.
Klucharev, A. N., and Sepman, V. Yu. (1975). *Opt. Spect.* **38**, 712.
Klucharev, A. N., Sepman, V., and Vujnovic, Y. (1977a). *Opt. Spect.* **42**, 336.
Klucharev, A. N., Sepman, V., and Vujnovic, Y. (1977b). *J. Phys. B* **10**, 715.
Klucharev, A. N., Lazarenko, A. V., and Vujnovic, V. (1980). *J. Phys. B* **13**, 1143.
Kobrin, P. H., Rosenberg, R. A., Becker, V., Southworth, S., Truesdale, C. M. Lindle, D. W., Thornton, G., White, M. G., Poliakoff, E. D., and Shirley, D. A. (1983). *J. Phys. B* **16**, 4339.
Koch, E. E. and Sonntag, B. F. (1979). "Topics in Current Physics: Synchrotron Radiation Techniques and Applications." Springer-Verlag, Berlin.
Koch, P. M. (1982). "X-Ray and Atomic Inner Shell Physics." (B. Crasemann, ed.). AIP Conference Proc. No. **94**, p. 645. Am. Inst. of Physics, New York.
Kollath, K. J. (1980). *J. Phys. B* **13**, 2901.
Kopystynska, A., and Kowalczyk, P. (1978). *Opt. Comm.* **25**, 351.
Kowalczyk, P. (1979). *Chem. Phys. Lett.* **68**, 203.
Kowalczyk, P. (1984). *J. Phys. B* **17**, 817.
Krause, M. O. (1980). *In* "Synchrotron Radiation Research." (H. Winick and S. Doniach, eds.), p. 101-151. Plenum, New York.
Krummacher, S., Schmidt, V., Bizau, J. M., Ederer, D. L., Dhez, P., Wuilleumier, F. (1982). *J. Phys. B* **15**, 4363.
Krebs, D. J. and Schearer, L. D. (1981). *J. Chem. Phys.* **75**, 3340.
Kushawaha, V. S. and Leventhal, J. J. (1980). *Phys. Rev. A* **22**, 2468.

Kushawaha, V. S. and Leventhal, J. J. (1982a). *Phys. Rev. A* **25**, 246.
Kushawaha, V. S. and Leventhal, J. J. (1982b). *Phys. Rev. A* **25**, 570.
Lahiri, J. (1981). "Workshop on Photoionization of Excited Atoms and Molecules," JILA, unpublished.
Lahiri, J. and Manson, S. T. (1982). *Phys. Rev. Lett.* **48**, 614.
Lahiri, J. and Manson, S. T. (1984). Unpublished.
Lambropoulos, P. (1976). *Adv. At. and Mol. Phys.* **12**, 87.
Larssen, P. K., van Beers, W. A. M., Bizau, J. M., Wuilleumier, F., Krummacher, S., Schmidt, V., and Ederer, D. L. (1982). *Nucl. Instr. Meth.* **195**, 245.
Laughlin, C. (1978). *J. Phys. B* **11**, 1399.
LeGouët, J. L., Picqué, J. L., Wuilleumier, F. J., Bizau, J. M., Dhez, P., Koch, P., and Ederer, D. L. (1982). *Phys. Rev. Lett.* **48**, 600.
Leuchs, G., Smith, S. J., Khawaja, E. E., and Walter, H. (1979). *Opt. Comm.* **31**, 313.
Leutwyler, S., Hofmann, M., Harri, H. P., and Schumacher, E. (1981). *Chem. Phys. Lett.*, **77**, 257.
Lucatorto, T. B. (1985). Private communication.
Lucatorto, T. B. and McIlrath, T. J. (1976). *Phys. Rev. Lett.* **37**, 428.
Lucatorto, T. B. and McIlrath, T. J. (1980). *Appl. Opt.* **19**, 3948.
Lucatorto, T. B., McIlrath, T. J., Sugar, J., and Younger, S. M. (1981). *Phys. Rev. Lett.* **47**, 1124.
Luke, T. M. (1982). *J. Phys. B* **15**, L1217.
Martin, G. A. and Wiese, W. L. (1976). *Phys. Rev. A* **13**, 699.
McGuire, E. J. (1970). Research Report SC-RR-721, Sandia Laboratories.
McIlrath, T. J. (1969). *Appl. Phys. Lett.* **15**, 41.
McIlrath, T. J. and Lucatorto, T. B. (1977). *Phys. Rev. Lett.* **38**, 1390.
Measures, R. M. (1970). *J. Quant. Spectrosc. Radiat. Transfer.* **10**, 107.
Measures, R. M. (1977). *J. Appl. Phys.* **48**, 2673.
Measures, R. M. and Cardinal, P. G. (1981). *Phys. Rev. A* **23**, 804.
Measures, R. M., Drewell, N., and Cardinal, P. (1979). *J. Appl. Phys.* **50**, 2662.
Mehlman, G. and Esteva, J. M. (1974). *Astrophys. J.* **188**, 191.
Moores, D. L., Norcross, D. W., and Sheorey, V. B. (1974). *J. Phys. B* **7**, 371.
Mohler, F. L. (1929). *Rev. Mod. Phys.* **1**, 216.
Mohler, F. L. (1933). *J. Res. Nat. Bur. Std. (U.S.)* **10**, 771.
Mohler, F. L. and Boeckner, C. (1929). *J. Res. Nat. Bureau Std. (U.S.)* **2**, 489.
Montagnani, P., Riani, P., Salvetti, O. (1983). *Chem. Phys. Lett.* **102**, 571.
Moskvin, Yu. (1963). *Optics Spectrosc.* **15**, 316.
Msezane, A. Z. and Manson, S. T. (1975). *Phys. Rev. Lett.* **35**, 364.
Msezane, A. Z. and Manson, S. T. (1984a). *Phys. Rev. A* **30**, 1795.
Msezane, A. Z., Manson, S. T. (1984b). *Phys. Rev. A* **29**, 1594.
Nayfeh, M. H. (1980). *Opt. Eng.* **19**, 57.
Nienhuis, G. (1982). *Phys. Rev. A* **26**, 3137.
Norcross, D. W. (1971). *J. Phys. B* **4**, 652.
Norcross, D. W. and Stone, P. M. (1966). *J. Quant. Spectrosc. Radiat. Transfer* **6**, 277.
Norcross, D. W. (1973). *Phys. Rev. A* **7**, 606.
Nygaard, K., Hebner, R. E., Jr., Jones, J. D., Corbin, R. J. (1975). *Phys. Rev. A* **12**, 1440.
Nygaard, K. J., Corbin, R. J., and Jones, J. D. (1978). *Phys. Rev. A* **17**, 1543.
Nunneman, A., Prescher, Th., Richter, M., Schmidt, M., Sonntag, B., Wetzel, H. E., Zimmerman, P. (1985). *J. Phys. B* **18**, L337.
Parr, A. C., Southworth, S. H., Dehmer, J. L., and Holland, D. M. P. (1984). *Nuc. Inst. Meth.* **222**, 221.
Pratt, S. T., Dehmer, P. M., and Dehmer, J. L. (1983). *J. Chem. Phys.* **78**, 4315.

Preses, J. M., Burkhardt, C. E., Corey, R. L., Erson, D. L., Daulton, T. L., Garver, W. P., Leventhal, J. J., Msezane, A. Z., Manson, S. T. (1985). *Phys. Rev. A* **32**, 1264.
Radler, K. and Berkowitz, J. (1979). *J. Chem. Phys.* **70**, 216.
Register, D. F., Trajmar, S., Jensen, S. W., and Poe, R. T. (1978). *Phys. Rev. Lett.* **41**, 749.
Roth, D. L. (1969). *J. Quant. Spectrosc. Radiat. Transfer* **9**, 49; ibid, (1971). **11**, 355.
Rosenberg, R. A., White, M. S., Thornton, G., Shirley, D. A. (1979). *Phys. Rev. Lett.* **43**, 1384.
Roussel, F., Breger, P., Spiess, G., Manus, C., and Geltman, S. (1980). *J. Phys. B* **13**, L631.
Rudkjobing, M. (1940). *Publ. Kbh. Obs.* **18**, 1.
Rundel, R. D., Dunning, F. B., Goldwire, Jr., M. C., and Stebbings, R. F. (1975). *J. Oct. Soc. Am.* **65**, 628.
Salzmann, D. (1983). Private communication.
Salzmann, D. and Pratt, R. H. (1984). *Phys. Rev. A* **30**, 2757.
Seaton, M. J. (1966). *Proc. Phys. Soc. London* **88**, 801.
Shevelko, V. P. (1970). *Preprint Fiz. Inst. Akad. Nauk SSSR.* No. 1 (Moscow).
Shorer, P. (1979). *Phys. Rev. A* **20**, 642.
Siegbahn, H. and Karlsson, L. (1982). "Handbuch der Physik, Vol. XXXI, Corpuscles and Radiation in Matter." (W. Mehlhorn, ed.), p. 215 and refs. Springer-Verlag, Berlin.
Siegel, A., Ganz, J., Bussert, W., and Hotop, H. (1983). *J. Phys. B* **16**, 2945.
Skinner, C. H. (1980). *J. Phys. B* **13**, 55.
Smith, A. V., Goldsmith, J. E. M., Nitz, D. E., and Smith, S. J. (1980). *Phys. Rev. A* **22**, 577.
Spitzer, L. (1968). "Diffuse Matter in Space." Interscience Publishers, New York, NY.
Stacewicz, T. (1980). *Opt. Comm.* **35**, 239.
Stacewicz, T. and Krasinski, J. (1981). *Opt. Comm.* **39**, 35.
Starace, A. F. (1982). "Handbüch der Physik Vol. 1. XXXI, *Corpuscles and Radiation in Matter* (W. Mehlnorn, ed.), p. 1 and refs. Springer-Verlag, Berlin.
Stebbings, R. F., Dunning, F. B., Tittel, F. K., and Rundel, R. D. (1973). *Phys. Rev. Lett.* **30**, 815.
Sugar, J., Lucatorto, T. B., McIlrath, T. J., and Weiss, A. W. (1979). *Opt. Lett.* **4**, 109.
Theodosiou, C. (1986). *Phys. Rev. A* **33**, 2164.
Valance, A. and Nguyen, Tuan, Q. (1981). *Phys. Lett.* 82A, 116.
Wang, M. X., de Vries, M. S., Keller, J., and Weiner, J. (1985). *Phys. Rev. A* **32**, 681.
Weiner, J. (1985). *J. Physique* **45**, C1-173.
Weiner, J. and Polak-Dingels, P. (1981). *J. Chem. Phys.* **14**, 508.
Weisheit, J. C. (1972). *J. Quant. Spectrosc. Radiat. Transfer* **12**, 1241.
Wendin, G. and Starace, A. F. (1978). *J. Phys. B* **11**, 4119.
Wendin, G. (1985). Private communication.
White, M. G., Seaver, M., Chupka, W. A., and Colson, S. D. (1982). *Phys. Rev. Lett.* **49**, 28.
Willison, J. R., Falcone, R. W., Wang, J. C., Young, J. F., and Harris, S. E. (1980). *Phys. Rev. Lett.* **44**, 1125.
Wolff, H. W., Radler, K., Sonntag, B., and Haensel, R. (1972). *Z. Physik* **257**, 353.
Wuilleumier, F. J. (1976). Ed., "Photoionization and Other Probes of Atomic Interactions." Plenum, New York.
Wuilleumier, F. (1980). "Electronic and Atomic Collisions." (N. Oda and K. Takayanagi, eds.), p. 55. North-Holland, Amsterdam.
Wuilleumier, F. J. (1981a). *Atomic Physics* **7**, 482.
Wuilleumier, F. J. (1981b). "Inner Shell and X-Ray Physics of Atoms." (D. J. Fabian *et al.*, eds.), p. 395. Plenum, New York.
Wuilleumier, F. J. (1982a). *J. Physique* **43**, C2 p. 347.
Wuilleumier, F. J. (1982b). "X-ray and Atomic Inner-Shell Physics." (B. Crasemann, ed.). AIP Conference Proc. No. 94, p. 615. Am. Inst. of Physics, New York.

Wuilleumier, F. J. (1984). "Laser Techniques in the Extreme Ultraviolet." (S. E. Harris and T. B. Lucatorto, eds.), AIP Conf. Proc. No. 119, p. 220. Am. Inst. of Physics, New York.
Wuilleumier, F. J. and Krause, M. O. (1979). *J. Elett. Spect.* **15**, 15.
Wuilleumier, F. J., Adam, M. Y., Dhez, P., Sandner, N., Schmidt, V., and Mehlhorn, V. (1977). *Phys. Rev. A* **16**, 661.
Wynne, J. J. and Armstrong, J. A. (1978). *IBM J. Res. Rev.* **23**, 490.
Yang, C. N. (1948). *Phys. Rev.* **74**, 764.
Ya'akobi, B. (1957). *Proc. Phys. Soc.* **92**, 108.

INDEX

A

Ab initio calculation, 92–93
Absorption arrays, 151–154
   variances in, 154
Absorption spectroscopy, 8–15
   in photoionization, 205, 206
AC Stark effect, 3, 5
Atomic orbitals
   electron correlation in, 94
   as expression of molecular orbitals, 90, 92
   gaussian-type functions in, 93
   polarization of atoms in, 93–94
   Slater-type, 93
Atomic site charges
   calculation of, 96–100
   dependent, 96
   grid derivation of, 100
   net, 96, 100
Aufbau principle, 46
Axial systems, for grid generation, 100

B

Basis sets
   computing time needed for, 94, 102
   in electric potential calculation, 93–94
   finite, *see* Finite basis sets
   gaussian, 93–94
   in population analysis, 88–89
   scaling relationships in, 102
   and Schrodinger limit, 68
   sensitivity of atomic site charges to, 102
   size of, 64, 67, 102, 104
Binding energies, of inner shell electrons, 252
Bond site models
   for analysis of fluorocarbons, 117–122
   improvement in fit with, 121–122
   see also PD-BC; PD-EC
Boundary conditions
   as $r \to \infty$, 43–44
   and basis set size, 67
   and Coulomb potential, 45
   exclusion of continuum-type orbitals, 55
   in extended nuclear charge models, 45–46
   and pairing of basis functions, 68
   at $r = \theta$, 44–46
   in relativistic quantum theory, 43–46
Bound states, 65–66
Breit interaction, 70

C

CI methods
   in open shell problem, 56–57
Collisional ionization, 261–278
   associative, 271–274, 275
   atomic beam collisions in, 264–265
   and electron production, 263
   and electron seeding, 262–263
   electron spectrometry in, 266–267
   energy pooling collisions in, 275–276
   excited state populations in, 267
   laser excitation in, 256–266
   Penning, 271, 274–275
   real time energy calibration in, 267
   of sodium, 268–278
   superelastic collisions in, 276–278
   synchrotron radiation in, 267–268
Collisional-radiative model of transition arrays, 178
Configuration states
   absorbent zones of, 169
   centers of gravity in, 166–167
   computation of moments in, 138
   coupling in, 169–171

emissive zones of, 166–169
energy distribution in, 137–142
energy variance in, 138
J-file sum rule in, 143
as matrix basis states, 137
mixing of, 171–174
and second quantization method, 139–141
stochasticity in, 178
subconfiguration mixing, 175
subconfiguration states, 141–142
widths of active zones in, 167
Coulombic energy, intermolecular, 126
Coulomb potential
  and boundary conditions, 45
  in relativistic quantum theory, 61–62
Coulomb's law
  effect on molecular packing, 121–122
  in electric potential, 90
  and intermolecular energy, 95
  model potential from, 97
Coupling
  in de-excitation, 169–171
  intermediate, 184
  and line strength, 183–185
Crystal structure
  modelled vs. static, 116
  shifts in, 121
cw dye lasers
  in collisional ionization, 265
  in photoabsorption experiments, 229–230
  and synchrotron radiation, 208, 228, 231–238
  see also Laser-synchrotron radiation

D

De-excitation, coupling in, 169–171
Dimer site charge polarization, 124
Dipole or quadrupole moments
  calculated vs. observed, 104–105
  misestimation of, 104
  scaling of, 105, 121
Dipoles
  in crystal structure energy calculation, 115
  point vs. distributed, 104
Dirac–Fock equations, finite difference methods for, 59
Dirac–Fock matrix, 69–71
  block structure of, 69

Dirac operator, 39–43
  analytic behavior of eigensolutions, 39
  and nonrelativistic limit, 64–65
  unboundedness of, 38
Dirac's wave equation, 39, 40
  locally square integrable solutions, 43–44
  plane wave solutions of, 40
  scattering-type solutions, 42
Distribution moments
  computation of, 134–137
  and computation of configuration states, 138
  levels vs. states in, 179
  line strengths in, 164
  uses of, 186
D values, vs. MCY, PD-AC, and PD-SC, 127

E

Electric potential
  ab initio calculation, 92–93
  basis sets in, 93–94
  calculation of, 89–94
  contour maps of, 94–95
  definition of, 90, 91
  and determination of point charge, 89
  envelope, 99–100
  of ethylene, 94–96, 107–108
  as expectation value, 92
  grid for evaluation of, 94, 100
  interaction of, 88, 94–95
  modelling by atomic site charges, 96–100
  vs. net atomic point charge, 89
  and quantum mechanics, 91, 96–98
  relative values of, 92–93
  represented by multipoles, 89
  scaling of values, 94
  and van der Waals envelope, 87
  of water, 94–95, 97, 98, 108
  and wavefunction, 92, 94
Electron bombardment, in pumping excitation, 205
Electron cloud
  division of overlap in, 88–89
  limit of extension in, 87
  polarization of, 101
Electron spectrometry
  in associative ionization, 273
  in collisional ionization, 266–267

INDEX 289

in laser ionization, 262
in Penning ionization, 275
and synchrotron radiation, 267
Emissive zones, 166-169
  de-excitation in, 169-171
  and perturbation, 187
  in transition arrays, 164-171
Excitation spectroscopy, 8-15
  vs. fluoresence spectroscopy, 15
  frequency-selective, 11-15
  nonselective, 8-11
  role configuration interaction and electron correlation in spectra, 251
Excited state atoms, 201-202
  see also Photoionization

F

Feynman diagrams
  and perturbation expansion, 49-53
  radiative corrections in, 62-63
Finite basis sets
  and Breit interaction, 69-71
  choice of, 71-73
  comparison of methods, 73-81
  and Dirac-Fock matrix, 64, 69-71
  even-tempered, 72
  expansions with, 67
  pairing of functions in, 66-67, 68
  principles of calculation in, 65-69
  problems with, 64-65
  in relativistic quantum theory, 64-81
  Slater and gaussian type, 66-67, 68
  and spectrum prediction, 64
  variational collapse in, 38, 64, 65, 69
Fluorescence
  measurements in excitation spectroscopy, 10, 11, 15
  multiphoton, 3, 4
  in two-step excitation spectroscopy, 25-28
Fluorescence spectroscopy, 15-22
  vs. excitation spectroscopy, 15
  lifetime measurements, 18, 20-22
  time resolved, 16, 18, 19
  wavelength selective, 15-17
Fock space field equations, 47
Furry bound interaction picture, and relativistic quantum theory, 38

G

Gaussian basis functions
  for modelling of atomic orbitals, 93-94
  in relativistic quantum theory, 66

H

Hamiltonians
  bound state solutions in, 37-38
  Dirac-Coulomb, 54
  effective, 51
  standard model of, 52, 54
Harmonic generation, 5
Hartree-Fock approximation
  and photoionization cross section, 221-222
  and relativistic quantum theory, 54-56

I

Ionization, 3, 4, 23
  see also Photoionization
Ionization spectroscopy, 23-24

J

J-file sum rule
  in computation of average transition energy, 143
  generalization of, 165-166

L

Lamb shift, of electronic bound levels, 51
Laser excitation
  in collisional ionization, 265-266
  one-photon-two-electron vs. two-photon, 236
  two-photon, 238
  see also Photoionization
Laser ionization, 261-263
  stages of, 262
  superelastic collisions in, 277
Laser spectroscopy
  high-resolution, 6-8

multiphoton, 31–32
VUV, 1–32
Laser-synchrotron radiation, 231–261
  and 5d photoionization cross section, 254–256
  and 6s–5d resonant photoemission, 257–261
  experimental conditions for, 231
  oscillator strength in, 238–241
  and two-photon experiments, 237
  in VUV photoelectron spectroscopy, 251–254
LCAO/MO (linear combination of atomic orbitals as
  expression of molecular orbital), 90–92
Least-squares fitting method, 89
  see also PD/LSF point-charge model
Level emissivity, strength of, 164–166
Level statistics and configuration states, 180–181
Lifetime variations in singlet and triplet states, 20–22
Lines
  amplitude of, 182
  definition of, 137
  intensity in transition array, 176–179
Line strength
  and coupling, 183–185
  distribution function in, 179
  vs. line intensity, 176–177
  statistics of, 179, 181–185
  in transition arrays, 164–166
Lone-pair electron sites, 123
  see also PD-LP
Lorentz transformation, and destruction of factorization, 41

M

MCDF methods, 57–59
MCY
  basis set in, 127
  defined, 124
  vs. PD-AC and PD-SC, 127
Molecular boundary, see van der Waals envelope
Molecular orbital, 90, 92
  see also LCAO/MO
MPA (molecular packing analysis)
  calculation of intermolecular energy in, 114
  and large site charges, 111, 114
  objective of, 115
  vs. PA, DC–AC, and PD–LP, 116, 117
  for prediction of azabenzene crystal structure, 110, 115–116
Multiphoton spectroscopy, 2–6
  fluorescence and ionization in, 3, 4
  quantitative interpretation of, 3, 5
  VUV, 31–32

N

NC (no-charge), vs. MPA, PA, DC–AC, and PD–LP, 116
Nuclear motion in relativistic quantum theory, 59–61

O

Open shell problem
  CI methods, 56–57
  GRASP code, 59
  MCDF methods, 57–59
Open shell systems, vs. closed shell systems, 59
Open subshell electrons, in transition arrays, 169
Orbitals
  "core" vs. "virtual," 55, 59
  definition by local mean potential, 53
  energy differences in, 53
  expansion as linear combinations of basis functions, 38
  spread away from nucleus, 88–89
Oscillator strength
  applications from, 200
  and electron spectrum, 242
  for inner shell excitations, 238–251

P

Particle-hole formalism
  in closed shell systems, 55–56
  in open shell systems, 59
PD–AC (PD/LSF atomic site charges)
  and basis sets, 102

in calculation of multipole moments, 104–105
effects of polarization on, 101, 106
from fitting to crystal structures and multipole moments, 106–107
vs. quantum mechanical potential, 107–109
scale factors in, 105
for small molecules, 102–103
for water dimer, 124, 126–128
for water monomer, 123
PD-BC (bond charge site), vs. PD-AC and PD-EC, 120
PD-EC (extension of C-F bond axis), vs. PD-AC and PD-BC, 120
PD-LP (lone-pair electron sites), 110–117
vs. multipole models, 114
vs. PA and PD-AC, 114, 116, 117
root-mean-square fit in, 114
variability of fit with distance, 111
PD/LSF point-charge model
and analysis of fluorocarbon crystals, 117–122
calculation of charges in, 94–101
for calculation of electric integration, 95
charges for small molecules, 102–103
definition of point charge in, 89
degrees of freedom in, 118
establishment of, 90
non-atomic charge sites in, 101
polarization in, 101
scaling in MPA calculations, 121
selection of points in, 99
PD-SC, vs. PD-AC, 126–128
PD-SC (four charge site), 123
Perturbation expansion, 47, 49–52
CI method, 56–57
truncation of, 53
Photoabsorption experiments, 229–230
Photodissociation spectroscopy, vs. VUV lasers, 22–23
Photoelectron spectroscopy, see VUV photoelectron spectroscopy
Photoionization of laser-excited atoms, 198–279
from $2p$ state, 237
from $3p$ state, 219, 221, 237, see also Resonant $3p$ cross section
from $5d$ state, 254–256
of argon, 216–217
of barium, 227–228, 251–254, 257–261
of cesium, 223–226
continuum proton sources in, 205
and dipole selective rules, 199
direct vs. resonant, 248
of excited alkaline earths, 226–228
experimental techniques of, 201–209
ground state vs. excited state cross sections, 240
Hartree–Fock approximation in, 221–222
of helium, 210–212
ionization saturation, 202–204
of krypton, 216–217
of lithium, 217–218
long-lived triplet states in, 226
multiproton, 227
of neon, 212–215
one-photon cross section, 210
of outer electron, 210–218
photoabsorption, 202–209, see also Photoabsorption
of potassium, 222
pump and ionization, 204–209
recombinant radiation, 201–202
of rubidium, 223
of sodium, 218–222, 238–241
synchrotron radiation in, 205
theoretical calculations of, 221
time delayed, 18
two-photon cross section, 209
uses of, 198–199
of xenon, 216–217
Point-charge model, 88–89
see also PD/LSF point-charge model
Point charges
atomic, 96–100
non-atomic, 101
transferable from small to large molecules, 93
Polarization
estimation of, 126, 127
in modelling of atomic orbitals, 93–94
Population analysis (PA)
vs. PD-AC and PD-LP values, 114, 116, 117
in point-charge model, 88–89
Potential-derived (PD) point charge
definition of, 89
see also PD/LSF point-charge model
Pump and ionization, for excited state atoms, 204–209.

Q

Quantum mechanical potential
 calculation of, 96–98
 vs. PD/AC potential, 107–109

R

Radiative corrections, in relativistic quantum theory, 62–63
Relativistic corrections, in transition array calculation, 176
Relativistic quantum theory, 37–83, 46–63
 Breit interaction in, 69–71
 corrections to, 59–63
 Coulomb potential corrections in, 61–62
 criticisms of, 38, 54
 Dirac–Fock matrix in, 64, 69–71
 finite basis sets in, 64–81, see also Finite basis sets
 Furry bound interaction picture in, 47–48
 and Hartree–Fock approximation, 54–56
 nuclear motion corrections in, 59–61
 open shell problem in, 56–59
 perturbation theory in, 49–52
 problems with, 37–38
 and quantum electrodynamics, 46–63
 radiative corrections in, 62–63
 standard model of, 51–52, 54–56
 two-body interaction kernel in, 52–54
Resonance
 changes with atomic size and nuclear charge, 213–214
 ionization mass spectrometry, 227
 resonant photoemission of laser excited barium atoms, 257–261
Resonant $5d$ cross section, 254–256
 resonance in, 258
Resonant $3p$ cross section
 and oscillator strengths, 238–241
Root-mean-square fits, in atomic point-charge model, 99, 100

S

SCF equations, 54
Site charges
 changes in, 124
 and coulombic energy, 124, 126
 net, 104
 non-atomic, 101
 see also PD/LP
 in PD-AC, PD-SC, and MCY, 126, 127
 shifting of, 126
 in water molecule, 123–124
Skewness
 effect on gaussian shape, 158, 160–161
 and energy values, 189
 as function of Z, 161
 and higher moments, 189–190
 incompleteness of formula, 187
 in transition arrays, 148–150
Spinor basis functions, in relativistic quantum theory, 68
Spin-orbit-split arrays, 150–151, 161–163
 and accuracy of moments, 161
 moments of subarrays, 150–151
 splitting of, 150
 and subarray intensity distribution, 161
Sum frequency mixing, 5–6
 in gases, 6–8
 nonlinear media in, 7
 in VUV laser spectroscopy, 6–8
 wavelength calibration in, 8, 9
Synchrotron radiation, 205, 207–209
 circularly polarized, 221
 in collisional ionization, 267–268
 in collision experiments, 263
 and electron production, 263
 and laser experiments, 208, 228–229, 231–261
 and photoelectron spectroscopy, 207
 spectral range of, 228
 see also Laser-synchrotron radiation

T

Time-delayed ionization, 18, 23
Transition arrays, 131, 191
 absorption arrays, 151–154
 average energies of, 142–146
 collisional-radiative model, 178
 and collision strengths, 155
 definition of, 132
 dipole–velocity formulation in, 187
 electric dipole, 154–155
 extension of theory, 188–189
 global properties of, 142, 171–172

intensity distribution of, 134, 155
line intensities in, 176–179
mean energy and width of, 189
as narrow band emission source, 190–191
one-gaussian model, 156–159
and perturbation theory, 188
in plasma diagnostics, 191
relativistic corrections in, 176
resolvability of, 190
skewed gaussian model, 158, 160–161
skewness in, 148–150
spectrum identification with, 189–190
spin-orbit-split arrays, 150–151, 161–163
theory vs. model, 135, 155–156
unresolved, 156–157
validity of theory, 186
variance of, 146–148
Transitions, definition of, 137
Tunable lasers, spectral range of, 205
Two-body interaction kernel, 52–54
Two-step excitation spectroscopy, 24–30
analysis of perturbations in, 26–27
collision-induced transitions in, 29, 30
description of, 24–25
fine structure measurements in, 31
fluorescence in, 25–28
predissociation in, 27–28, 30

V

van der Waals envelope, 37, 39
Variance
in absorption arrays, 154
in configuration states, 138
dependence on level energies, 148
in preferential de-excitation, 171
of transition array energies, 146–148
VUV
electronic sources, 6
tunable sources, 6–8, 22
VUV laser spectroscopy, 1–32
high-resolution, 6–8
multiphoton, 31–32
VUV photoabsorption spectra, 205
VUV photoelectron spectroscopy, 208, 209
of laser-excited barium, 251–254
spectra in, 251–252

W

Water monomer and dimer
charge sites in, *see* Site charges
PD/LSF analysis of, 122–128
wavefunction calculation in, 124
Wavefunctions
in calculation of multipole moments, 104, 105
directional dependence of, 38
electron correlation and, 94
in evaluation of electric potential, 94
factorization of, 41
sensitivity of intensities to, 175
for water monomer and dimer, 124

RAYMOND H. FOGLER LIBRARY